# Encyclopedia of Aircraft Technology

# Encyclopedia of Aircraft Technology

Edited by **Natalie Spagner**

CLANRYE
INTERNATIONAL

New Jersey

Published by Clanrye International,
55 Van Reypen Street,
Jersey City, NJ 07306, USA
www.clanryeinternational.com

**Encyclopedia of Aircraft Technology**
Edited by Natalie Spagner

International Standard Book Number: 978-1-63240-171-7 (Hardback)

Printed in the United States of America.

# Contents

# Preface

Aircraft technology is a dynamic field. Substantial information regarding advancements in aircraft technology has been provided in this insightful book. It elucidates the most recent developments in technologies for several areas of aircraft systems. Specifically, it covers a broad range of topics in aircraft structures and modern materials, and control systems. The authors are leading veterans in their fields. This book should appeal to both the students as well as researchers.

This book is a result of research of several months to collate the most relevant data in the field.

When I was approached with the idea of this book and the proposal to edit it, I was overwhelmed. It gave me an opportunity to reach out to all those who share a common interest with me in this field. I had 3 main parameters for editing this text:

1. Accuracy – The data and information provided in this book should be up-to-date and valuable to the readers.

2. Structure – The data must be presented in a structured format for easy understanding and better grasping of the readers.

3. Universal Approach – This book not only targets students but also experts and innovators in the field, thus my aim was to present topics which are of use to all.

Thus, it took me a couple of months to finish the editing of this book.

I would like to make a special mention of my publisher who considered me worthy of this opportunity and also supported me throughout the editing process. I would also like to thank the editing team at the back-end who extended their help whenever required.

**Editor**

# Part 1

# Aircraft Structures and Advanced Materials

# Study of Advanced Materials for Aircraft Jet Engines Using Quantitative Metallography

Juraj Belan
*University of Žilina, Faculty of Mechanical Engineering,*
*Department of Materials Engineering, Žilina*
*Slovak Republic*

## 1. Introduction

The aerospace industry is one of the biggest consumers of advanced materials because of its unique combination of mechanical and physical properties and chemical stability. Highly alloyed stainless steel, titanium alloys and nickel based superalloys are mostly used for aerospace applications. High alloyed stainless steel is used for the shafts of aero engine turbines, titanium alloys for compressor blades and finally nickel base superalloys are used for the most stressed parts of the jet engine – the turbine blades. Nickel base superalloys were used in various structural modifications: as cast polycrystalline, a directionally solidified, single crystal and in last year's materials which were produced by powder metallurgy.

So what exactly is a superalloy? Let us have a closer look to its definition. An interesting thing about it is that there is no standard definition of what constitutes a superalloy. The definitions which are provided in the various handbooks and reference books, although somewhat vague, are typically based on the service conditions in which superalloys are utilised. The most concise definition might be that provided by Sims et al. (1987): "...superalloys are alloys based on Group VIII-A base elements developed for elevated-temperature service, which demonstrate combined mechanical strength and surface stability." Superalloys are typically used at service temperatures above 540 C° (1000 F°), and within a wide range of fields and applications, such as components in turbine engines, nuclear reactors, chemical processing equipment and biomedical devices; by volume, its predominant use is in aerospace applications. Superalloys are processed by a wide range of techniques, such as investment casting, forging and forming, and powder metallurgy.

The superalloys are often divided into three classes based on the major alloying constituent: iron-nickel-based, nickel-based and cobalt-based. The iron-nickel-based superalloys are considered to have developed as an extension of stainless steel technology. Superalloys are highly alloyed, and a wide range of alloying elements are used to enhance specific microstructural features (and - therefore - mechanical properties). Superalloys can be further divided into three additional groups based on their primary strengthening mechanism:

- solid-solution strengthened;

- precipitation strengthened;
- oxide dispersion strengthened (ODS) alloys.

Solid-solution strengthening results from lattice distortions caused by solute atoms. These solute atoms produce a strain field which interacts with the strain field associated with the dislocations and acts to impede the dislocation motion. In precipitation strengthened alloys, coherent precipitates resist dislocation motion. At small precipitate sizes, strengthening occurs by the dislocation cutting of the precipitates, while at larger precipitate sizes strengthening occurs through Orowan looping. Oxide dispersion strengthened alloys are produced by mechanical alloying and contain fine incoherent oxide particles which are harder than the matrix phase and which inhibit dislocation motion by Orowan looping (MacSleyne 2008).

Figure 1. provides a representation of the alloy and process development which has occurred since the first superalloys began to appear in the 1940s; the data relates to the materials and processes used in turbine blading, such that the creep performance is a suitable measure for the progress which has been made. Various points emerge from a study of the figure. First, one can see that - for the blading application - cast rather than wrought materials are now preferred since the very best creep performance is then conferred. However, the first aerofoils were produced in wrought form. During this time, alloy development work – which saw the development of the first Nimonic alloys - enabled the performance of blading to be improved considerably; the vacuum introduction casting technologies which were introduced in the 1950s helped with this since the quality and cleanliness of the alloy were dramatically improved.

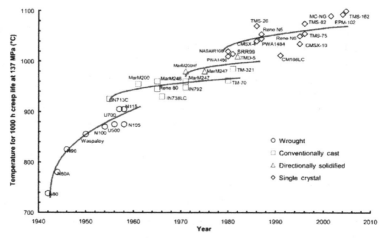

Fig. 1. Evolution of the high–temperature capability of superalloys over a 70 year period, since their emergence in the 1940s (Reed 2006).

Second, the introduction of improved casting methods and - later - the introduction of processing by directional solidification enabled significant improvements to be made; this was due to the columnar microstructures that were produced in which the transverse grain boundaries were absent (see Figure 2.)

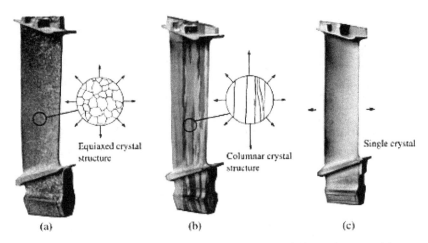

Fig. 2. Turbine blading in the (a) equiaxed-, (b) columnar- and (c) single–crystal forms.

Once this development had occurred, it was quite natural to remove the grain boundaries completely such that monocrystalline (single-crystal) superalloys were produced. This allowed, in turn, the removal of grain boundary strengthening elements such as boron and carbon which had traditionally been added, thereby enabling better heat treatments to reduce microsegregation and induced eutectic content, whilst avoiding incipient melting during heat treatment. The fatigue life is then improved.

Nowadays, single–crystal superalloys are being used in increasing quantities in the gas turbine engine; if the very best creep properties are required, then the turbine engineers turn to them (although it should be recognised that the use of castings in the columnar and equiaxed forms is still practiced in many instances).

In this chapter, a problem of polycrystalline (equiaxed) nickel base superalloy turbine blades - such as the most stressed parts of the aero jet engine - will be discussed.

The structure of polycrystalline Ni–based superalloys - depending on heat–treatment - consists of a solid solution of elements in Ni ($\gamma$-phase, an austenitic fcc matrix phase) and inter-metallic strengthening precipitate $Ni_3(Al, Ti)$ ($\gamma'$-phase, which is an ordered coherent precipitate phase with a Ll2 structure). A schematic showing representative microstructures of both a wrought and a cast nickel-base superalloy is shown in Figure 3. The $\gamma'$ precipitates in precipitate strengthened nickel-base superalloys remain coherent up to large precipitate sizes due to the small lattice mismatch between the matrix phase $\gamma$ and the $\gamma'$ precipitates. The $\gamma'$ precipitates are usually present in volume fractions in the range of 20-60%, depending on the alloy (Sims et al. 1987), with typical shapes from the spherical at small sizes to cuboid at larger sizes, although more complex dendritic shapes are also observed in some cases (see Figure 4). The alignment of $\gamma'$ precipitates along the elastically soft (100) directions is frequently observed. Nickel based superalloys are precipitation hardened, with a typical precipitate size of 0.25-0.5 µm for high temperature applications (Sims et al. 1987).

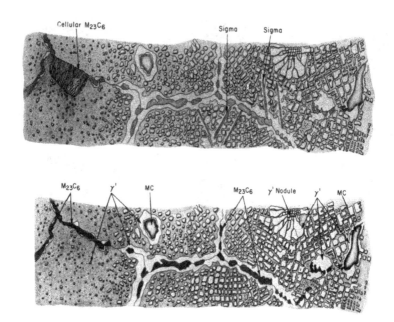

Fig. 3. Structure of a wrought and a cast nickel-base superalloy (M. J. Donachie & S. J. Donachie 2002).

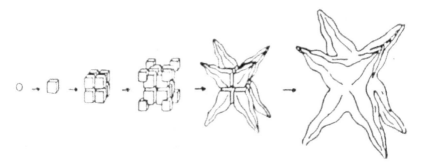

Fig. 4. Schematic showing the evolution of $\gamma'$ morphology during continuous cooling. Sphere → cube → ogdoadically diced cubes → octodendrite → dendrite (Durand–Charre 1997).

In niobium-strengthened nickel-base superalloys - such as IN-718 - the principal strengthening phase is $\gamma''$ (Ni$_3$Nb), which has a bct ordered DO22 structure. When $\gamma''$ precipitates are observed, they form as disk-shaped precipitates on {100} planes with a thickness of 5-9 nm and an average diameter of 60 nm (Durand–Charre 1997).

The next structural components are MC type primary carbides (created by such elements as Cr and Ti) and M$_{23}$C$_6$ type secondary carbides (created by such elements as Cr, Co, Mo and W). However, except of these structural components, "unwanted" TCP (Topologically

Close-Packed) phases are also presented, such as σ-phase $A_xB_y$ (Cr, Mo)$_x$(Fe, Ni)$_y$, μ–phase $A_7B_6$ (Co, Fe, Ni)$_7$(Mo, W, Cr)$_6$, Laves phases $A_2B$ (Fe, Cr, Mn, Si)$_2$(Mo, Ti, Nb) and $A_3B$ phases (π Ni$_3$(AlTa), η Ni$_3$Ti, δ Ni$_3$Ta and ε (NiFeCo)$_3$(NbTi)). The shape and size of these structural components have a significant influence on final the mechanical properties of alloys and - mainly - on creep rupture life.

Although alloy-specific heat treatments are generally proprietary, the typical heat treatment of nickel-base superalloys consists of a solution treatment followed by an aging step (precipitation and coarsening). For additional details on alloy-specific heat treatments, see Sims et al. (1987) and M. J. Donachie & S. J. Donachie (2002). Nickel-base superalloys are highly-alloyed: because of the complexity which this adds, many experimental studies use binary or ternary alloys as model alloy systems. The nickel-rich region of the binary Ni-Al alloy system is frequently used as a model alloy system. The Al-Ni phase diagram is shown in Figure 5, and we will use it to consider the typical heat treatments of nickel-base superalloys.

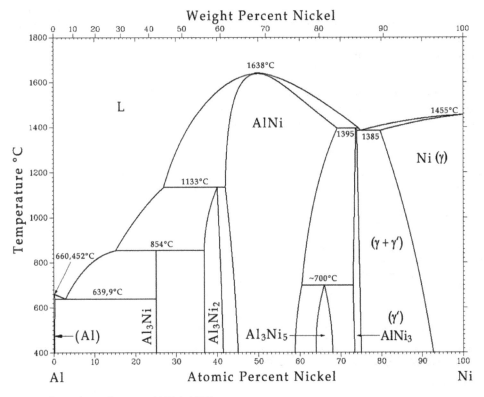

Fig. 5. Al-Ni phase diagram (ASM, 1992).

The solution treatment occurs above the γ′ solvus and is required for the ordered γ′ to go into the solution. The γ′ solvus separates the γ + γ′ and γ regions in Figure 5. This is usually followed by a quench (air, water or oil, depending upon the alloy) to room temperature. The aging occurs at a temperature below the solvus temperature and allows for the homogeneous nucleation, growth and coarsening of $\gamma'$, followed by air or furnace cooling to room temperature. Although heterogeneous nucleation is observed on grain boundaries and dislocations - for example - nucleation is primarily homogeneous. The temperature and duration of the aging treatment are selected so as to optimise the morphology, alignment and size distribution of $\gamma'$ precipitates. The resulting microstructure, in addition to its dependence on heat-treatment parameters, is also dependent on the physical properties of the alloy (and their isotropic or anisotropic nature) such as the lattice mismatch, the coherent $\gamma'$ interface energy, the volume fraction of $\gamma'$ and the elastic properties of the matrix and precipitate.

Polycrystalline turbine blades typically work within a temperature range from 705°C up to 800°C. As such, they must be protected from heat by various heat-proof layers; for example an alitise layer, MCrAlY coating or TBC (Thermal Barrier Coating). For this reason, dendrite arm-spacing, carbide size and distribution, morphology, the number and value of the $\gamma'$-phase and protective layer degradation are very important structural characteristics for the prediction of a blade's lifetime as well as the aero engine itself. In this chapter, the methods of quantitative metallography (Image Analyzer software NIS – Elements for carbide evaluation, the measurement of secondary dendrite arm-spacing and a coherent testing grid for $\gamma'$-phase evaluation) are used for the evaluation of the structural characteristics mentioned above on experimental material – Ni base superalloy ŽS6K.

For instance, a precipitate $\gamma'$ size greater than 0.8 µm significantly decreases the creep rupture life of superalloys and a carbide size greater than 5 µm is not desirable because of the initiation of fatigue cracks (Cetel, A. D. & Duhl, D. N. 1988).

For this reason, the needs of new methods of the evaluation of non–conventional structure parameters were developed. Quantitative metallography, deep etching and colour-contrast belong to the basic methods. The analysis of quantitative metallography has a statistical nature. The elementary tasks of quantitative metallography are:

- Dendrite arm-spacing evaluation;
- Carbide size and distribution;
- Volume ratio of evaluated gamma prime phase;
- Number ratio of evaluated gamma prime phase;
- Size of evaluated gamma prime phase;
- Protective alitise layer degradation.

The application of quantitative metallography and colour contrast on the ŽS6K Ni–base superalloy are the main objectives discussed in this chapter.

## 2. Description of experimental methods and experimental material

### 2.1 Experimental methods

For the evaluation of structural characteristics the following methods of quantitative metallography were used:

- Carbide distribution and average size was evaluated by the software NIS-Elements;
- Secondary dendrite arm-spacing measurement;

- For the number of γ'-phase particles, a coherent testing grid with 9 square shape area probes was used;
- For the volume of γ'-phase particles, a coherent testing grid with 50 dot probes made from backslash crossing was used.

Secondary dendrite arm spacing was evaluated according to Figure 6 and calculated with formula (1). The changing of the distance between the secondary dendrite arms "d" is an important characteristic because of base material, matrix γ, degradation via the equalising of chemical heterogeneity and also grain size growing.

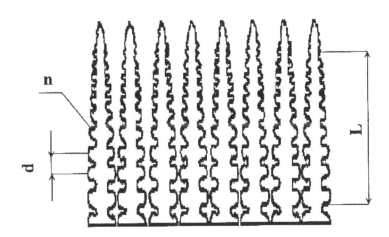

Fig. 6. Scheme for the evaluation of secondary dendrite arm-spacing.

$$d = \frac{L}{n} \cdot \frac{1}{z} \cdot 1000 \quad (\mu m) \tag{1}$$

- where "L" is a selected distance on which secondary arms are calculated (the distance is usually chosen with the same value as used magnification "z" – the reason why this is so is in order to simplify the equation), "n" is the number of secondary dendrite arms and "z" is the magnification used.

For the evaluation of the γ- and γ'-phases the method of coherent testing grid was used, and the number of γ' "N" was evaluated by a grid with 9 square-shaped area probes (Figure 7a) and the volume of γ' "V" was evaluated by grid with 50 dot probes (Figure 7b). Afterwards, measurement of the values was calculated with formulas (2) and (3). For a detailed description of the methods used, see (Skočovský & Vaško 2007, Tillová & Panušková 2008, Tillová et al. 2011). The size of γ' is also important from the point of view of creep rupture life. A precipitate with a size higher than 0.8 μm can be considered to be heavily degraded and as causing decreasing mechanical strength at higher temperatures.

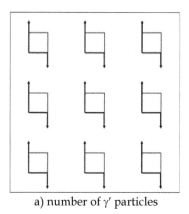

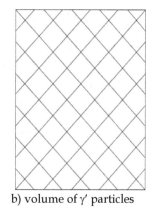

a) number of γ′ particles               b) volume of γ′ particles

Fig. 7. Coherent testing grid for γ′ evaluation.

$$N = 1{,}11 \cdot z^2 \cdot x_{str} \cdot 10^{-9} \quad \left( \mu m^{-2} \right)$$

(2)

- where "N" is a number of γ′ particles, "z" is the magnification used, "$x_{str}$" is the medium value of γ′-phase measurements.

$$V = \frac{n_s}{n} \cdot 100 \quad (\%)$$

(3)

- where "V" is a volume of γ′ particles, "$n_s$" is the medium value of γ′-phase measurement and "n" is a number of dot probes (when using a testing grid with 50 dot probes, the equation become more simple: $V = 2n_s$).

## 2.2 Experimental material

The cast Ni–base superalloy ŽS6K was used as an experimental material. Alloy ŽS6K is a former USSR superalloy which was used in the DV–2 jet engine. It is used for turbine rotor blades and whole-cast small-sized rotors with a working temperature of up to 800 ÷ 1050°C. The alloy is made in vacuum furnaces. Parts are made by the method of precise casting. The temperature of the liquid at casting in a vacuum to form is 1500 ÷ 1600°C, depending on the part's shape and its quantity. The cast ability of this alloy is very good, with only 2 ÷ 2.5% of shrinkage. Blades made of this alloy are also protected against hot corrosion, with a protective heat-proof alitise layer, and so they are able to work at temperatures of up to 750°C for 500 flying hours.

This alloy was evaluated at the starting stage, the stage with normal heat treatment after 600, 1000, 1500 and 2000 hours of regular working (for these evaluations, real ŽS6K turbine blades with a protective alitise layer were used as an experimental material), and different samples made from the same experimental material ŽS6K after annealing at 800 °C/ 10 and 800 °C/15 hours. This was followed by cooling at various rates in water, oil and air. The chemical composition in wt % is presented in Table 1.

A typical microstructure of the ŽS6K Ni–base superalloy as cast is shown by Figures 8 and 9. The microstructure of the as–cast superalloy consists of significant dendritic segregation

| C | Ni | Co | Ti | Cr | Al | W | Mo | Fe | Mn |
|---|----|----|-----|-----|-----|-----|-----|-----|-----|
| 0.13 ÷0.2 | Bal. | 4.0 ÷ 5.5 | 2.5 ÷ 3.2 | 9.5 ÷ 12 | 5.0 ÷ 6.0 | 4.5 ÷ 5.5 | 3.5 ÷ 4.8 | 2 | 0.4 |
| Adulterants | | | | | | | | | |
| P | | S | | Pb | | | Bi | | |
| 0.015 | | 0.015 | | 0.001 | | | 0.0005 | | |

Table 1. Experimental alloy's chemical composition.

caused by chemical heterogeneity (Fig. 8a) and particles of primary MC and secondary $M_{23}C_6$ carbides (Fig. 8b). Primary carbides MC (where M is (Ti, Mo and W)) are presented as block-shaped particles, mainly inside grains. Secondary carbides are presented by "Chinese" script-shaped particles on grain boundaries.

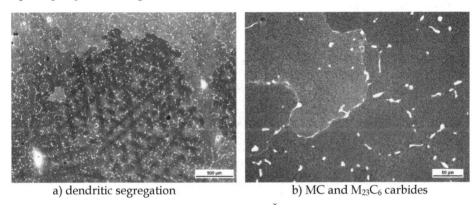

a) dendritic segregation          b) MC and $M_{23}C_6$ carbides

Fig. 8. Microstructure of as–cast Ni–base superalloy ŽS6K, Beraha III.

However, the microstructure also contains a solid solution of elements in the nickel matrix – the so-called γ-phase (Ni (Cr, Co and Fe)) and strengthening-phase, which is a product of artificial age–hardening and has a significant influence on mechanical properties and creep rupture life – so-called γ'-phase (gamma prime, Ni3 (Al and Ti)), Fig. 9a. Of course, both of these phases - γ (gamma) and γ' (gamma prime) - create an eutectic γ/γ', Fig. 9b.

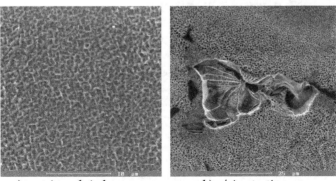

a) matrix and γ' phases          b) γ/γ' eutectic

Fig. 9. Ni–base superalloy ŽS6K microstructure, as–cast.

## 3. Experimental results and discussion

### 3.1 Carbide evaluation

Polycrystalline and columnar grain alloys contain carbon additions to help improve grain–boundary strength and ductility. While the addition of carbon is beneficial to grain boundary ductility, the large carbides that form can adversely affect fatigue life. Both low-and high-cycle fatigue-cracking have been observed to initiate with the large (lengths greater than 0.005 mm) carbides presented in these alloys. When polycrystalline alloys were cast in a single crystal form, it was determined that carbides did not impart any beneficial strengthening effects in the absence of grain boundaries, and thus could be eliminated. Producing essentially carbon–free single crystal alloys led to significant improvements in fatigue life as large carbide colonies were no-longer present to initiate fatigue cracks (Cetel, A. D. & Duhl, D. N. 1988).

The first characteristic were carbide size and its distribution evaluated. Specimens made of the ŽS6K superalloy were compared at the starting stage (non-heat-treated, as-cast) after 800°C/10 hrs and 800°C/15 hrs. The cooling rate depends on the cooling medium; in our case these were air, oil and water. The results for the ratio of carbide particles in the observed area are in Figure 10 and the results on the average carbide size are in Figure 11.

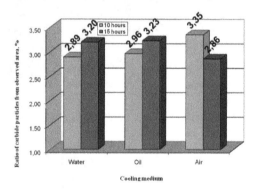

Fig. 10. The ratio of carbide particles from the observed area.

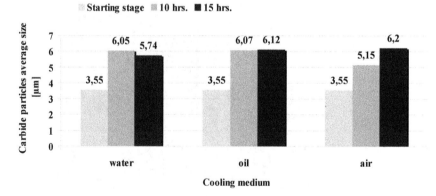

Fig. 11. Average carbide size [µm].

From the relations presented (Figure 11) it is obvious that the holding time on various temperatures for annealing and cooling in selected mediums does not have a significant influence on carbide particle size. More significant, the influence on the ratio of carbide particles has a cooling rate (Figure 10). With increasing speed of cooling and a longer holding time on the annealing temperature, the carbide particles' ratio decreases.

Generally, we can suppose that carbide particles are partially dissolved with the temperature of annealing and elements, which are consider as an carbide creators (in this case mainly Ti) have create a new particles of $\gamma'$ phase. This phenomenon has an influence on decreasing the segregated carbide percentage ratio. With an increase of the cooling rate (water, oil), an amount of the $\gamma'$-phase decreases and the carbides percentage ratio is higher. At slow cooling and a longer time of holding is higher amount of $\gamma'$ segregate and, therefore, the ratio of carbides decreases. It is all happen according to scheme:

$$MC + \gamma \rightarrow M_{23}C_6 + \gamma'$$

The microstructures which are equivalent to these evaluations are in Figures 12 and 13. For carbide evaluation, etching is not necessary. All of the micrographs are non-etched.

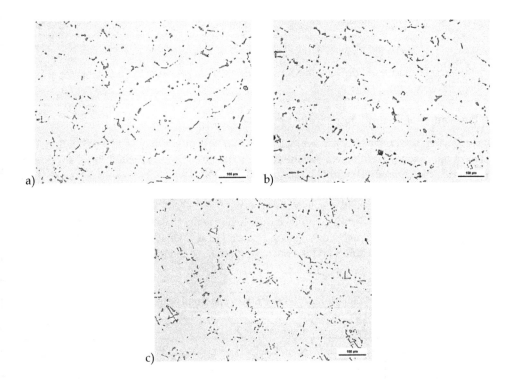

Fig. 12. Microstructure of ŽS6K, carbides ratio after 800°C annealing/10 hrs: a) water cooling; b) oil cooling; c) air cooling.

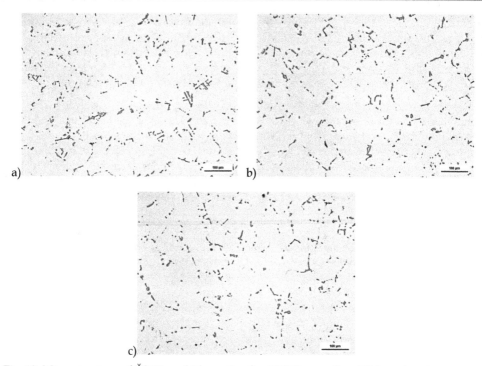

Fig. 13. Microstructure of ŽS6K, carbides ratio after 800°C annealing/15 hrs: a) water cooling; b) oil cooling; c) air cooling.

## 3.2 Evaluation of secondary dendrite arm-spacing

The second characteristic which is evaluated is dendrite arm-spacing. In this evaluation, two different approaches were taken. For the first evaluation, non-heat treated ŽS6K specimens were used and compared with loading at 800°C/10(15) hrs. The results of these first evaluations can be seen in Table 2 and Figures 14, 15 and 16. The second evaluation was performed on ŽS6K turbine blades used in the DV-2 (LPT – Low Pressure Turbine and HPT – High Pressure Turbine) aero jet engine at the starting stage (basic heat treatment) and after an engine exposition (at real working temperatures) for 600, 1000, 1500 and 2000 hours. Again, the results are in Table 3 and the microstructures are in Figure 17.

| Secondary dendrite arm spacing [µm] | | | |
|---|---|---|---|
| ŽS6K – starting stage | 185.19 | | |
| Cooling medium | | | |
| | Water | Oil | Air |
| ŽS6K/10hrs. | 126.58 | 131.58 | 138.89 |
| ŽS6K/15hrs. | 113.64 | 131.58 | 156.25 |

Table 2. Results from secondary dendrite arm-spacing for a non-heat treated ŽS6K alloy

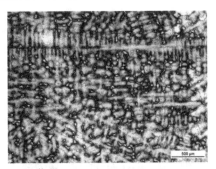

Fig. 14. Dendritic segregation of ŽS6K, starting stage, Marble etchant.

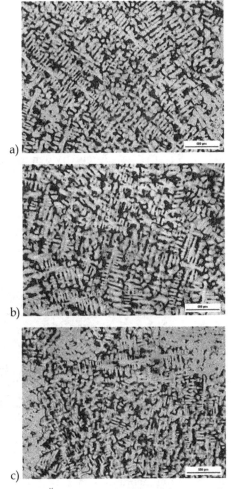

Fig. 15. Dendritic segregation of ŽS6K, 800°C/10 hrs: a) water cooling; b) oil cooling,; c) air cooling, Marble etchant.

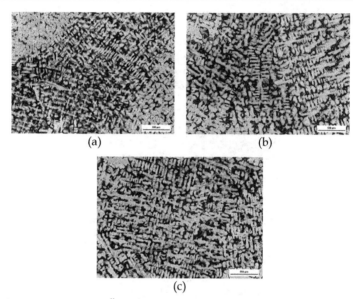

Fig. 16. Dendritic segregation of ŽS6K, 800°C/15 hrs.: a) water cooling, b) oil cooling, c) air cooling, Marble etchant.

| Type of blade | Secondary dendrite arm-spacing [µm] |
|---|---|
| Blade of 1°LPT – starting stage | 24.38 |
| Blade of HPT - 600 hrs. | 24.78 |
| Blade of HPT - 1000 hrs. | 27.98 |
| Blade of HPT - 1500 hrs. | 48.73 |
| Blade of HPT - 2000 hrs. | 66.66 |

Table 3. Results from secondary dendrite arm spacing for real turbine blades, heat-treated ŽS6K alloy.

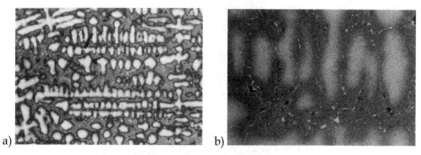

Fig. 17. Dendritic segregation of ŽS6K turbine blades: a) 1°LPT – starting stage; b) HPT – after 1500 hrs of work, Marble etchant.

The cast materials are characterised by dendritic segregation, which is caused by chemical heterogeneity. With the influence of holding at an annealing temperature, chemical heterogeneity decreases. This means that the distance between secondary dendrite arms increases (the dendrites are growing). From the results mentioned above (Table 2), it is clear to see that with a higher cooling rate comes a slowing of the diffusion processes and the dendrite arm-spacing decreases in comparison with the starting stage (Figure 14). All of these changes are also obvious in Figures 15 and 16. The ŽS6K dendrite arm-spacing increases in relation to the annealing time, with an annealing temperature and cooling medium of between 113.64 and 156.25 µm.

The same phenomena can be observed with heat-treated turbine blades after various working times. Of course, the secondary dendrite arm-spacing is smaller, but again it has a tendency to growth. So, this confirms the results from Table 2: that a longer time of exposure has a significant influence on dendrite and grain size.

### 3.3 Evaluation of γ′ morphology

Since the advanced high-strength nickel–base alloys owe their exceptional high temperature properties to the high volume fraction of the ordered γ′-phase that they contain, it should not be surprising that control of precipitate distribution and morphology can profoundly affect their properties. The post-casting processing of these alloys - especially solution heat treatment - can radically affect microstructure.

The high-strength alloys typically contain about $55 \div 75$ % of the γ′ precipitates which, in the cast condition, are coarse ($0.4 \div 1.0$ µm) and irregularly-shaped cuboid particles (see Figure 9).

The evaluation of the γ′-phase is also divided into two parts, just as the dendrite evaluation was. Firstly, the γ′-phase was evaluated on the cast stage, and secondly on turbine blades. The characteristics of γ′-phase morphology were also measured using the coherent testing grid methods. As was mentioned above, the number and volume of the γ′-phase have a significant influence on the mechanical properties of this alloy, especially on creep rupture life. The average satisfactory size of the γ′-phase is about 0.35–0.45 µm (Figure 18) and also the carbide size should not exceed a size of 5 µm because of fatigue crack initiation (M. J. Donachie & S. J. Donachie 2002). Another risk in using high temperature loading (or annealing) is the creation of TCP phases - such σ-phase or Laves-phase - within the temperature range of 750 °C–800 °C. The results of first evaluation are in Table 4. The microstructures related to this evaluation are in Figures 19 and 20.

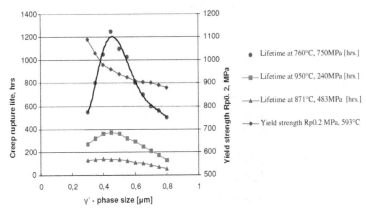

Fig. 18. Influence of γ′-phase size on the lifetime and mechanical properties of Ni superalloy.

| Cooling medium | Number of γ′ - phase N [μm⁻²] | Volume of γ′ - phase V [%] | Average size of γ′ - phase u [μm] |
|---|---|---|---|
| Start. stage | 2.47 | 39.4 | 0.61 |
| 10h water | 1.95 | 56.2 | 0.54 |
| 10h oil | 1.60 | 63 | 0.63 |
| 10h air | 1.50 | 72.4 | 0.69 |
| 15h water | 1.90 | 66.8 | 0.59 |
| 15h oil | 1.59 | 71.8 | 0.67 |
| 15h air | 1.49 | 76.6 | 0.72 |

Table 4. Results from γ′-phase evaluation at the cast stage at 800°C/10 (15) hrs.

With exposure for 10 hours at an annealing temperature, the volume of γ′-phase was increased by about 16.8–33% when compared with the starting stage (Figure 19). The significant increase of the γ′-phase was observed at a holding time of 15 hours (Figure 20), and cooling on air, where volume of γ′-phase is 76.6 %.

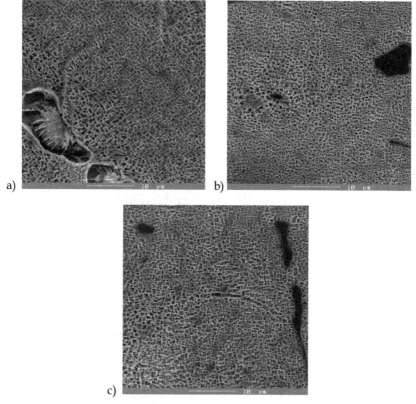

Fig. 19. Morphology of γ′-phase after 800°C/10 hrs: a) air cooling; b) oil cooling; c) water cooling, Marble etchant, SEM.

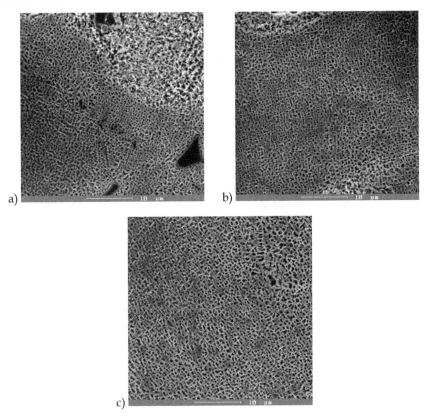

Fig. 20. Morphology of γ'-phase after 800°C/15 hrs: a) air cooling; b) oil cooling; c) water cooling, Marble etchant, SEM.

The highly alloyed nickel–base alloys solidify dendritically and, due to the effects of chemical segregation across the dendrites, a higher concentration of the γ'-phase forms elements such as aluminium and titanium which are more present in the inter-dendritic areas than in the dendrite core. This results in the γ' solvus (the temperature at which γ' first precipitates upon cooling) being lower in the core region than the inter-dendritically region.

Varying the cooling rate from the solution heat treatment temperature can significantly affect the γ' particle size, as rapid rates do not allow sufficient time for the particles to coarsen as they precipitate upon cooling below the γ' solvus temperature. Increasing the cooling rate of the solution heat treatment temperature from 30 to 120°C/minute results in an average particle size refinement of more than 30% (Figure 21) (Cetel, A. D. & Duhl, D. N. 1988).

By controlling both the solution heat treatment and the cooling rate, both the volume fraction of the fine particles as well as their size can be controlled. Heat treating an alloy close to its γ' solvus temperature and completely dissolving its coarse γ' particles can produce consistently high-elevated temperature creep–rupture strength.

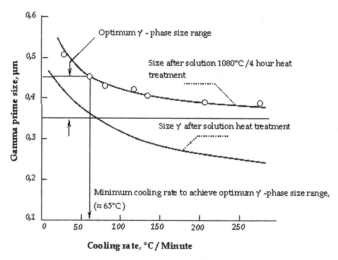

Fig. 21. Optimum γ′ size achieved by rapid cooling of the solution temperature combined with post-solution heat treatment.

Work performed by (Nathal et al. 1987) indicates that the optimum γ′-phase size for an alloy is dependent on the lattice mismatch between the γ- and γ′-phases (Figure 22), which is composition dependent.

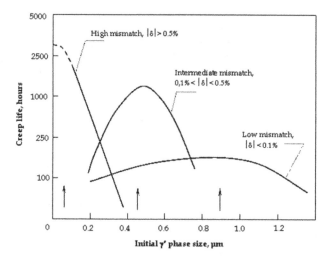

Fig. 22. The optimum γ′-phase size to maximize creep strength is dependent on mismatch between the γ- and γ′-phases (Nathal et al. 1987).

The second evaluation of the γ′-phase was provided on heat-treated turbine blades of a DV-2 aero jet engine after various working times. The results obtained are shown by Table 5. For the evaluation a coherent testing grid was used - the same procedure as in the first evaluation. The microstructures related to this evaluation are shown in Figures 23 and 24.

| Time of work [hours] | Number of γ'-phase N [μm⁻²] | Volume of γ'-phase V [%] | Average size of γ'-phase u [μm²] |
|---|---|---|---|
| 0 | 0.98542 | 67.2 | 0.6819 |
| 600 | 1.1242 | 67.6 | 0.60131 |
| 1000 | 1.1004 | 59 | 0.53615 |
| 1500 | 0.81938 | 57.4 | 0.7005 |
| 2000 | 0.6968 | 40.6 | 0.5826 |

Table 5. Results from the γ'-phase evaluation on heat-treated turbine blades at various working times.

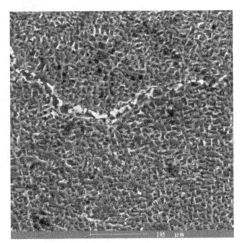

Fig. 23. Morphology of the γ'-phase heat-treated turbine blade, starting stage (0 hours), HCl + $H_2O_2$ etchant.

The morphology of the γ'-phase at the starting stage is cuboid and distributed equally in the base γ matrix (Figure 23.). With an increase of the hours of work at a temperature of up to 750°C, the γ'-phase morphology changes. The particles of the γ'-phase gradually coarsen (time of work to 1000 hours, Figure 24 a, b), which confirms the results of a number of γ'-phase evaluations "N" (see Table 5). A decrease of this value at a longer duration of work (1500 and 2000 hours) is caused by reprecipitation of new, fine γ'-phase particles in the area between the primal γ'-phase (Figure 24 c, d). From the results in Table 5, it is obvious that the γ'-phase of the ŽS6K alloy coarsen uniformly and increase its volume ratio in the structure after up to 1000 hours of exposition (regular work of a jet engine). However, after longer durations of work (1500 or 2000 hours) there occurs the reprecipitation of new, fine particles of the γ'-phase in the free space of matrix and which has caused structural heterogeneity.

In terms of structure degradation and the prediction of the life time of turbine blades - as well as the jet engine itself – and according to the results in Table 5, after up to 1000 hours of exposition the structure (with "N" = 1.1004, "V" = 59 and average size "u" = 0.53615) is at the "edge" of use because of its mechanical properties, as shown by Figure 18. However, the γ'-phase size is not the only parameter influencing the life time. In addition, the number "N" and volume "V" is important from the point of view of dislocation hardening. When "N" and "V" are smaller, this means that the distances between single particles are greater and that fact causes a decrease of the dislocation hardening effect. On the other hand, $M_{23}C_6$ carbides form a carbide net on the grain boundary which also decreases the creep rupture life by developing brittle grain boundaries. For a comparison of the increasing distance between $\gamma'$-phase particles see Figure 25.

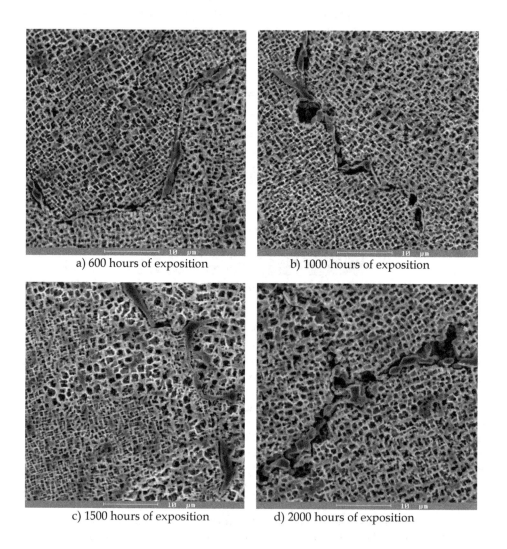

a) 600 hours of exposition    b) 1000 hours of exposition

c) 1500 hours of exposition    d) 2000 hours of exposition

Fig. 24. Morphology of the $\gamma'$-phase, heat treated turbine blade made of the ŽS6K alloy, after various exposition, $HCl + H_2O_2$ etchant.

|  |  |
|---|---|
| a) starting stage | b) 600 hours of exposition |
| c) 1500 hours of exposition | d) 2000 hours of exposition |

Fig. 25. Detail of the ŽS6K alloy's γ'-phase showing the increasing distance between γ' particles as an affect of working exposition - at normal working loading of a jet engine - which has a significant influence on the dislocation hardening effect.

### 3.4 Evaluation of the Al–Si protective layer

To improve the lifetime of turbine blades made from the ŽS6K alloy against a hot corrosion environment, an alitise Al–Si protective layer is used. What is important about this kind of layer is that it does not improve the high temperature properties of the base alloy but only its hot corrosion resistance.

An alitise layer is used for the protection of HPT blades (Figure 26) and only 1° of LPT blades, which means that Al-Si suspension is applied on to the surface of the blades. Silicon is added due to its ability to increase resistance to corrosion in sulphide and sea environments. Generally, an alitise layer AS-2 type is used for corrosion protection of aero gas turbine parts which work at temperatures of up to 950 °C; type AS-1 up to a temperature of 1100 °C (DV–2–I–62: Company standard). The standard procedure of applying Al–Si protective coating is in Table 6.

| Heat – treatment | Conditions |
|---|---|
| Homogenization annealing | In vacuum, temperature 1225 °C, holding 4 hrs, cooling with argon to 900 °C per 10 min. |
| Alitise AS2 | 1. Spraying of AS2 layer (AS2 – koloxylin solution 350 ml, Al – powder 112 g, Si – powder 112 g) |
|  | 2. Diffusion annealing temperature 1000 °C, 3 hrs, slowly cooled in retort |

Table 6. Steps for protective Al–Si coating as applied on to a ŽS6K turbine blade.

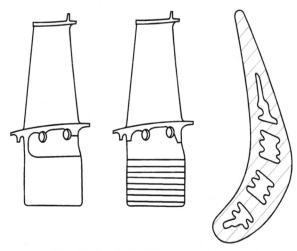

Fig. 26. A high pressure turbine blade of a DV–2 aero jet engine, left-side, right-side and cross-section with cooling chambers.

The Al–Si layer consists of two layers at the starting stage. The upper layer (the aluminium-rich layer) is created by aluminides - a complex compound of Si, Cr and Mo - and by carbides. The lower part of layer (the silicon rich layer) is created mainly by silicon and titanium carbides and the $\gamma$ matrix. The average thickness of the layer is 0.04 mm. According to the evaluation by metallography, the alitise layer is equally distributed across the whole blade surface at the starting stage (Figure 27 a, b).

In cases of overheating (here, at 1000°C - the normal working temperature is 705°C ÷ 750°C) the alitise layer is significantly degraded (Figure 27 c, d). The layer is non-homogeneous, with a rough surface and in place of the flow edge in the area of the flap pantile is a layer which is evenly broken (Figure 27d). Layer degradation is connected with the diffusion of elementary elements - such as Cr, Ti, Ni and Al - from the base material into the surface area (Table 7.). Where Cr and Ti creates carbides, Ni and Al form fine $\gamma'$ particles and Al as itself also creates NiAl ($\beta$–phase) and $Al_2O_3$ oxides on the surface of the layer. With decreasing of the layer's heat resistance, the base material is impoverished, which leads to the growth of $\gamma'$ particles and decreasing of its volume.

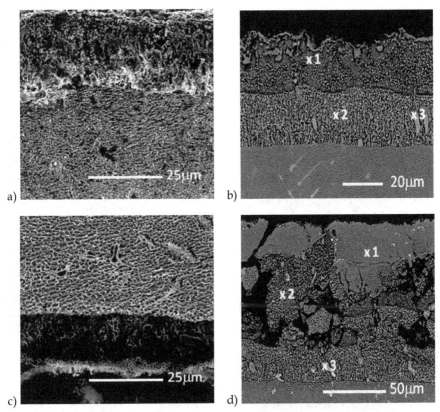

Fig. 27. Alitise layer a, b) starting stage; c, d) after overheating at 1000°C, SEM.

| Sample | Marked spots | AlK | SiK | MoK | TiK | CrK | CoK | NiK | Wk |
|---|---|---|---|---|---|---|---|---|---|
| | 1 | 19.7 | 2.21 | 0.23 | 0.83 | 6.54 | 4.14 | 65.5 | 0.34 |
| Starting stage, Fig. 27b | 2 | 15.01 | 3.09 | 2.56 | 2.28 | 5.16 | 4.06 | 65 | 2.27 |
| | 3 | 1.57 | 7.57 | 17.51 | 5.82 | 13.53 | 3.01 | 28.45 | 21.94 |
| | 1 | 9.04 | 5.94 | 7.42 | 0.92 | 9.25 | 4.15 | 52.6 | 10.6 |
| Overheating at 1000°C, Fig. 27d | 2 | 18.2 | 3.55 | 3.56 | 0.73 | 6.18 | 3.81 | 58.2 | 5.7 |
| | 3 | - | 9.54 | 13.5 | 9.5 | 11.3 | 3.25 | 33.1 | 19.6 |

Table 7. Spot analysis of selected particles. The marked spots (in wt%) correspond with Figure 27b, d.

The alitise layer on the blades which have worked at regular conditions is also degraded, which is represented by the changing of the layer thickness and the surface relief. Changes in layer thickness are caused by heterogeneity of the temperature field along the blade and the abrasive and erosive effect of gases and exhaust gases. The level of layer degradation varies, depending on the area of blade. From a metallographic point of view, the highest degradation is in the flap pantile region close to the flow edge, in the case of blades after

1500 and 2000 hours of work (Figure 28 c, d). In region close to the Si sub-layer, needle particles (probably a special form of Cr base carbides) are created which grow depending upon the time of work (compare Figures 27 a, b and 28). These needle particles start to form after 600 hours of loading, which means that after 600 (Figure 28 a, b) hours of work and aero jet engine should to be taken in for overhauling and the old alitise layer replaced by a new one. However, when it comes to the local overheating of the turbine blades, all of the degradation processes are much faster.

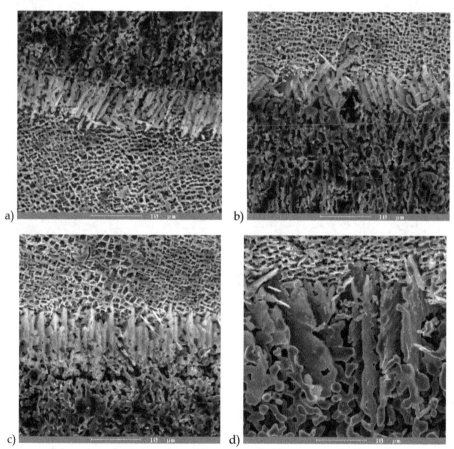

Fig. 28. Creation of needle particles in the region under the Si sub-layer: a) 600 hours; b) 1000 hours; c) 1500 hours; d) 2000 hours of regular work, SEM, Marble etchant.

## 4. Conclusion

As cast Ni–base ŽS6K superalloy was used as an experimental material, the structural characteristics were evaluated from the starting stage of the sample, after annealing at 800 °C/10 and 800 °C/15 hrs and after various working times in real jet engines with the use of the methods of quantitative metallography. The results are as follows:

- The structure of the samples is characterised by dendritic segregation. In dendritic areas, fine $\gamma'$-phase is segregated. In inter dendritic areas, eutectic cells $\gamma/\gamma'$ and carbides are segregated.
- The holding time (10–15 hrs.) has a significant influence on the carbide particles' size. The size of the carbides is under a critical level for the initiation of fatigue crack only at the starting stage. An increase in the rate of cooling has a significant effect on the carbide particles' ratio.
- The chemical heterogeneity of the samples with a longer holding time decreases. This is a reason of the fact that there is sufficient time for the diffusion mechanism, which is confirmed by the measurement results of secondary dendrite arm-spacing.
- The volume of the $\gamma'$-phase with a longer holding time increases and the $\gamma'$-phase size grows. With a higher rate of cooling the $\gamma'$ particles become finer.
- There was no evidence of the presence of TCP phase even at a high annealing temperature.
- Cooling rate also has an influence on the hardness. At a lower rate of cooling, the internal stresses are relaxed, which causes hardness to increase – a changing of the dislocation structure.

The cooling rates, represented by various cooling mediums, have a significant influence on the diffusion processes which are operating within the structure. These diffusion processes are the main mechanisms for the formation and segregation of carbide particles, the equalising of chemical heterogeneity (represented by dendrite arm-spacing) and segregation of the $\gamma'$-phase; they are also responsible for structural degradation of such alloys.

Air - as a cooling medium - provides sufficient time for the realisation of diffusion reactions and it leads to a decrease of chemical heterogeneity, which is presented by an increase of secondary dendrite arm-spacing. Also, this "slow" cooling rate has a positive effect on carbides' segregation and on the morphology, number and volume of the strength precipitate $\gamma'$ (the precipitate has a greater diameter and its volume increases).

Water is the most intensive cooling medium, which breaks diffusion processes and which leads to an increase of carbide particles in the observed area; the precipitate $\gamma'$ is smaller, increasing the hardness and at least also increasing the strength.

From a general point of view we can perform cooling in oil, which might be consider as a medium point between cooling in air and cooling in water.

For the turbine blades, which have been worked at normal loading and for various durations (600, 1000, 1500 and 2000 hours), the following results were achieved:

- The medium distance of secondary dendrite arms "d" grows in dependence on the time of work, caused by changes of the grain size of the $\gamma$-matrix.
- The gradual dissolving of primary carbides rests and the reprecipitation of secondary carbides on grain boundary. After longer durations of work (1000–2000 hours) it changes its chain morphology onto the carbide net, which has a significant influence on lowering the mechanical properties of the alloy.
- The inter-metallic phase-$\gamma'$ was evaluated with the methods of quantitative metallography; this evaluation shows gradual morphology changes of the $\gamma'$-phase – coarsening, spheroidisation and reprecipitation.

- The alitise layer degradation was expressed by a changing thickness and needle-like Cr carbide segregation at the sub-layer region, which has a negative influence on the layer's lifetime. There is strong recommendation for overhauling after every 500 hours of regular work.

## 5. Acknowledgment

The authors acknowledge the financial support of the projects VEGA No. 1/0841/11 and No. 1/0460/11; KEGA No. 220-009ŽU-4/2010 and European Union - the Project *"Systematization of advanced technologies and knowledge transfer between industry and universities (ITMS 26110230004)"*.

## 6. References

ASM. (1992). *ASM Handbook Volume 3: Alloy Phase Diagrams* (10th edition), ASM International, ISBN 0–871–70381–5, USA.

Cetel, A. D. & Duhl, D. N. (1988). Microstructure – Property Relationships In Advanced Nickel Base Superalloy Airfoil Castings, *2nd International SAMPE Metals Conference*, pp. 37–48, USA, August 2–4, 1988.

Donachie, M. J. & Donachie, S. J. (2002). *Superalloys – A technical Guide* (2nd edition), ASM International, ISBN 0–87170–749–7, USA.

Durand–Chare, M. (1997). *The Microstructure of Superalloys*, Gordon & Breach Science Publishers, ISBN 90–5699–097–7, Amsterdam, Netherlands.

DV–2–I–62: Company standard, Považské machine industry, Division of Aircraft Engine DV–2, Považská Bystrica, Slovakia, 1989.

MacSleyne, J. P. (2008). Moment invariants for two-dimensional and three-dimensional characterization of the morphology of gamma-prime precipitates in nickel-base superalloys, In: *Doctoral Thesis / Dissertation*, n.d., Available from: <http://www.grin.com/en/doc/263761/moment-invariants-for-two-dimensional-and-three-dimensional-characterization>

Nathal, M. V. (1987) *Met. Trans.*, Vol. 18 A, pp. 1961–1970.

Reed, R. C. (2006). *The superalloys: fundamentals and applications*, Cambridge University Press, ISBN 0–521–85904–2, New York, USA.

Sims, Ch. T., Stoloff, N. S. & Hagel, W. C. (1987). *Superalloys II* (2nd edition), Wiley-Interscience, ISBN 0–471–01147–9, USA.

Skočovský, P. & Vaško, A. (2007). *The quantitative evaluation of cast iron structure* (1st edition), EDIS, ISBN 978-80-8070-748-4, Žilina, Slovak Republic.

Tillová, E. & Panuškova, M. (2008). Effect of Solution Treatment on Intermetallic Phase's Morphology in AlSi9Cu3 Cast Alloy. *Mettalurgija/METABK*, No. 47, pp. 133-137, 1-4, ISSN 0543-5846.

Tillová, E., Chalupová, M., Hurtalová, L., Bonek, M., & Dobrzanski, L. A. (2011). Structural analysis of heat treated automotive cast alloy. *Journal of Achievements in Materials and Manufacturing Engineering/JAMME*, Vol. 47, No. 1, (July 2011), pp. 19-25, ISSN 1734-8412.

# One Dimensional Morphing Structures for Advanced Aircraft

Robert D. Vocke III[1], Curt S. Kothera[2], Benjamin K.S. Woods[1],
Edward A. Bubert[1] and Norman M. Wereley[1]
[1]*University of Maryland, College Park, MD*
[2]*Techno-Sciences, Inc., Beltsville, MD,*
*USA*

## 1. Introduction

Since the Wright Brothers' first flight, the idea of "morphing" an airplane's characteristics through continuous, rather than discrete, movable aerodynamic surfaces has held the promise of more efficient flight control. While the Wrights used a technique known as wing warping, or twisting the wings to control the roll of the aircraft (Wright and Wright, 1906), any number of possible morphological changes could be undertaken to modify an aircraft's flight path or overall performance. Some notable examples include the Parker Variable Camber Wing used for increased forward speed (Parker, 1920), the impact of a variable dihedral wing on aircraft stability (Munk, 1924), the high speed dash/low speed cruise abilities associated with wings of varying sweep (Buseman, 1935), and the multiple benefits of cruise/dash performance and efficient roll control gained through telescopic wingspan changes (Sarh, 1991; Gevers, 1997; Samuel and Pines, 2007).

While the aforementioned concepts focused on large-scale, manned aircraft, morphing technology is certainly not limited to vehicles of this size. In fact, the development of a new generation of unmanned aerial vehicles (UAVs), combined with advances in actuator and materials technology, has spawned renewed interest in radical morphing configurations capable of matching multiple mission profiles through shape change – this class has come to be referred to as "morphing aircraft" (Barbarino *et al.*, 2011). Gomez and Garcia (2011) presented a comprehensive review of morphing UAVs. Contemporary research is primarily dedicated to various conformal changes, namely, twist, camber, span, and sweep. It has been shown that morphing adjustments in the planform of a wing without hinged surfaces lead to improved roll performance, which can expand the flight envelope of an aircraft (Gern *et al.*, 2002), and more specifically, morphing to increase the span of a wing results in a reduction in induced drag, allowing for increased range or endurance (Bae *et al.*, 2005). The work presented here is intended for just such a one dimensional (1-D) span-morphing application, for example a UAV with span-morphing wingtips depicted in Figure 1. By achieving large deformations in the span dimension over a small section of wing, the wingspan can be altered during flight to optimize aspect ratio for different roles. Furthermore, differential span change between wingtips can generate a roll moment, replacing the use of ailerons on the aircraft (Hetrick *et al.*, 2007). This one dimensional morphing could also be used in the chordwise direction, and is not limited in application to fixed-wing aircraft, as rotorcraft would also benefit from a variable diameter or chord rotor.

Fig. 1. Illustration of span-morphing UAV showing 1-D morphing wingtips.

A key challenge in developing a one dimensional morphing structure is the development of a useful morphing skin, defined here as a continuous layer of material that would stretch over the morphing structure and mechanism to form a smooth aerodynamic skin surface. For a span-morphing wingtip in particular, the necessity of a high degree of surface area change, large strain capability in the span direction, and little to no strain in the chordwise direction all impose difficult requirements on any proposed morphing skin. The goal of this effort was a 100% increase in both the span and area of a morphing wingtip, or "morphing cell."

Reviews of contemporary morphing skin technology (Thill *et al.*, 2008; Wereley and Gandhi, 2010) yield three major areas of research being pursued: compliant structures, shape memory polymers, and anisotropic elastomeric skins. Compliant structures, such as the FlexSys Inc. Mission Adaptive Compliant Wing (MACW), rely on a highly tailored internal structure and a conventional skin material to allow small amounts of trailing edge camber change (Perkins *et al.*, 2004). Due to the large geometrical changes required for a span-morphing wingtip as envisioned here, metal or resin-matrix-composite skin materials are unsuitable because they are simply unable to achieve the desired goal of 100% increases in morphing cell span and area.

Shape memory polymer (SMP) skin materials are relatively new and have recently received attention for morphing aircraft concepts. They may at first glance seem highly suited to a span-morphing wingtip: shape memory polymers made by Cornerstone Research Group exhibit an order of magnitude change in modulus and up to 200% strain capability when heated past a transition temperature, yet return to their original modulus upon cooling. There have been attempts to capitalize on the capabilities of SMP skins, such as Lockheed Martin's Z-wing morphing UAV concept (Bye and McClure, 2007) and a reconfigurable segmented variable stiffness skin composed of rigid disks and shape memory polymer proposed by McKnight *et al.* (2010). However, electrical heating of the SMP skin to reach transition temperature proved difficult to implement in the wind tunnel test article and the

SMP skin was abandoned as a high-risk option. Additionally, the state-of-the-art of SMP technology does not appear to be well-suited for dynamic control morphing objectives.

With maximum strains above 100%, low stiffness, and a lower degree of risk due to their passive operation, elastomeric materials are ideal candidates for a morphing skin. Isotropic elastomer morphing skins have been successfully implemented on the MFX-1 UAV (Flanagan et al., 2007). This UAV employed a mechanized sliding spar wing structure capable of altering the sweep, wing area, and aspect ratio during flight. Sheets of silicone elastomer connect rigid leading and trailing edge spars, forming the upper and lower surfaces of the wing. The elastomer skin is reinforced against out-of-plane loads by ribbons stretched taught immediately underneath the skin, which proved effective for wind tunnel testing and flight testing. Morphing sandwich structures capable of high global strains have also been investigated (Joo et al., 2009; Bubert et. al., 2010; Olympio et al., 2010). However, suitable improvements over these structures, such as anisotropic fiber reinforcement and a better developed substructure for out-of-plane reinforcement, are desired for a fully functional morphing skin.

The present research therefore focuses on the development of a passive anisotropic elastomer composite skin with potential for use in a 1-D span-morphing UAV wingtip. The skin should be capable of sustaining 100% active strain with negligible major axis Poisson's ratio effects, giving a 100% change in surface area, and should also be able to withstand typical aerodynamic loads, assumed to range up to 200 psf (9.58 kPa) for a maneuvering flight surface, with minimal out-of-plane deflection. The following will describe the process of designing, building, and testing a morphing skin with these goals in mind, and will compare the performance of the final article to the initial design objectives.

## 2. Conceptual development

The primary challenge in developing a morphing skin suitable as an aerodynamic surface is balancing the competing goals of low in-plane actuation requirements and high out-of-plane stiffness. In order to make the skin viable, actuation requirements must be low enough that a reasonable actuation system within the aircraft can stretch the skin to the desired shape and hold it for the required morphing duration. At the same time, the skin must withstand typical aerodynamic loads without deforming excessively (e.g., rippling or bowing), which would result in degradation to the aerodynamic characteristics of the airfoil surface.

To achieve these design goals, a soft, thin silicone elastomer sheet with highly anisotropic carbon fiber reinforcement, called an elastomeric matrix composite (EMC), would be oriented such that the fiber-dominated direction runs chordwise at the wingtip, and the matrix-dominated direction runs spanwise (Figure 2a). Reinforcing carbon fibers controlling the major axis Poisson's ratio of the sheet would limit the EMC to 1-D spanwise shape change (Figure 2b). For a given skin stiffness, actuation requirements will increase in proportion to the skin thickness, $t_s$, while out-of-plane stiffness will be proportional to $t_s^3$ by the second moment of the area. To alleviate these competing factors, a flexible substructure is desired (Figure 2c) that would be capable of handling out-of-plane loads without greatly adding to the in-plane stiffness. This allows a thinner skin which, in turn, reduces actuation requirements. The combined EMC sheet and substructure form a continuous span-morphing skin.

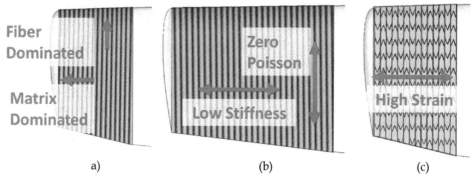

a)                                          (b)                                          (c)

Fig. 2. Design concept as a span morphing wingtip. (Bubert *et al.*, 2010)

To motivate the goal of low in-plane stiffness for this research, the skin prototype was designed to be actuated by a span-morphing pneumatic artificial muscle (PAM) scissor mechanism described separately by Wereley and Kothera (2007). The PAM scissor mechanism shown in Figure 3 was designed to transform contraction of the PAM actuator into extensile force necessary in a span-morphing wing. Based upon the maximum performance of the PAM and the kinematics of the scissor frame, the maximum force output of the actuation system was predicted and a skin stiffness goal was determined such that 100% active strain could be achieved, with the skin simplified as having linear stiffness. A margin of 15% was added to the 100% strain goal to account for anticipated losses due to friction or manufacturing shortcomings in the skin or actuation system.

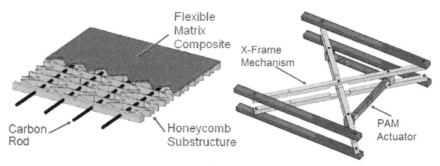

Fig. 3. Morphing skin demonstrator including PAM actuation system.

In addition, minimal out-of-plane deflection of the skin surface under aerodynamic loading was desired. No specific out-of-plane deflection goal was set or designed for, but out-of-plane stiffness of the substructure was kept in mind during the design process. Deflection due to distributed loads was included as a final test to ensure that the aerodynamic shape of a UAV wing morphing structure could be maintained during flight.

## 3. Skin development

The primary phase of the morphing skin development was to fabricate the EMC sheet that would make up the skin or face sheet. A number of design variables were available for tailoring the EMC to the application, including elastomer stiffness, durometer, ease of handling during manufacturing, and the quantity, thickness, and angle of carbon fiber reinforcement.

## 3.1 Elastomer selection

Initially, a large number of silicone elastomers were tested for viability as matrix material. Desired properties included maximum elongation well over 100%, a low stiffness to minimize actuation forces, moderate durometer to avoid having too soft a skin surface, and good working properties. Workability became a primary challenge to overcome, as two-part elastomers with high viscosities or very short work times would not fully wet out the carbon fiber layers. While over a dozen candidate elastomer samples were examined, only four were selected for further testing. Table 1 details the silicone elastomers tested as matrix candidates.

| Material | Modulus (kPa) | Viscosity (cP) | % Elongation at Break | Comments |
|---|---|---|---|---|
| DC 3-4207 | 130 | 430 | 100+ | difficult to demold |
| Sylgard-186 | 410 | 65,000 | 100+ | too viscous |
| V-330, CA-45 | 570 | 10,000 | 500 | excellent workability |
| V-330, CA-35 | 330 | 10,000 | 510 | excellent workability |

Table 1. Elastomer properties.

The most promising compositions tested were Dow Corning 3-4207 series and the Rhodorsil V-330 series. Both exhibited the desired low stiffness and greater than 100% elongation, but DC 3-4207 suffered from poor working qualities and lower maximum elongation and was not down-selected. Rhodorsil's V-330 series two-part room temperature vulcanization (RTV) silicone elastomer had the desired combination of low viscosity, long working time, and easy demolding to enable effective EMC manufacture, and also demonstrated very high maximum elongation and tear strength. V-330 with CA-35 had the lowest stiffness of the two V-330 elastomers tested. This led to selecting V-330, CA-35 for use in test article fabrication.

## 3.2 CLPT predictions and validation

Concurrently, using classical laminated plate theory (CLPT), a simple model of the EMC laminate was developed to study the effects of changing composite configuration on performance. The skin lay-up shown in Figure 4a was examined: two silicone elastomer face sheets sandwiching two symmetric unidirectional carbon fiber/elastomer composite laminae. The unidirectional fiber layers are offset by an angle $\theta_f$ from the 1-axis, which corresponds to the chordwise direction. Orienting the fiber-dominated direction along the wing chord controls minor Poisson's ratio effects while retaining low stiffness and high strain capability in the 2-axis, which corresponds to the spanwise direction.

In order to determine directional properties of the EMC laminate, directional properties of each lamina must first be found. The following micromechanics derivation comes from Agarwal et al. (2006). For a unidirectional sheet with the material longitudinal (L) and transverse (T) axes oriented along the fiber direction as shown in Figure 4b, we assume that

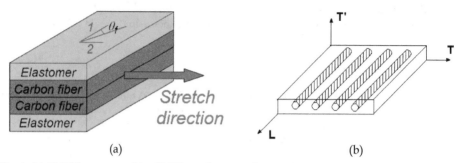

(a)                                                                    (b)

Fig. 4. (a) EMC lay-up used in CLPT predictions. (b) Unidirectional composite layer showing fiber orientation.

perfect bonding occurs between the fiber and matrix material such that equal strain is experienced by both fiber and matrix in the **L** direction. Based upon these assumptions, the longitudinal elastic modulus is given by the rule of mixtures:

$$E_{\mathrm{L}} = E_f V_f + E_m (1 - V_f) \qquad (1)$$

Here $E_{\mathrm{L}}$ is the longitudinal elastic modulus for the layer, $E_f$ is the fiber elastic modulus, $E_m$ is the matrix elastic modulus, and $V_f$ is the fiber volume fraction. To find the elastic modulus in the transverse direction, it is assumed that stress is uniform through the matrix and fiber. The equation for the transverse modulus, $E_T$, is:

$$E_{\mathrm{T}} = 1 / (V_f / E_f + (1 - V_f) / E_m) \qquad (2)$$

Calculations based on these micromechanics assumptions supported the intuitive conclusion that thinner EMC skins would have a lower in-plane stiffness modulus in the spanwise direction, $E_2$. Predictions for the transverse elastic modulus and the minor Poisson's ratio are plotted versus fiber offset angle in Figure 5a and Figure 5b, respectively, as solid lines. In order to provide some validation for the CLPT predictions, three EMC sample coupons were manufactured, consisting of 0.5 mm elastomer face sheets sandwiching two 0.2-0.3 mm composite lamina with a fiber volume fraction of 0.7. Nominal fiber axis offset angles of 0°, 10°, and 20° were used. The measured transverse modulus and minor Poisson's ratio are plotted as circles in their respective figure. As expected, increasing fiber offset angle increases the in-plane stiffness of the EMC, requiring greater actuation forces. Also, it is noteworthy that the inclusion of unidirectional fiber reinforcement at 0° offset angle nearly eliminates minor Poisson's ratio effects as predicted by CLPT theory.

It is of critical importance to note that, according to the assumptions used in deriving the lamina transverse modulus in Eq. (2), the transverse modulus has a lower bound equal to the matrix modulus. This lower bound is shown in Figure 5a as a horizontal black line at $E_2/E_m = 1$. However, the experimental data is close to this lower bound for the 10° and 20° samples, and the modulus is actually below the lower bound for the 0° case. Clearly in this case there is a problem in the micromechanics from which the transverse modulus prediction was derived.

Recall it was assumed that perfect bonding between fiber and matrix occurred, as illustrated in Figure 6a. This implies stress was equally shared between matrix and fiber under transverse loading. Close visual examination of the EMC samples during testing revealed that the fiber/matrix bond was actually very poor, and the matrix pulled away from individual fibers under transverse loading as illustrated in Figure 6b. Thus, the fibers carry no stress in the transverse direction, and the effective cross-sectional area of matrix left to carry transverse force in the lamina is reduced by the fiber volume fraction. For the case of poor transverse bonding exhibited in the fiber laminae, the transverse modulus in Eq. (2) can thus be simplified to:

$$E_T = E_m / (1 - V_f)$$ (3)

Using Eq. (3) to calculate transverse modulus for the fiber laminae, new CLPT predictions for EMC non-dimensionalized transverse modulus and minor Poisson's ratio are also plotted in Figure 5a and 5b, respectively. Much better agreement is seen between the analytical and experimental values for $E_2/E_m$. In spite of the poor bond between fiber and matrix material in the EMCs, the fiber stiffness still appears to contribute to the transverse stiffness at higher fiber offset angles. The minor Poisson's ratio is also influenced by the fiber offset angle. The EMC's longitudinal modulus, not shown, also remains high. These findings clearly indicate the fiber continues to contribute to the longitudinal stiffness of the fiber laminae even when bonding between matrix and fiber is poor.

To explain this contribution, it is hypothesized that friction between fiber and matrix help share load between the two materials in the longitudinal direction, while the matrix is free to pull away from the fiber in the transverse direction. This would explain the stiffening effect seen in the transverse modulus at increased offset angles and the controlling effect the fiber appears to have on Poisson's ratio at very low offset angles.

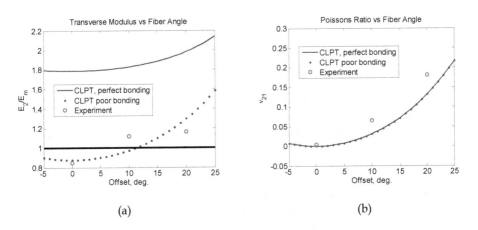

(a)                                                      (b)

Fig. 5. Comparison of CLPT predictions with experimental data for three different fiber angles (a) non-dimensionalized transverse elastic modulus $E_2/E_m$, (b) minor Poisson's ratio $\nu_{21}$.

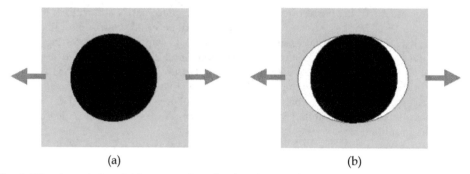

(a)                                                 (b)

Fig. 6. Fiber/matrix bond (a) assumed perfect bonding and equal transverse stress sharing in CLPT, (b) actual condition with poor fiber/matrix bond and no fiber stress under transverse loading.

Based upon these CLPT results, a fiber offset angle of 0° was selected to minimize transverse stiffness and also to minimize the minor Poisson's ratio. As the analytical and experimental results in Figure 5b indicate, a 0° fiber offset angle can resist chordwise shape change during spanwise morphing. While this conclusion appears obvious, the results demonstrate that with the appropriate correction to micromechanics assumptions in the transverse direction, simple CLPT analysis can be more confidently used to predict EMC directional properties. This simplifies the morphing skin design procedure by allowing in-plane EMC stiffness to be predicted by analytical methods.

### 3.3 EMC fabrication and testing

A key issue in this study was developing a dependable and repeatable skin manufacturing process. The final manufacturing process involved a multi-step lay-up process, building the skin up through its thickness (Figure 7). First, a sheet of elastomer was cast between two aluminum caul plates using shim stock to enforce the desired thickness. Secondly, unidirectional carbon fiber was applied to the cured elastomer sheet, with particular attention paid to the alignment of the fibers to ensure that they maintained their uniform spacing and unidirectional orientation (or angular displacement, depending on the sample). Enough additional liquid elastomer was then spread on top of the carbon to wet out all of the fibers. An aluminum caul plate was placed on top of the lay-up, compressing the carbon/elastomer layer while the elastomer cured. The third and final step in the skin lay-up process was to build the skin up to its final thickness. The bottom sheet of skin with attached carbon fiber was laid out on a caul plate. As in the first step, shim stock was used to enforce the desired thickness (now the full thickness of the skin) and liquid elastomer was poured over the existing sheet. A caul plate was then placed on top of this uncured elastomer and left for at least 4 hours. Once cured, the completed skin was removed from the plates, trimmed of excess material, and inspected for flaws. A successfully manufactured skin had a consistent cross-section and no air bubbles or visible flaws.

Several EMC sheets were originally manufactured in an effort to experimentally test the effect of fiber thickness and orientation on in-plane and out-of-plane characteristics and to attempt to optimize both. Table 2 describes the nominal dimensions and fiber angle values

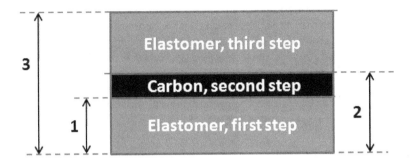

Fig. 7. Progression of skin manufacturing process.

for the three EMC samples. EMC #3 was not intended to be used in the final morphing skin demonstrator, but instead was an academic exercise intended to increase out-of-plane stiffness at the expense of in-plane stiffness.

|  | Sheet thickness (mm) | Fiber orientation (deg) | Fiber layer thickness (mm) | Total Thickness (mm) |
|---|---|---|---|---|
| EMC #1 | 0.5 | 0 | 0.4 | 1.4 |
| EMC #2 | 0.5 | 0 | 0.7 | 1.7 |
| EMC #3 | 0.5 | +15, 0, -15 deg | 0.8 | 1.8 |

Table 2. Summary of EMC sample properties.

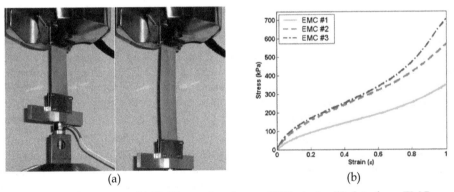

(a)　　　　　　　　　　　　　　　　(b)

Fig. 8. In-plane skin testing, (a) EMC sample taken to 100% strain; (b) data from EMC samples.

Sample strips measuring 51 mm x 152 mm were cut from the three EMCs and tested on a Material Test System (MTS) machine. Each sample was strained to 100% of its original length and then returned to its resting position. The test setup is depicted in Figure 8a and data from these tests are presented in Figure 8b. Notice the visibly low Poisson's ratio effects as the EMC is stretched to 100% strain in Figure 8a – there is little measurable reduction in width. It is also important to note that the stress-strain curves measured for each EMC reflect not only the impact of their lay-ups on stiffness, but also improvements in manufacturing ability. Thus, due to improved control of carbon fiber angles and the thickness of elastomer matrix, EMC #3 has roughly the same stiffness as EMC #2, in spite of the larger amount of carbon fiber present and higher fiber angles. EMC #1 exhibited high quality control and linearity of fiber arrangement and has the lowest stiffness of all, regardless of its nominal similarity to EMC #2. Based upon these tests, EMC #1 and EMC #2 were selected for incorporation into integrated test articles. EMC #1 displayed the lowest in-plane stiffness, while EMC #2 had the second lowest stiffness, making them the most attractive candidates for a useful morphing skin.

## 4. Substructure development

The most challenging aspect of the morphing skin to design was the substructure. Structural requirements necessitated high out-of-plane stiffness to help support the aerodynamic pressure load while still maintaining low in-plane stiffness and high strain capability.

### 4.1 Honeycomb design

The substructure concept originally evolved from the use of honeycomb core reinforcement in composite structures such as rotor blades. Honeycomb structures are naturally suited for high out-of-plane stiffness, and if properly designed can have tailored in-plane stiffness as well (Gibson and Ashby, 1988). By modification of the arrangement of a cellular structure, the desired shape change properties can be incorporated.

In order to create a honeycomb structure with a Poisson's ratio of zero, a negative Poisson's ratio cellular design presented by Chavez et al. (2003), or so-called auxetic structure (Evans et al., 1991), was rearranged to resemble a series of v-shaped members connecting parallel rib-like members, as seen in Figure 9. This arrangement gives large strains in one direction with no deflection at all in the other by means of extending or compressing the v-shaped members, which essentially act as spring elements. The chordwise rib members act as ribs in a conventional airplane wing by defining the shape of the EMC face sheet and supporting against out-of-plane loads. The v-shaped members connect the ribs into a single deformable substructure which can then be bonded to the EMC face sheet as a unit, with the v-shaped bending members controlling the rib spacing.

For a standard honeycomb, Gibson and Ashby (1988) describe the in-plane stiffness as a ratio of in-plane modulus to material modulus, given in terms of the geometric properties of the honeycomb cells. By modifying this standard equation, it is possible to describe the in-plane stiffness of a zero-Poisson honeycomb structure with cell geometric properties as illustrated in Figure 10a. Here $t$ is the thickness of the bending (v-shaped) members, $\ell$ is the length of the bending members, $h$ is the cell height, $c$ is the cell width, and $\theta$ is the angle between the rib members and the bending members. Note that in the figure the cell is being stretched vertically and $F$ is the force carried by a bending member under tension. Also note that the depth of the cell, denoted as $b$, is not represented in Figure 10.

With the geometry of the cell defined, an expression can be found for the honeycomb's equivalent of a stress-strain relationship. For small deflections, the bending member between points **1** and **2** can be considered an Euler-Bernoulli beam as shown in Figure 10b, with the forces causing a second mode deflection similar to a pure moment. From Euler-

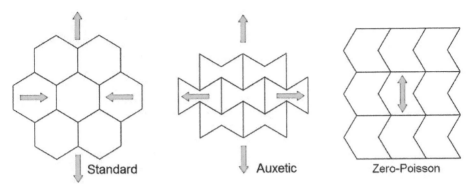

Fig. 9. Comparison of standard, auxetic, and modified zero-Poisson cellular structures showing strain relationships.

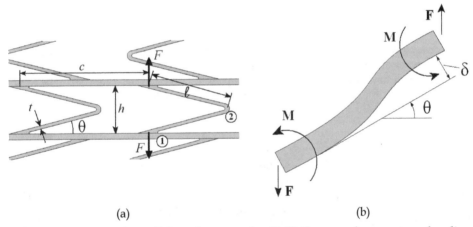

(a)                                                          (b)

Fig. 10. (a) Geometry of zero-Poisson honeycomb cell, (b) Forces and moments on bending member leg.

Bernoulli theory, the cosine component of the force $F$ will cause a bending deflection $\delta$ (Shigley *et al.*, 2004):

$$\delta = \frac{F\cos\theta\ell^3}{12E_0I} \tag{4}$$

Here $E_0$ is the Young's Modulus of the honeycomb material and $I$ is the second moment of the area of the bending member; in this case $I = bt^3/12$. In order to determine an effective tensile modulus for the honeycomb substructure, the relationship in Eq. (4) between force and displacement needs to be transformed into an equivalent stress-strain relationship. The equivalent stress through one cell can be found by using the cell width $c$ and honeycomb depth $b$ to establish a reference area, and the global equivalent strain is determined by non-dimensionalizing the v-shaped member's bending deflection $2\delta$ by the cell height $h$. These equivalent stresses and strains are used to determine a transverse stiffness modulus for the honeycomb, $E_2$:

$$\sigma_2 = \frac{F}{cb},$$
(5)

$$\varepsilon_2 = \frac{\delta \cos\theta}{h/2},$$
(6)

$$E_2 = \frac{\sigma_2}{\varepsilon_2}$$
(7)

Substituting Eqns. (5) through (7) into Eq. (4) and simplifying yields the following expression for the stiffness of the overall honeycomb relative to the material modulus:

$$\frac{E_2}{E_0} = \left(\frac{t}{l}\right)^3 \frac{\sin\theta}{\frac{c}{l}\cos^2\theta}$$
(8)

Because this modified Gibson-Ashby model assumes the bending member legs to be beams with low deflection angles and low local strains, Eq. (8) should only be valid for global strains that result in small local deflections. However, it will be shown that due to the nature of the honeycomb design, relatively large global strains are achievable with only small local strains.

With this fairly simple equation, the cell design parameters can easily be varied and their effect on the overall in-plane stiffness of the structure can be studied. For fixed values of $t$, $h$, $c$, and $b$, the modulus ratio of the structure, $E_2/E_0$, increases with the angle $\theta$. Noting the definitions in Figure 10a, it can be seen that decreasing $\theta$ consequently affects the bending member length $l$, as the upper and lower ends must meet to form a viable structure. Thus, for a given cell height $h$, minimum stiffness limitations are introduced into the design from a practicality standpoint in that the bending members must connect to the structure and cannot intersect one another. Lower in-plane stiffness can be achieved by increasing cell width to accommodate lower bending member angles.

In Figure 11a, an example is given of a zero-Poisson substructure designed in a commercial CAD software and produced on a rapid prototyping machine out of a photocure polymer. Using this method, a large number of samples could be fabricated with variations in bending member angle, $\theta$. By testing these structures on an MTS machine (Figure 11b), a comparison could be made between the predicted effect of bending member angle on in-plane stiffness and the actual observed effect.

The stress-strain test data from a series of rapid prototyped honeycombs is presented in Figure 12a. Each honeycomb was tested over the intended operating range, starting at a reference length of 67% of resting length (pre-compressed) and extending to 133% of resting length to achieve 100% total length change. To test the validity of the modified Gibson-Ashby model, comparisons of experimental data and analytical predictions were made. The stiffness modulus of each experimentally tested honeycomb was determined by applying a linear least squares regression to the data in Figure 12a. The resulting stiffnesses were then plotted with the analytical predictions from Eq. (8) in Figure 12b.

The strong correlation between the analytical predictions and measured behaviour suggests the assumptions made in the modified Gibson-Ashby equation are accurate over the intended operating range of the honeycomb substructure, and local strains are indeed relatively low. Having low local strain is a benefit as it will increase the fatigue life of the substructure. The low local strains were verified with a finite element analysis that predicted a maximum local strain of 1.5% while undergoing 30% compression globally, a 20:1 ratio. This offers hope that a honeycomb substructure capable of high global strains with a long fatigue life can be designed by minimizing local strain, an area which should be a topic of further research. Further details regarding this structure can be found by consulting Kothera *et al.* (2011).

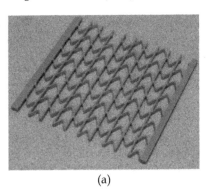 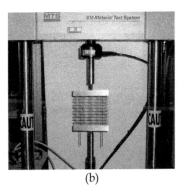

(a)                                          (b)

Fig. 11. (a) Example of Objet PolyJet rapid-prototyped zero-Poisson honeycomb, (b) morphing substructure on MTS machine.

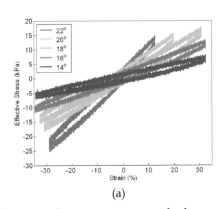 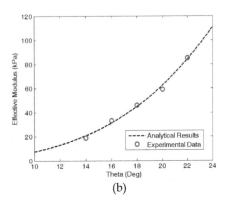

(a)                                          (b)

Fig. 12. (a) Stress-strain curves of substructures of various interior angles, (b) In-plane substructure stiffness, analytical versus experiment.

To minimize the in-plane stiffness of the substructure, the lowest manufacturable bending member angle, 14°, was selected for integration into complete morphing skin prototypes.

Furthermore, this testing demonstrated the usefulness of the modified Gibson-Ashby equation for future honeycomb substructure design efforts. The in-plane stiffness of zero-Poisson honeycomb structures can be predicted.

## 5. One dimensional morphing demonstrator

### 5.1 Carbon fiber stringers

One unfortunate aspect of the zero-Poisson honeycomb described above is the lack of bending stiffness about the in-plane axis perpendicular to the rib members. Another structural element is needed to reinforce the substructure for out-of-plane loads. In order to reinforce the substructure, carbon fiber "stringers" were added perpendicular to the rib members. Simply comprised of carbon fiber rods sliding into holes in the substructure, the stringers reinforce the honeycomb against bending about the transverse axis.

The impact of the stringers on the in-plane stiffness of the combined skin was imperceptible. Fit of the stringer through the holes in the substructure was loose and thus the assembly had low friction. Additionally, the EMC sheet and bending members of the substructure kept the substructure ribs stable and vertical, preventing any binding while sliding along the stringers.

### 5.2 EMC/substructure adhesive

In order to integrate the EMC face sheets with the honeycomb substructure and carry in-plane loads, a suitable bonding agent was necessary. The desired adhesive was required to bond the silicone EMC to the plastic rapid-prototyped honeycomb sufficiently to withstand the shear forces generated while deforming the structure. In addition, the adhesive also needed to be capable of high strain levels in order to match the local strain of the EMC at the bond site. Loads imposed on the adhesive by distributed loads (such as aerodynamic loads on the upper surface of a wing) were not taken into account in this preliminary study.

Due to the fact that the substructure, and not the EMC itself, would be attached to the actuation mechanism, the adhesive was required to transfer all the force necessary to strain the EMC sheet. Based upon the known stiffness of the EMCs selected for integration into the morphing skin prototype, the adhesive was required to withstand up to 10.5 N/cm of skin width for 100% area change. The adhesive was to bond the EMC along a strip of plastic 2.54 cm deep, so the equivalent shear strength required was 41.4 kPa. A couple silicone-based candidate adhesives were selected for lap shear evaluation, all of which were capable of high levels of strain. Test results indicated that Dow Corning (DC) 700, Industrial Grade Silicone Sealant, a one-part silicone rubber that is resistant to weathering and withstands temperature extremes, was most capable of bonding the EMC skin to the substructure, as it had a safety factor of 2.

### 5.3 Morphing structure assembly

A 152 mm x 152 mm morphing skin sample was fabricated from EMC #1. A 14° angle honeycomb was used for the substructure, and DC 700 adhesive was used to bond the EMC to the honeycomb substructure. To assist in the attachment, the rib members of the honeycomb core were designed with raised edges on one side, as shown in Figure 13a. This figure shows a side view of the zero-Poisson honeycomb, where it can be seen that the top surface has the ribs extended taller than the bending members. Therefore, the bonding layer can be applied to the raised rib surfaces and pressed onto the EMC without bonding the bending members to the EMC. A sectional side view of a single honeycomb cell, shown in Figure 13b, illustrates conceptually how the bonded morphing skin looks. A thin layer of adhesive is shown between the EMC and the ribs of the honeycomb, but it does not affect the movement of the bending members. The outermost two ribs on the substructure were each 26 mm wide, providing large bonding areas to carry the load of the skin under strain. This left 100 mm of active length capable of undergoing high strain deformation.

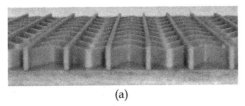
(a)

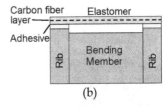
(b)

Fig. 13. EMC-structure bonding method – (a) honeycomb core; (b) single cell diagram.

The configuration of the morphing skin design is summarized in Table 3. The assembled morphing skin sample was used to assess in-plane and out-of-plane stiffness before fabricating a final 165 mm x 330 mm full scale test article for combination and evaluation with the PAM actuation system described in Section 2.

| EMC | Honeycomb | Adhesive | Active Length |
|---|---|---|---|
| EMC #1, 1.4 mm thick, two CF layers at 0° | 14° zero-Poisson prototyped VeroBlue | rapid DC-700 | 100 mm |

Table 3. Morphing skin configuration.

### 5.4 In-plane testing

The morphing skin sample was tested on an MTS machine to 50% strain. The level of strain was limited in order to prevent unforeseen damage to the morphing skin before it could be tested for out-of-plane stiffness as well. In Figure 14a, the morphing skin is shown undergoing in-plane testing, with results presented in Figure 14b. Note that the test procedure strained the specimen incrementally to measure quasi-static stiffness, holding the position briefly before starting with the next stage. Relaxation of the EMC sheet is the cause for the dips in force seen in the figure.

Based upon the individually measured stiffnesses of the EMC and substructure components used in the morphing skin and the stiffness of the skin overall, the energy required to strain each structural element can be determined (Figure 15), with the adhesive strain energy found by subtracting the strain energy of the other two components from the total for the morphing skin. The strain energy contribution of each element is broken down in energy per unit width required to strain the sample from 10 cm to 20 cm.

It can be seen that the adhesive had a considerable strain energy requirement, more than double that of the honeycomb substructure. When designing future morphing skins, the energy to strain the adhesive layer must be taken into account to ensure sufficient actuation force is available to meet strain requirements. More careful attention to minimizing the amount of adhesive used to bond the skin and substructure would also likely reduce the in-plane stiffness of the morphing skin by a non-trivial amount.

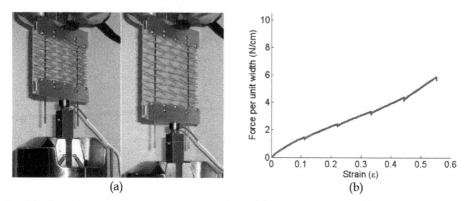

(a)                                                                    (b)

Fig. 14. Morphing skin sample in-plane testing – (a) Skin #1 on MTS; (b) Data from morphing skin in-plane testing.

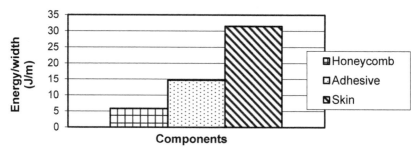

Fig. 15. Contributions to morphing skin strain energy.

### 5.5 Out-of-plane testing

The final phase of evaluation for morphing skin sample required measuring out-of-plane deflection under distributed loadings, approximating aerodynamic forces. A number of testing protocols were investigated, including ASTM standard D 6416/D 6416M for testing simply supported composite plates subject to a distributed load. This particular test protocol is intended for very stiff composites, not flexible or membrane-like composites. A simpler approach to the problem was adopted wherein acrylic retaining walls were placed above the morphing skin sample into which a distributed load of lead shot and sand could be poured. The final configuration of the out-of-plane deflection testing apparatus can be seen in Figure 16a. A set of lead-screws stretched the morphing skin sample from rest to 100% strain. The acrylic retaining walls could be adjusted to match the active skin area, and were tall enough to contain lead shot equivalent to a distributed load of 200 psf (9.58 kPa). By applying a thin layer of sand directly to the surface of the skin, the weight of the lead shot was distributed relatively evenly over the surface of the EMC. Moreover, as the skin deflected under load, the sand would adjust to conform to the surface and continue to spread the weight of the lead. A single-point laser position sensor was also placed underneath to measure the maximum deflections at the center of the skin, between the rib members.

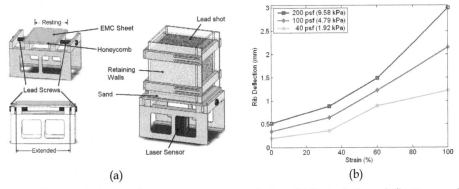

(a)    (b)

Fig. 16. (a) Out-of-plane deflection test apparatus design. (b) Out-of-plane deflection results as measured on the center rib.

The test procedure for each morphing skin covered the full range of operation, from resting (neutral position) to 100% area change. Lead-screws were used to set the skin to a nominal strain condition between 0% and 100% of the resting length. The laser position sensor shown below the skin in the figure was positioned in the center of a honeycomb cell at the center of the morphing skin, where the greatest deflection is seen. This positioning was achieved using a small two-axis adjustable table. The laser was zeroed on the under-surface of the EMC, and the relative distance to the bottom of an adjacent rib was measured. This established a zero measurement for rib deflection as well. A layer of sand with known weight was poured onto the surface of the EMC, and lead shot sufficient to load the skin to one of the three desired distributed loads was added to the top of the sand. Wing loadings of 40 psf (1.92 kPa), 100 psf (4.79 kPa), and 200 psf (1.92 kPa) were simulated. Once the load had been applied, measurements were taken at the same points on the EMC and the adjacent rib to determine deflection. These measurements were repeated for four different strain conditions (0, 25%, 66%, 100%) and the three different noted distributed loads.

Experimental results from the morphing skins is provided in Figure 16b. It was observed that, relative to the rib deflections, the EMC sheet itself deflected very little (less than 0.25 mm). The results therefore ignore the small EMC deflections and show only the maximum deflection measured on the rib at the midpoint of each morphing skin. Overall, the morphing skin deflections show that as the skin is strained and unsupported length increases, the out-of-plane deflection increases. Naturally, the deflection increases with load as well. Based on observation and on these results, the EMC sheets appeared to carry a greater out-of-plane load than expected, probably due to tension in the skin. EMC deflections between ribs remained low at all loading and strain conditions, while the substructure experienced deflections an order of magnitude greater. Future iterations of morphing skins will require stiffer substructures to withstand out-of-plane loads.

## 5.6 Full scale integration and evaluation

After proving capable of reaching over 100% strain with largely acceptable out-of-plane performance, the morphing skin sample from the previous subsection was used as the basis for a larger test article. A 34.3 cm x 14 cm morphing skin, nominally identical to the morphing skin sample in configuration, was fabricated and attached to the actuation assembly. The actuation assembly, honeycomb substructure, and completed morphing cell can be seen in Figure 17. Individual components of the system are pictured in Figure 17a, while the assembled morphing skin test article appears in Figure 17b. The active region stretches from 9.1 cm to 18.3 cm with no transverse contraction, thus, producing a 1-D, 100% increase in surface area with zero Poisson's ratio.

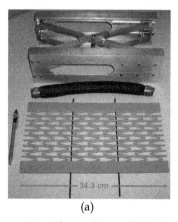

(a)                                                                    (b)

Fig. 17. Integration of morphing cell – (a) actuation and substructure components; (b) complete morphing cell exhibiting 100% area change.

To characterize the static performance of the morphing cell, input pressure to the PAM actuators was increased incrementally and the strain of the active region was recorded at each input pressure, and a load cell in line with one PAM recorded actuator force for comparison to predicted values. This measurement process was repeated three times, recording strain, input pressure, and actuator force at each point. Note that the entire upper surface of the EMC is not the active region: each of the fixed-length ends of the honeycomb was designed and manufactured with 25.4 mm of excess material to allow adequate EMC bonding area and an attachment point to the mechanism. This inactive region can be seen on the top and bottom of the honeycomb shown in Figure 17a. The two extremities of the arrows in Figure 17b also account for the inactive region at both ends of the morphing skin.

The static strain response to input actuator pressure is displayed in Figure 18. Strain is seen to level off with increasing pressure due to a combination of mechanism kinematics and the PAM actuator characteristics, but the system was measured to achieve 100% strain with the PAMs pressurized to slightly over 620 kPa.

The measured system performance matches analytical predictions very closely. The previously mentioned analytical predictions and associated experimental data for the actuation system and skin performance are also repeated in this figure. The morphing cell performance data, while not perfectly linear, approximately matches the slope of the experimental skin stiffness and intersects the actuation system experimental data near 100% extension. Furthermore, although the performance data falls roughly 15% short of original predictions, the morphing skin meets the design goal, validating the analytical design process. Losses were not included in the original system predictions. However, the margin of error included in the original design for friction, increased skin stiffness, and other losses enabled the final morphing cell prototype to achieve 100% strain. It should also be noted that 100% area increases could be achieved repeatedly at 1 Hz using manual actuator pressurization.

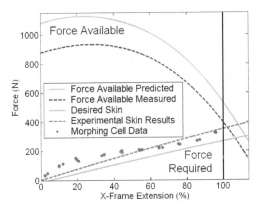

Fig. 18. Morphing cell data comparison with predictions.

## 6. Wind tunnel prototype

Building on the success of the 1-D morphing demonstrator, a wind tunnel-ready morphing wing was designed and tested. A key technical issue addressed here was determining the scalability of the skin and substructure manufacturing processes for use on a real UAV. Thus, the prototype airfoil system was designed such that future integration with a candidate UAV is feasible, and experimentally evaluated as a wind tunnel prototype. Nominal design parameters for the prototype are a 30.5 cm chord wing section capable of 100% span extension over a 61.0 cm active morphing section with less than 2.54 mm of out-of-plane deflection between ribs due to dynamic pressures consistent with a 130 kph maximum speed.

### 6.1 Structure development

Initially, the planar core design was extruded and cut into the form of a NACA $63_3$-618 airfoil with a chord of approximately 30.5 cm and span of 91.4 cm. A segment of the resulting morphing airfoil core appears in Figure 19a. While this morphing structure is capable of achieving greater than 100% length change itself, it has insufficient spanwise bending and torsional stiffness and so does not constitute a viable wing structure. The structure was therefore augmented with continuous sliding spars. Additionally, the center of the wing structure was hollowed out to potentially accommodate an actuation system for the span extension.

The final form of the morphing airfoil core is shown in Figure 19b. This figure shows a shell-like section mostly around the center of the airfoil, where an actuator could be located. Both the leading and trailing edges feature circular cut-outs to accommodate the carbon fiber spars, and near the trailing edge is a solid thickness airfoil shape for more rigidity where the airfoil is thinnest. The spars were sized using simple Euler-Bernoulli beam approximations and a desired tip deflection of less than 6.4 mm at full extension.

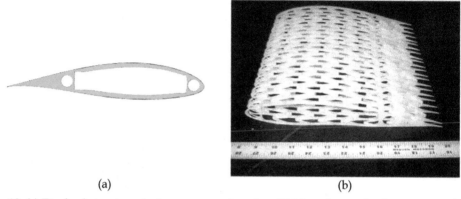

(a)                                                         (b)

Fig. 19. (a) Final substructure design, cross-section view (b) Manufactured substructure, side view.

Due to the complex geometry of the morphing core and the desire for rapid part turn around, a stereo lithographic rapid-prototyping machine was again used to manufacture the morphing core sections from an acrylic-based photopolymer. The viability of this approach for flyable aircraft applications would have to be studied, but the material/manufacturing approach was sufficient for this proof-of-concept structural demonstrator. Other fabrication techniques such as investment casting, electrical discharge machining, etc. could be considered when fabricating this structure to meet full scale aircraft requirements. It should also be noted that the prototype will feature three of the core segments shown in Figure 19b. They will be pre-compressed when the EMC skin is bonded to allow for more expansion capability and introduce a nominal amount of tension in the EMC skin.

Figure 20a shows the core sections together between two aluminum end plates, with the leading edge and trailing edge support spars. The end plates were sized to provide a suitably large bonding surface for attaching the skin on the tip and root of the morphing section. In this configuration, the core sections are initially contracted such that the active span length is 61.0 cm. In terms of the aircraft, this contracted state will be considered the neutral, resting state because the EMC skin will not be stretched here and a potential actuation system would not be engaged. Hence, this is the condition in which the skin would be bonded to the core. Also shown in Figure 20b is the same arrangement in the fully extended (100% span increase) state with a span of 122.0 cm. The figure shows that the spacing between each of the rib-like members has nearly doubled from what was shown in the contracted state. This figure helps illustrate the large area morphing potential of this technological development in a way that could not be seen once the skin was attached.

Spanwise bending and torsional stiffness was provided by two 1.91 cm diameter carbon fiber spars. The spars were anchored at the leftmost outboard portion of the wing but were free to slide through the inboard end plates, thus allowing the wing to extend while maintaining structural integrity. The spars were sized in bending to deflect less than 2.54 cm at 100% extension under the maximum expected aerodynamic loads. Note that the spars are also capable of resolving torsional pitching moments, but as the express purpose of the present work was to demonstrate the feasibility of a span morphing wing, these torsional properties were not directly evaluated.

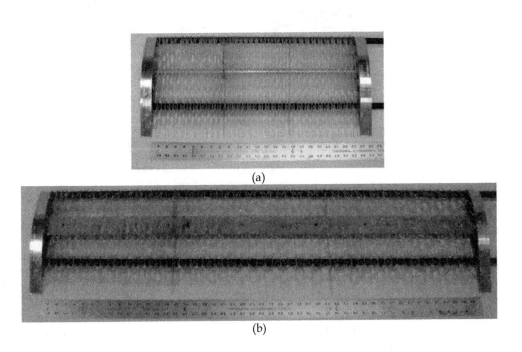

(a)

(b)

Fig. 20. Assembled core with spars and end plates – (a) contracted state; (b) extended state.

## 6.2 Prototype integration

The skin was bonded to the morphing substructure using DC-700. The skin was attached to each rib member, but not to the v-shaped bending members. Particular caution was used when bonding the skin to the end plates, as all of the tensile stress in the skin was resolved through its connection to the end plates.

At the resting condition with no elastic energy stored in the skin, Figure 21 shows the 0% morphing state with a 61.0 cm span. Increasing the span by another 61.0 cm highlights the full potential of this morphing system as the prototype wing section doubles its initial span, which has gone from 61.0 cm to 122 cm to show the 100% morphing capability (Figure 21b). Recall from the design that the wing section chord stays constant during these span extensions, so the morphing percentages indicated (e.g., 100%) are consistent with the increase in wing area. As a fixed point of reference in each of these figures, note that the length of the white poster board underneath the prototype wing section does not change. Note also that this demonstration will use fixed-length internal spreader bars to hold the structure in different morphing lengths. Actuation was achieved by manually stretching the skin/core structure and then attaching the appropriate spreader bar to maintain the stretched distance.

(a)                                                         (b)

Fig. 21. Prototype morphing wing demonstration – (a) resting length, 0% morphing; (b) 61.0 cm span extension, 100% morphing.

## 6.3 Wind tunnel testing

Having shown that the prototype morphing wing section could achieve the goal of 100% span morphing for a total 100% wing area increase, the final test that was performed placed the wing section in a wind tunnel. The purpose of this test was to ensure that the EMC skin and core could maintain a viable airfoil shape at different morphing states under true aerodynamic loading, with minimal out-of-plane deflection between ribs. An open circuit wind tunnel at the University of Maryland with a 50.8 cm tall, 71.1 cm wide test section was used in this test. An overall view of the test section is shown in Figure 22a, with the wing at its extended length, and a close-up view of the test section is shown in Figure 22b looking upstream from the trailing edge.

With only a 50.8 cm tall test section in the wind tunnel, where only this span length of the prototype morphing wing would be placed in the wind flow, while the remaining span and support structure was below the tunnel. This is illustrated in Figure 22a, where the full extension condition (100% morphing) is shown. It should also be noted that while only a 50.8 cm span section of the wing is in the air flow, this is sufficient to determine whether or not the skin and core can maintain a viable airfoil shape in the presence of representative aerodynamic conditions, which was the primary goal of this test. That is, the morphing core motion and skin stretching is consistent and substantially uniform across the span of the prototype, so any characteristics seen in one small section of the wing could similarly be seen or expected elsewhere in the wing, making this 50.8 cm span "sampling" a reasonable measure of system performance.

Both the cruise (105 kph) and maximum (130 kph) rated speeds of the candidate UAV were tested. Three angles-of-attack (0°, 2°, 4°) and three wing span morphing conditions (0%, 50%, 100%) were also included in the test matrix. Tests were performed by first setting the morphing condition of the wing section, then positioning the wing for the desired angle-of-attack (AOA). With these values fixed, the tunnel was turned on and the speed was increased incrementally, stopping at the two noted test speeds while experimental observations were made. Tests were completed at each of the conditions in the table indicated with an x-mark. Note that tests were not performed at two of the angles-of-attack at the 100% morphing condition. This was because the skin began to debond near the trailing edge at one of the end plates. This occurred over a section approximately 7.6 cm in span at the 100% morphing condition, though the majority of the prototype remained intact. After removing the wing section from the wind tunnel and inspecting the debonded corner, it was discovered that very little adhesive was on the skin, core, and end plate. Thus, the likely cause for this particular debonding was inconsistent surface preparation, which can easily be rectified in future refinements. Note that the upper surface of the trailing edge experiences relatively small dynamic pressures compared to the rest of the wing, so that this debonding was most likely unrelated to the wind tunnel test. Rather, it was the result of manufacturing inconsistency.

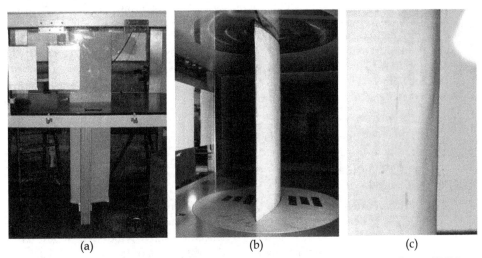

(a)  (b)  (c)

Fig. 22. Wind tunnel test setup – (a) Overall wind tunnel setup at 100% morphing; (b) Wing installed in wind tunnel – from trailing edge; (c) Picture of wing section leading edge at 130 kph, 100% morphing.

During execution of the test matrix, digital photographs (Figure 22c) were taken of the leading edge at each test point to determine the amount of skin deflection (e.g., dimpling) that resulted from the dynamic pressure. The leading edge location was chosen as the point to measure because the pressure is highest at the stagnation point. Pictures were taken perpendicular to the air flow direction and angled from the trailing edge, looking forward on the upper skin surface. Grids were taped to the outside of the transparent wall on the opposite side of the test section to provide reference lengths for processing. The grids form 12.7 mm squares and are located 35.6 cm behind the airfoil in the frame of view, which is also 35.6 cm from the camera lens. These can be seen in Figure 22c. Using image processing, the maximum error in the measurements was determined to be ±7%. This error can be attributed to vibration of the wind tunnel wall, which the camera lens was pressed against, or deviations in the focus of the pictures. In all the data processed, the maximum discernible out-of-plane deflection was approximately 0.51 ± 0.04 mm, which is well within the goal of less than 2.54 mm. In reference to the 30.5 cm chord and 5.49 cm thickness, this deflection accounts for only 0.17% and 0.93%, respectively. Additionally, in observing this experiment, it can be qualitatively stated the morphing wing held its shape remarkably well under all tested conditions. This can be confirmed through visual inspection of the figures, as well.

## 7. Conclusions

This work explored the development of a continuous one dimensional morphing structure. For an aircraft, continuous morphing wing surfaces have the capability to improve efficiency in multiple flight regimes. However, material limitations and excessive complexity have generally prevented morphing concepts from being practical. Thus, the goal of the present work was to design a simple morphing system capable of being scaled to UAV or full scale aircraft. To this end, a passive 1-D morphing skin was designed, consisting of an elastomer matrix composite (EMC) skin with a zero-Poisson honeycomb substructure intended to support out-of-plane loads. In-plane stiffness was controlled to match the capabilities of an actuator by careful design and testing of each separate skin component. Complete morphing skins were tested for in-plane and out-of-plane performance and integrated with the actuator to validate the design process on a small-scale morphing cell section.

Design goals of 100% global strain and 100% area change were demonstrated on a laboratory prototype using the combined morphing skin and actuation mechanism. The morphing skin strained smoothly and exhibited a very low in-plane Poisson's ratio. Actuation frequencies of roughly 1 Hz were achieved.

This work was then extended to a full morphing UAV-scale wing suitable for testing in a wind tunnel. The system was assembled as designed and demonstrated its ability to increase span by 100% while maintaining a constant chord. Wind tunnel tests were conducted at cruise (105 kph) and maximum speed (130 kph) conditions of a candidate UAV test platform, at 0º, 2º, and 4º angles-of-attack, and at 0%, 50%, and 100% extensions. At each test point, image processing was used to determine the maximum out-of-plane deflection of the skin between ribs. Across all tests, the maximum discernable out-of-plane deflection was little more than 0.5 mm, indicating that a viable aerodynamic surface was maintained throughout the tested conditions.

## 8. Acknowledgement

This work was sponsored by the Air Force Research Laboratory (AFRL) through a Phase I STTR (contract number FA9550-06-C-0132), and also by a Phase I SBIR project from NASA Langley Research Center (contract number NNX09CF06P).

# 9. References

Agarwal, B. D., Broutman, L. J., and Chandrashekhara, K. (2006). *Analysis and Performance of Fiber Composites*, John Wiley & Sons, Hoboken.

Bae, J.S., Seigler, T.M. and Inman, D.J. (2005). "Aerodynamic and Static Aeroelastic Characteristics of a Variable-Span Morphing Wing," *Journal of Aircraft*, 42(2): 528-534. doi: 10.2514/1.4397

Barbarino, S., Bilgen, O., Ajaj, R.M., Friswell, M.I., and Inman, D.J. (2011). "A Review of Morphing Aircraft," *Journal of Intelligent Material Systems and Structures*, 22: 823-877. doi:10.1177/1045389X11414084.

Bubert, E.A., Woods, B.K.S., Lee, K., Kothera, C.S., and Wereley, N.M. (2010). "Design and Fabrication of a Passive 1D Morphing Aircraft Skin," *Journal of Intelligent Material Systems and Structures*, 21(17):1699-1717 doi: 10.1177/1045389X10378777

Buseman, A. (1935) "Aerodynamic Lift at Supersonic Speeds," Ae. Techl. 1201, Report No. 2844 (British ARC, February 3, 1937), Bd. 12, Nr. 6: 210-220.

Bye, D.R. and McClure, P.D. (2007). "Design of a Morphing Vehicle," *48th AIAA Structures, Structural Dynamics, and Materials Conference*, 23-26 April, Honolulu, HI, Paper No. AIAA-2007-1728.

Chaves, F. D., Avila, J., and Avila, A. F. (2003). "A morphological study on cellular composites with negative Poisson's ratios,"*44th AIAA Structures, Structural Dynamics, and Materials Conference*, Norfolk, VA, Paper No. AIAA 2003-1951.

Evans, K.E., Nkansah, M.A., Hutchinson, I.J., and Rogers, S.C. (1991). "Molecular network design," *Nature*, 353: 124.

Flanagan, J.S., Strutzenberg, R.C., Myers, R.B., and Rodrian, J.E. (2007). "Development and Flight Testing of a Morphing Aircraft, the NextGen MFX-1,"*AIAA Structures, Structural Dynamics and Materials Conference*, Honolulu, HI. Paper No. AIAA-2007-1707.

Gern, F.H., Inman, D.J., and Kapania, R.K. (2002). "Structural and Aeroelastic Modeling of General Planform Wings with Morphing Airfoils," *AIAA Journal*, 40(4): 628-637. doi: 10.2514/2.1719

Gevers, D.E. (1997). "*Multi-purpose Aircraft*," US Patent No. 5,645,250.

Gibson, L. J. and Ashby, M. F. (1988). *Cellular Solids: Structure and Properties*, Pergamon Press, Oxford.

Gomez, J. C., and Garcia, E. (2011). Morphing unmanned aerial vehicles. *Smart Materials and Structures*, 20(10):103001. doi:10.1088/0964-1726/20/10/103001

Hetrick, J. A., Osborn, R. F., Kota, S., Flick, P. M., and Paul, D. B. (2007). "Flight Testing of Mission Adaptive Compliant Wing,"*48th AIAA Structures, Structural Dynamics, and Materials Conference*, Honolulu, HI, Paper No. AIAA 2007-1709.

Joo, J.J. Reich, G.W. and Westfall , J.T. (2009). "Flexible Skin Development for Morphing Aircraft Applications Via Topology Optimization." *Journal of Intelligent Material Systems and Structures*, 20(16):1969-1985.

Kothera, C.S., Woods, B.K.S., Bubert, E.A., Wereley, N.M., and Chen, P.C. (2011). "Cellular Support Structures for Controlled Actuation of Fluid Contact Surfaces." U.S. Patent 7,931,240. Filed: 16 Feb 2007. Issued: 26 Apr 2011.

McKnight, G. Doty, R. Keefe, Herrera, A.G. and Henry, C. (2010). "Segmented Reinforcement Variable Stiffness Materials for Reconfigurable Surfaces." *Journal of*

*Intelligent   Material   Systems   and   Structures,*   21:1783-1793, doi:10.1177/ 1045389X10386399

Munk, M. M. (1924). "Note on the relative Effect of the Dihedral and the Sweep Back of Airplane Wings," NACA Technical Note 177.

Olympio, K.R., and Gandhi, F. (2010). "Flexible Skins for Morphing Aircraft Using Cellular Honeycomb Cores," *Journal of Intelligent Material Systems and Structures,* 21:1719-1735, doi:10.1177/1045389X09350331

Parker, H.J. (1920). "The Parker Variable Camber Wing," Report #77 Fifth Annual Report, *National Advisory Committee for Aeronautics,* Washington, D.C.

Perkins, D. A., Reed, J. L., and Havens, E. (2004). "Morphing Wing Structures for Loitering Air Vehicles," *45th AIAA Structures, Structural Dynamics & Materials Conference,* Palm Springs, CA, Paper No. AIAA 2004-1888.

Sarh, B., (1991). "*Convertible Fixed Wing Aircraft,*" US Patent No. 4,986,493.

Samuel, J.B. and Pines, D.J. (2007). "Design and Testing of a Pneumatic Telescopic Wing for Unmanned Aerial Vehicles," *Journal of Aircraft,* 44(4) DOI: 10.2514/1.22205

Shigley, J., Mishke, C., and Budynas, R. (2004). *Mechanical Engineering Design,* McGraw-Hill, New York.

Thill, C., Etches, J., Bond, I., Potter, K., and Weaver, P. (2008). "Morphing Skins," *The Aeronautical Journal,* 112(1129):117-139.

Wereley, N. and Gandhi, F. (2010). "Flexible Skins for Morphing Aircraft." *Journal of Intelligent   Material   Systems   and   Structures,*   21:   1697-1698, doi:10.1177/1045389X10393157.

Wereley, N. M. and Kothera, C. S. (2007). "Morphing Aircraft Using Fluidic Artificial Muscles," *International Conference on Adaptive Structures and Technologies,* Ottawa, ON, Paper ID 171.

Wright, O. and Wright, W. (1906). "Flying-Machine" U.S. Patent 821,393. Filed: 23 Mar 1903. Issued: 22 May, 1906.

# 3

# A Probabilistic Approach to Fatigue Design of Aerospace Components by Using the Risk Assessment Evaluation

Giorgio Cavallini and Roberta Lazzeri

*University of Pisa-Department of Aerospace Engineering*
*Italy*

## 1. Introduction

Fatigue design of aerospace metallic components is carried out by using two methodologies: damage tolerance and safe-life. At present, regulations mainly recommend the use of the former, which entrusts safety to a suitable inspections plan. Indeed, a crack or a flaw is supposed to have been present in the component since the beginning of its operative life, and it must remain not critical, i.e. it must not cause a catastrophic failure in the life period between two following inspections, (Federal Aviation Administration, 1998; Joint Aviation Authorities, 1994; US Department of Defence, 1998). When a crack is detected, the component is repaired or substituted and the structural integrity is so restored.

If the damage tolerance criterion cannot be applied, the regulations state the safe-life criterion should be used, i.e. components must remain free of crack for their whole operative life and, at their ends, components must be in any case substituted.

Therefore, both methodologies have deterministic bases and a single value (usually the mean value) is associated to each parameter that can influence the fatigue phenomenon, which on the contrary has a deep stochastic behaviour.

To take these items into account and in order to protect against unexpected events, it is necessary to introduce safety factors in the fatigue life design (generally equal to 2 or 3 for damage tolerance and equal to 4 or even more for safe-life). They usually produce heavy or expensive structures and, in the past, they were not always able to protect against catastrophic failures, because the real risk level is in any case unknown. On the one hand, indeed, the inspected structures or the substituted components may be still undamaged, with high costs; on the other hand, highly insidious phenomena, such as Multiple Site Damage and Widespread Fatigue Damage (which are typical of ageing aircrafts) cannot be taken into account very well and in the past they were the causes of some catastrophic accidents.

For these reasons, researchers are hypothesizing the possibility of facing fatigue design in a new way, by using the risk evaluation from a probabilistic point of view. Indeed, the parameters that affect the phenomenon have a statistical behaviour, and this can be described by means of statistical distributions.

In such a way, by using a statistical method, such as the Monte Carlo Method (Besuner, 1987; Hammersley & Handscomb, 1983), all the parameter distributions can be managed and each simulated 'event' can be considered as a possible 'event'. So, the computer simulation of the fatigue life of a big amount of components and the evaluation of the real risk level are possible, making this approach extremely useful.

The Authorities are interested in this approach but, before allowing the use of it as a design criterion, they require impartial evidence, first of all about the reliability of the analytical models used for fatigue simulation and for parameter distribution evaluation.

This paper intends to show a computer code – PISA, Probabilistic Investigation for Safe Aircrafts – and how it can be applied to the fatigue design of typical aerospace components, such as riveted joints, making it possible to integrate damage tolerance and the evaluation of the real risk level connected to the chosen inspection plan, (Cavallini et al., 1997; Cavallini & R. Lazzeri, 2007).

## 2. The parameters that mainly affect the fatigue phenomenon

A metallic component subjected to repeated loads can fail due to the fatigue phenomenon, (Schijve, 2001). Many research activities on this subject are known from both the theoretical and the experimental points of view and it is well known that fatigue has a random behaviour, with a high number of parameters (mechanical behaviour of the material, loads, geometry, manufacturing technologies, etc.) that can affect it.

Almost all these parameters have a statistical behaviour, but some among them play a more important role compared to the others and must be taken into account with their distributions, while the others can be assumed to be constant.

In detail, we can assume four main parameters as statistically distributed:

- the Initial Fatigue Quality (IFQ), described by using the (Equivalent) Initial Flaw Size, (E)IFS, or the Time To Crack Initiation, TTCI, distribution, (Manning et al., 1987);
- the crack grow rate, CGR (constant C in the Paris law);
- the fracture toughness $K_{Ic}$, and
- the inspection reliability, i.e. the Probability of crack Detection, PoD.

They are described in the following.

### 2.1 The Initial Fatigue Quality (IFQ)

Structural components can have, until the end of the manufacturing process, defects due to metallurgical effects, scratches, roughness, inclusions, welding defects, etc.

So, the IFQ can be considered as a property linked to the material and the manufacturing process. Defects can be the starting point for fatigue cracks. For this reason it is extremely important to know their position and size, but, as they are very small, they are very difficult to be measured even if by using very sophisticated inspection methods.

As a consequence, this information can be reached only through an indirect evaluation by means of a 'draw-back' procedure starting from experimental data about detectable cracks.

Therefore, a 'tool' able to characterize the component initial condition is necessary, in order to predict the fatigue life. At present, two approaches are available, Fig. 1:

- the (Equivalent) Initial Flaw Size distribution,
- the Time To Crack Initiation distribution.

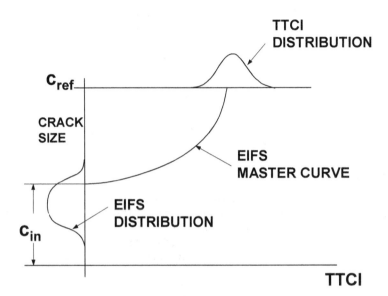

Fig. 1. TTCI and EIFS distributions.

### 2.1.1 The Time To Crack Initiation distribution (TTCI)

The TTCI model can be used to describe fatigue crack nucleation in metallic components. It can be defined as the time (in cycles, flights or flight hours) necessary for an initial defect to grow to a detectable reference crack, $c_{ref}$. In such a way, this method does not reveal the crack dimension during the early steps of the component life. The TTCI can be described by using a Weibull, (Manning et al., 1987), or a lognormal distribution (Liao & Komorowski, 2004). It must be noted that the TTCI distribution is not only a material property, as it is connected to the crack growth, and so to the loading fatigue spectrum.

### 2.1.2 The (Equivalent) Initial Flaw Size distribution (EIFS)

The EIFS is the (fictitious) dimension of a crack at time t=0. The use of the adjective 'equivalent' indicates that the initial flaw is not the actual one and its size is only 'equivalent' to it because it is very difficult to account for the influence of all the relevant parameters. The distribution can be numerically obtained starting from experimental crack size data by using a 'fictitious' backward integration of the crack growth. It is affected by the material properties and the stress distribution: cold-working, rivet interference, … have to be taken into account, too.

The EIFS can be described by using a lognormal or a Weibull (Manning et al., 1987) distribution. In the present paper, a lognormal distribution is assumed.

## 2.2 The Crack Growth Rate, CGR (C constant in the Paris law)

Different models are available to describe the crack growth law according to Linear Elastic Fracture Mechanics and the distribution of the involved parameters. We assumed to use the simple and effective Paris law $dc/dN=C(\Delta K)^m$. The parameter $m$ is assumed to be constant and all the scatter is consolidated in C, which is assumed to belong to a normal distribution.

## 2.3 The Fracture Toughness $K_{Ic}$

Fracture toughness is a very important material property because it identifies a failure criterion (crack instability). Unfortunately, few experimental data are available to characterize its distribution. Anyway, a normal distribution (Hovey et al., 1991), or a lognormal distribution (Johnson, 1983; Schutz, 1980) can be supposed. We assumed the fracture toughness can be described through a lognormal distribution.

## 2.4 The Probability of crack Detection (PoD)

Non destructive inspections are among the principal items of the damage tolerance methodology. Indeed, during inspection, it is supposed that cracks are detected and the component can be re-qualified for further use. This action depends on many parameters, included the human factor and so it can be described only by using a probabilistic approach. Usually, (Lincoln, 1998), we define the inspection capability as the 90% probability of crack detection with the 95% of confidence.

Many distributions have been proposed for the Probability of Detection, (Tong, 2001; Ratwani, 1996).

In the present work we assumed a three parameters Weibull distribution, [Lewis et al., 1978]:

$$PoD = 1 - e^{-\left[\frac{c-c_{min}}{\lambda-c_{min}}\right]^\beta} \tag{1}$$

where, $c_{min}$ is the minimum detectable crack size, $c$ is the actual crack size and $\lambda$ and $\beta$ are parameters connected to the chosen inspection method.

## 3. The Monte Carlo method

A tool is necessary to manage all the parameter distributions together and at the same time. Some reliable approaches are available - FORM (First Order Reliability Method), SORM (Second Order Reliability Method) and many others (Madsen et al., 1986) – but we decided to use the Monte Carlo method because of its simplicity and effectiveness, as it can handle high numbers of different distributions for the stochastic variables to simulate many different deterministic situations. In addition, the Monte Carlo method easily allows the introduction of the repairs, that is a non-continuous change in the crack size.

The Monte Carlo method is based on a very easy assumption: the probability of an event $p_f$ – in the present paper the component failure – is evaluated by using the analytical expression

$$p_f = n / N \qquad (2)$$

where $N$ is the total simulation number and $n$ is the number of positive results.

Each simulation reproduces only one deterministic event, in which, for each deterministic or random variable a value is assumed; for the stochastic parameters, the value is randomly obtained from its distribution.

After a high number of trials, the method converges to the solution.

The only disadvantage of this method is that it requires a high number of simulations to have a low probability of the event. As an example, if the required probability is $10^{-6}$, with a confidence level of 95%, at least (Grooteman, 2002)

$$N_{trials} > \frac{3}{10^{-6}} = 3x10^6 \qquad (3)$$

are necessary.

## 4. The PISA code and the simulation of the fatigue phenomenon

The PISA code (Cavallini et al., 1997; Cavallini & R. Lazzeri, 2007), developed at the Department of Aerospace Engineering of the University of Pisa, allows the simulation of the whole fatigue life of typical aerospace components, such as simple plane panels, riveted lap-joint panels, Fig. 2, or stiffened panels, subjected to constant amplitude fatigue loading.

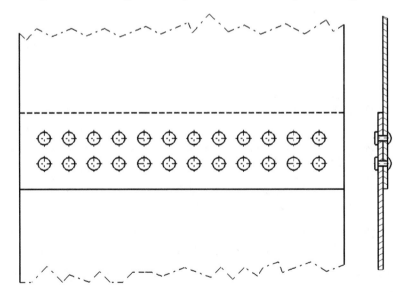

Fig. 2. Riveted lap-joint panel.

As far as the panel is concerned, the following hypotheses can be made:

| | |
|---|---|
| **Geometrical** | Plane panel |
| | Uniform thickness |
| | Constant rivet pitch |
| | Rivets with or without countersunk head |
| | Through cracks |
| | Cracks on one or both hole sides and orthogonal to the load direction |
| **Physical** | Uniform stress |
| | Plane stress |
| | Uniform pin load in the same row |
| | Rivets with extremely high stiffness |
| | Fretting and corrosion effects are negligible |

Table 1. Assumed geometrical and physical hypotheses.

The code can simulate crack nucleation, growth, inspection actions and failures in components subjected to uniform loading. The basic idea is that the damage process can be simulated as the continuous growth of an initial defect due to metallurgical effects and/or the manufacturing process, and/or other parameters.

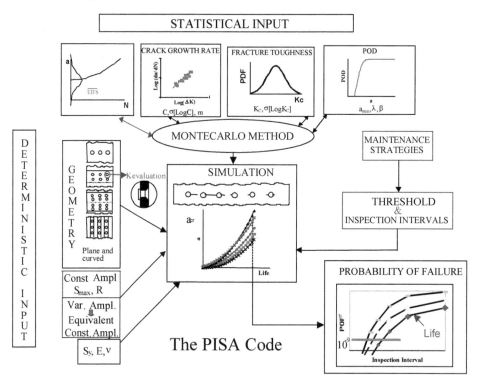

Fig. 3. Structure of the code.

Many deterministic simulations can be run and the risk assessment can be evaluated by using the Monte Carlo method, Fig. 3. In each simulation, the deterministic parameters are assigned by taking off a value from their own distributions.

The first phase – crack nucleation at holes – is simulated by using the EIFS distribution, which can be considered as an indication of the initial fatigue quality. In this context, this approach has to be preferred to the TTCI, as it allows to consider the whole life as the only propagation phase.

The second phase – crack growth – is simulated by using the simple, well-known Paris law $dc/dN = C(\Delta K)^m$

The core for the evaluation of the crack growth is the expression of the stress intensity factor $K$ from the beginning of the life to the final failure.

In the stress intensity factor evaluation, suitable corrective factors (Sampath & Broek, 1991; Kuo et al., 1986) have been used to take into account the different boundary conditions, and the load transfer inside the joints has been simulated by using the Broek & Sampath model, (Sampath & Broek, 1991).

In detail, the effect of different boundary conditions can be taken into account by using the composition approach (Kuo et al., 1986).

$K$ is analytically evaluated by means of a corrective coefficient which has been found by splitting the complex geometry into simple problems (open hole, finite width, …), whose solutions are known.

With regard to the open hole

$$K = K^R \cdot CR_1 \cdot CR_2 \cdot CR_n \tag{4}$$

$$K^R = S\sqrt{\pi(c-r)} \tag{5}$$

where $S$ is the uniform membrane stress and $(c-r)$ the crack length, Fig. 4.

Fig. 4. Crack at an open hole.

As to the rivet effect, the load $P$ on the hole has been approximated with a uniform pressure $p$ on the hole, (Kuo et al., 1986), Fig. 5.

$$P = t \cdot \int_0^\pi p \cdot sen\theta \cdot r \cdot d\theta = t \cdot p \cdot r \cdot \int_0^\pi sen\theta \cdot d\theta = 2 \cdot p \cdot r \cdot t \tag{6}$$

$$p = \frac{P}{2 \cdot r \cdot t} \qquad (7)$$

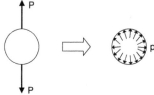

Fig. 5. Approximation of the $P$ load with a uniform pressure $p$.

$$K = K^P \cdot CP_1 \cdot CP_2 \cdots CP_n \qquad (8)$$

$$K^P = p\sqrt{\pi \, c} \qquad (9)$$

The main corrective factors for open holes and for filled holes taken into account are summarized in Table 2.

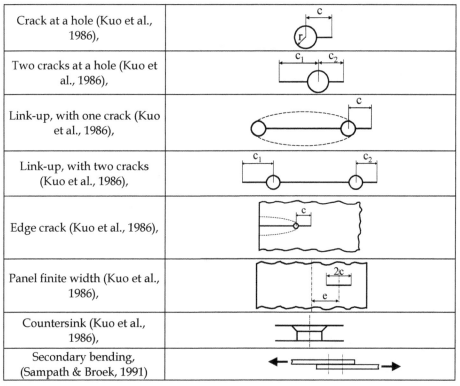

| | |
|---|---|
| Crack at a hole (Kuo et al., 1986), | |
| Two cracks at a hole (Kuo et al., 1986), | |
| Link-up, with one crack (Kuo et al., 1986), | |
| Link-up, with two cracks (Kuo et al., 1986), | |
| Edge crack (Kuo et al., 1986), | |
| Panel finite width (Kuo et al., 1986), | |
| Countersink (Kuo et al., 1986), | |
| Secondary bending, (Sampath & Broek, 1991) | |

Table 2. The main corrective factors for open holes and for filled holes taken into account in the PISA code.

The Broek & Sampath model joins the solutions related to the open hole and the loaded hole by using the superposition approach, Fig. 6.

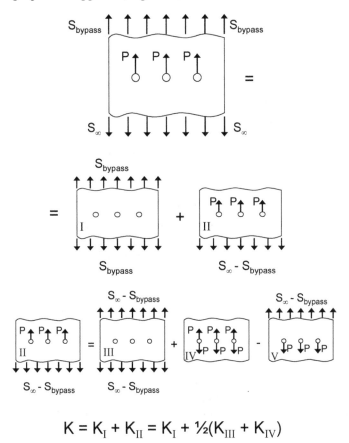

$$K = K_I + K_{II} = K_I + \tfrac{1}{2}(K_{III} + K_{IV})$$

Fig. 6. The Broek & Sampath model.

$$K = \frac{1}{2} \cdot CR \cdot \left( S_\infty + S_{bypass} \right) \cdot \sqrt{\pi \cdot c} + \frac{1}{2} \cdot CP \cdot p \cdot \sqrt{\pi \cdot c} \qquad (10)$$

$S_\infty$ is the membrane uniform stress and $S_{bypass}$ is the bypass stress, i.e. for each row, the stress not transferred by the rivets.

Also rivet interference introduces an additional stress distribution. Its main effect is that only a part of the applied load amplitude $S_{max}$-$S_{min}$ is effective for crack propagation. This effect can be taken into account by using the Wang model (Wang, 1988) for the evaluation of the lift-off stress $S_o$, corresponding to the separation of the rivet from the hole, and then by introducing in the simulations carried out with the PISA code only the effective amplitude $S_{max}$-$S_o$ for crack propagation.

Once the stress intensity factor is calculated, crack growth simulation can start.

Two collinear cracks are considered as linked according to the Swift criterion, i.e. when their plastic radii $r_p$ - evaluated by using the Irving expression - are tangential.

Inspections at planned intervals are simulated through the PoD distribution, applied at each crack at both hole sides. Though the repair of the hole has the same quality as the pristine panel, the repair itself of the detected crack is not immune from the possibility of having some tiny cracks.

The final failure can happen either for crack instability ($K_{max}$ higher than the fracture toughness, $K_{max} \geq K_{Ic}$) or for static failure ($S_{max}$ higher than the yield stress in the net section evaluated without the plastic zones, $S_{max} \geq S_{02}$)

## 5. Experimental activity as a support for the evaluation of the statistical distributions

To support this activity, all the parameter statistical distributions and the coefficients for the used analytical law (for example, the EIFS distribution, $C$ and $m$ coefficients for the Paris law, etc.) have to be experimentally evaluated. Of course, they cannot be obtained from tests on real components, but we have tested realistic simple specimens and we have demonstrated the applicability of the obtained results to the life evaluation of the actual components.

In this paper the activity carried out to find the distributions of the EIFS and the C constant in Paris law are shown. A similar approach can be used for the definition of the distributions for $K_{Ic}$ and PoD.

### 5.1 Equivalent Initial Flaw Size distribution evaluation

For the evaluation of the EIFS distribution it was necessary to use a 'fictitious' negative integration (draw-back) which, starting from the experimental crack data at assigned number of cycles, would be able to find the 'equivalent' initial size, i.e. the crack size at $N=0$.

To support this approach, a wide experimental activity was performed on 29 simple strip lap-joints, Fig. 7, in aluminum alloy 2024-T3 (Cavallini & R. Lazzeri, 2007). They were fatigue tested under a constant amplitude load spectrum with $S_{max}=120$ MPa and $R=0.1$. The tests were stopped at a set number of cycles, the specimens were statically broken and the crack dimensions were carefully measured. The tests confirmed an already well known result: all the cracks were found in the most critical location, i.e. in the first row, at the countersunk side.

At present, several numerical codes are available to simulate the growth of a single crack in the long crack range, but few can manage also the short crack range and none is able to carry out a direct negative integration that starting from the experimental crack data can find the initial dimension. For this reason, we decided to use the PISA code itself and the simplified model of a specimen with a through crack at a lap-joint, taking into account the effects of countersink of the hole, membrane stress, by-pass loading and pin load, Fig. 8, (Cavallini & R. Lazzeri, 2007).

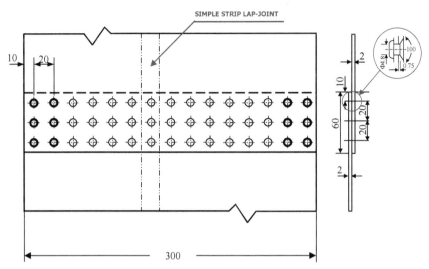

Fig. 7. Specimen geometry, all lengths in mm.

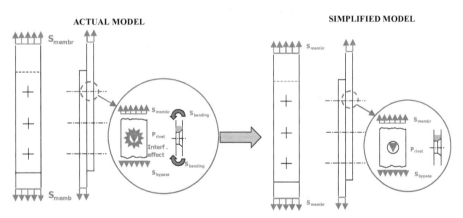

Fig. 8. Simplified model implemented inside the PISA code in order to evaluate the EIFS distribution.

In addition, an iterative positive integration was carried out starting from an initial tentative crack size value and stopping at the same number of cycles of the experimental result. The simulated final crack dimension was compared with the experimental one, and an iterative process was carried out to the required convergence.

In this way, a lognormal distribution for the EIFS was found, with $\mu[Log_{10}(c_0)]$=-2.88605, and $\sigma[Log_{10}(c_0)]$=0.28456, Fig. 9.

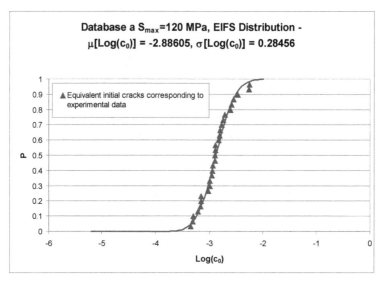

Fig. 9. EIFS distribution obtained by using the draw-back procedure.

## 5.2 The Crack Growth Rate (CGR) (C constant in the Paris law)

The crack growth law must be experimentally characterized, in order to evaluate the parameters involved in the selected crack growth law (Paris).

For their evaluation, a test campaign on a 2.1 mm thick Center Crack Tension (CCT) specimen with an open hole (4 mm in diameter) in aluminium alloy 2024-T3 was carried out (Imparato & Santini, 1997), Fig. 10.

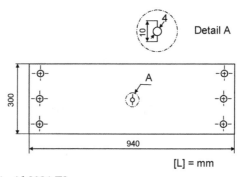

Fig. 10. CCT specimen in Al 2024-T3.

They were pre-cracked and the further crack propagation in the 5 to 40 mm range was investigated by using the Potential Drop Technique.

Tests were carried out under a constant amplitude (C.A.) spectrum, at the same $S_{max}$=68.7 MPa, with 4 different $R=S_{min}/S_{max}$ values ($R$=0.1, $R$=0.25, $R$=0.4, $R$=0.55). In Fig. 11 the experimental results for one test are shown.

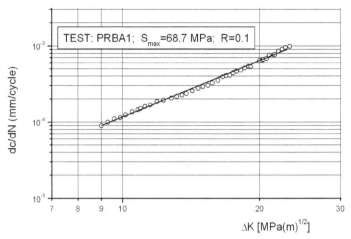

Fig. 11. Result for one CCT coupon test.

The test results were elaborated splitted for the different R values. For the characterization of the linear portion of the curve, we assumed $m$ as a deterministic parameter (equal for each test, Fig. 12), and we considered only $C$ as normal distributed.

In such a way, for $R=0.1$, we found that $m=2.555$ and $C$ is normal distributed with $\mu[(C)]=2.8834\times10^{-7}$, and $\sigma[(C)]=0.036792$.

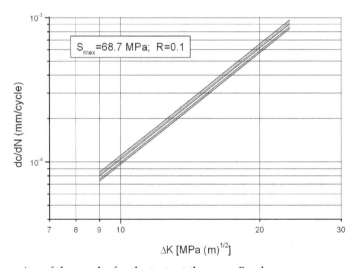

Fig. 12. Elaboration of the results for the tests at the same $R$ value.

## 5.3 The validation of the analytical models implemented inside the PISA code

To validate the approach and to verify the capability of the PISA code to simulate the fatigue behavior of aerospace structural components, further experimental tests were carried out on wide lap-joint panels, in the same aluminum alloy as the simple strips, Fig. 7, loaded under a constant amplitude spectrum ($S_{max}$ = 120 MPa, $R$=0.1).

A group of panels was fatigue tested for an assigned number of cycles (1 at 70,000 cycles, 4 at 75,000 cycles, 4 at 80,000 cycles, 3 at 85,000 cycles) and then statically broken to measure the sizes of the nucleated cracks. Also, in these panels the cracks were found only in the most critical row, i.e. the first one, at the countersunk side.

After having statically broken the tested panels, it was not possible, at all the hole sides, to detect a crack and they were considered as run-outs. The run-out effect has been introduced inside the crack size distribution by using the maximum likelihood method (Spindel & Haibach, 1979). Their crack dimensions have been supposed less than 0.1 mm. We supposed that the crack sizes (both with and without the run-outs) belong to two lognormal distributions.

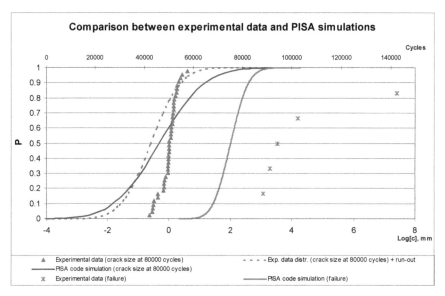

Fig. 13. Comparison between experimental data and PISA simulations (crack dimensions at 80,000 cycles and cycles to failure for the lap-joint panels).

Starting from the EIFS distribution obtained by the simple strip joints and by using the PISA code, the capability of the procedure has been verified by simulating the behavior of the cracks in the most critical row of the lap-joint panels.

The comparison has been made by generating 1000 runs, i.e. by simulating the crack sizes at different number of cycles in the most critical row of 1000 lap-joint panels similar to the tested lap-joint panels (i.e. 15 holes x 2 sides = 30 positions for each panel). In Figure 13 (Cavallini & R. Lazzeri, 2007) the comparison is shown between the predicted crack dimensions and the corresponding experimental results at 80,000 cycles.

The agreement between predictions and experimental results can be seen; in detail, the predicted distribution obtained by using PISA is included between the experimental data distribution with and without the run-outs, Fig. 13.

In addition, 5 further lap-joint panels were fatigue tested till failure. In Fig. 13 the comparison between PISA simulations and the experimental results is also shown. The agreement is good, though the simulation results are a little conservative.

## 6. Applications of the PISA code

The Pisa code is organized in such a way that all the information about geometry, material characterization, loads, inspection methods, failure criteria are collected in an input file.

The code can be used for the evaluation of the fatigue behaviour of only one component, starting from an initial known crack path, or for the generation of high numbers of deterministic simulations for the probabilistic approach.

### 6.1 The simulation of a single panel

In Fig. 14, the fatigue behaviour of a very simple panel, with only four open holes (diameter=4 mm), in Al 2024-T3, loaded under a constant amplitude load with $S_{max}$=100 MPa and $R$=0.1, is simulated. The initial crack path is extracted from the EIFS distribution, but it would be assigned also as an external input. The assigned life was 100,000 cycles. Inspections were planned every 25,000 cycles. For the probability of detection parameters, we assumed $c_{min}$=0.65 mm, $\lambda$=1.62 mm, $\beta$=1.35, (extrapolated from Ratwani, 1996).

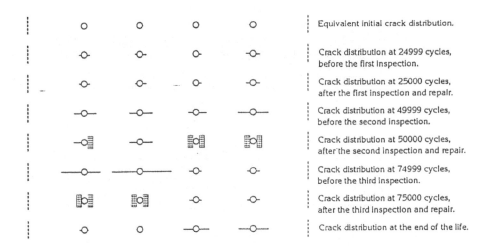

Fig. 14. Simulation of the fatigue life of an open hole panel (damage tolerance criterion).

As it can be seen in the Figure, during the first inspection (at 25,000 cycles) cracks were too small and were not detected. They grew till the second inspection (at 50,000 cycles), when five cracks were detected and the panel repaired. The simulation went on till the following inspection, when four more cracks were detected and repaired. With this inspection plan the panel could reach the target life.

## 6.2 Applications of the PISA code for the probabilistic risk evaluation

Before applying the Pisa code for the evaluation of the probability of failure of an aerospace component, an acceptable risk level must be identified. Indeed, one among the most debated items connected with the application of this methodology is the definition of the 'acceptable' risk level. Usually, 'risk' defines the probability of failure of some components within an assigned period.

Lincoln (Lincoln, 1998), says that for the USAF an acceptable global risk failure is $10^{-7}$ for flight, even if other authors suggest a safer $r(t) \le 10^{-9}$ per hour (Lundberg, 1959).

We fixed $r(t) \le 10^{-7}$. To reach a $10^{-7}$ probability of failure, at least $3 \times 10^{+7}$ simulations must be run.

Starting from a lap-joint in Al 2024-T3, Fig. 7, loaded at C.A. with $S_{max}$=120 MPa, $R$=0.1, our aims were the definition of a 'safe' maintenance plan, the comparison of the effects of the deterministic (safe-life and damage tolerance) and the probabilistic approaches, and the evaluation of their respective advantages and disadvantages, (Cavallini & R. Lazzeri, 2007).

We assumed the following distributions for the stochastic parameters:

- The EIFS fits a lognormal distribution, with $\mu[Log_{10}(c_0)]$=-2.88605, and $\sigma[Log_{10}(c_0)]$=0.28456, [$c_0$] in mm.
- The $C$ parameter in the Paris law is normal distributed, with $\mu[(C)]$=2.8834x$10^{-7}$, and $\sigma[(C)]$=0.036792. The corresponding $m$ value is $m$=2.555.
- The fracture toughness $K_{Ic}$ fits a lognormal distribution, with $\mu[Log_{10}(K_{Ic})]$=1.65 (Koolloons, 2002), and COV= $\sigma[Log_{10}(K_{Ic})]$/ $\mu[Log_{10}(K_{Ic})]$=0.14, [$K_{Ic}$]=MPa$(m)^{0.5}$ (Schutz, 1980),
- The probability of detection is expressed as (1), with $c_{min}$=0.65 mm, $\lambda$=1.62 mm, $\beta$=1.35, (extrapolated from Ratwani, 1996).

At first, we simulated the fatigue behaviour of $3 \times 10^{+7}$ lap-joint panels without any inspection actions. The number of cycles with probability of failure equal to $10^{-7}$ is 51,000 cycles, Fig. 15. So, this number of cycles can be fixed for the first inspection (threshold).

The second run was made after having fixed, for each panel, the first inspection at 51,000 cycles. In such a way, we obtained the new probability of failure curve and it was possible to fix the second inspection at 63,000 cycles, that is (63,000 - 51,000) = 12,000 cycles after the first one.

In Fig. 15 the probability of failures corresponding to the different deterministic approaches are also shown.

The safe life criterion requires the component replacement after a portion (as for example ¼) of its mean life. The mean life (probability of failure equal to 50%) corresponds to 74,550 cycles that, divided by four, gives 18,638 cycles. So, for the safe life criterion, after 18,638 cycles the component should be substituted, without any consideration about its real damage condition.

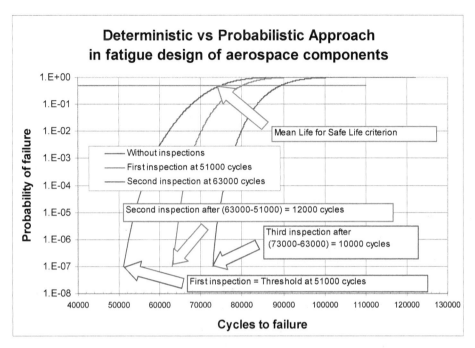

Fig. 15. Example of PISA simulation and maintenance strategy for a lap-joint panel.

The probability of failure at 18,638 cycles is extremely low, so the component could still be used without any loss in safety. In this situation, the component replacement only introduces costs.

As far as the damage tolerance criterion is concerned, the first inspection (threshold) is fixed by evaluating the number of cycles necessary for a crack of assigned dimension (regulations state a 1.27 mm size) to grow till the final failure. The inspections cited below are planned considering the propagation period of a sure visible crack (depending on the selected inspection method, in this case 6.35 mm) till the final failure. A safety factor equal to 2 at the threshold and 3 at the following period are additionally applied.

For this component, by analytical calculation or by using the PISA code itself, it can be found that the period necessary for a crack to grow from 1.27 mm to the final failure is equal to 57,700 cycles, and from 6.35 mm to the final failure is equal to 31,100 cycles.

Therefore, the first inspection will be carried out at 57,700/2=28,850 cycles and the next one after 31,100/3 = 10,367 cycles.

As it can be seen in Fig 15, the corresponding probability of failures is very low, and so the inspection plan, based on the deterministic damage tolerance approach, might be very expensive.

## 7. Conclusion

In this Chapter, a new possible and useful approach to fatigue design of aerospace metallic components is explained, founded on probabilistic bases, together with a tool – the PISA code - and the experimental test results used for the validation of the tool, and of the approach.

The validation analysis provided good results and therefore the PISA code can be used for the risk assessment analysis and to compare the effect of the deterministic approaches (damage tolerance and safe-life) with those of the probabilistic approach in the fatigue design of a wide lap-joint panel.

The advantages appeared to be very significant:

- the probability of failure can be defined as a design constraint (or goal), for instance $10^{-7}$; so the risk level is well defined. In deterministic approaches, this important element is not known and the assumption of conservative values of the inputs can produce uneconomical designs without benefits;
- in each design condition, it is possible to know the "distance" from the critical condition in terms of probability of failure;
- the Multiple Site Damage event can be handled in a logical way because it is one of the possible statistical events. The same problem, faced on deterministic bases, could bring to heavy and/or very expensive solutions;
- components are inspected or withdrawn and substituted only if really necessary, thus avoiding too early inspections or the substitution of intact components.

The comparison between the different approaches, applied to a lap joint panel, shows that a more economic inspection plan can be applied if the probabilistic approach is used, without loss of safety.

Of course, this new methodology can be safely applied only if reliable models for the crack growth are available, and the parameter distributions have been carefully obtained.

## 8. References

Besuner P.M. (1987). Probabilistic Fracture Mechanics, In: *Probabilistic fracture mechanics and reliability*, Provan Ed., pp. 387-436, Martinus Nijhoff Publ., ISBN 90-247-3334-0, Dordrecht (NL).

Cavallini G., Lanciotti A. & Lazzeri L. (1997). A Probabilistic Approach to Aircraft Structures Risk Assessment, Proceedings of the 19th ICAF Symposium, Edinburgh (UK), June 1997, pp. 421-440.

Cavallini G. & Lazzeri R. (2007). A probabilistic approach to fatigue risk assessment in aerospace components. *Eng. Fracture Mech.*, vol. 74, issue 18, (Dec. 2007), pp. 2964-2970.

Federal Aviation Administration (1998). Federal Aviation Regulations – Part 25. Airworthiness Standards: Transport Category Airplanes, Section 571, Damage-tolerance and fatigue evaluation of structures. Available from http://rgl.faa.gov/Regulatory_and_Guidance_Library/rgFAR.nsf/Frameset?OpenPage

Grooteman F. P. (2002). *WP4.4: Structural Reliability Solution Methods – Advanced Stochastic Method,* Admire Document N. ADMIRE-TR-4.4-03-3.1/NLR-CR-2002-544.

Hammersley J. M. & Handscomb D. C. (1983). *Monte Carlo Methods,* Chapman and Hall Publ., ISBN 0-412-15870-1, New York.

Hovey P. W., Berens A. P. & Skinn D. A. (1991). *Risk Analysis for Aging Aircraft Volume 1 – Analysis,* Flight Dynamics Directorate, Wright Laboratory, Wright-Patterson AFB, OH 45433-6553.

Imparato G. & Santini L. (1997). *Prove sperimentali sul comportamento a fatica di strutture con danneggiamento multiplo,* Thesis in Aeronautical Engineering, Department of Aerospace Engineering, University of Pisa.

Joint Aviation Authorities (1994). Joint Airworthiness Requirements, JAR-25, Large Aeroplanes, Section 1, Subpart D, JAR 25.571, Damage-tolerance and fatigue evaluation of structures.

Johnston G. O. (1983). Statistical scatter in fracture toughness and fatigue crack growth rates, In: *Probabilistic fracture Mechanics and Fatigue Methods: Applications for structural design and maintenance,* ASTM STP 798, pp. 42-66, Bloom J.M. & Ekvall J.C., American Society for Testing Materials.

Koolloons M. (2002). Details on Round Tobin Tests, ADMIRE Document, ADMIRE-TR-5.1-04-1.1/NLR.

Kuo, A., Yasgur, D. & Levy, M. (1986). Assessment of damage tolerance requirements and analyses - Task I report., *ICAF Doc. 1583,* AFVAL-TR-86-3003, vol. II, AFVAL Wright-Patterson Air Force Base, Dayton, Ohio.

Lewis W. H., Sproat W.H., Dodd B. D. & Hamilton J. M. (1978). *Reliability of nondestructive inspection-final report,* San Antonio Air Logistic Center, Rep. SA-ALC/MME 76-6-38-1.

Liao M. & Komorowski J. P. (2004). Corrosion risk assessment of aircraft structures. *Journal of ASTM International,* vol. 1, no. 8 (September 2004), pp. 183-198.

Lincoln J.W. (1998). Role of nondestructive inspection airworthiness assurance, *RTO AVT Workshop on Airframe Inspection Reliability under field/depot conditions,* Brussels, Belgium, May 1998.

Lundberg, B. (1959). The Quantitative Statistical Approach to the Aircraft Fatigue Problem, Full-Scale Fatigue Testing of Aircraft Structures, *Proceedings of the 1st ICAF Symposium,* Amsterdam, Netherlands, 1959, Pergamon Press, pp. 393-412 (1961).

Madsen H. O., Krenk S. & Lind N. C. (1986). *Methods of Structural Safety,* Prentice Hall, Inc., ISBN 0-13-579475-7, Englewood Cliffs, NJ.

Manning, S.D., Yang, J.N. & Rudd, J.L. (1987). Durability of Aircraft Structures, In: *Probabilistic Fracture Mechanics and Reliability,* Provan J.W. (ed.), pp. 213-267, Martinus Nijhoff.

Ratwani M. M. (1996). Visual and non-destructive inspection technologies, In: *Aging Combat Aircraft Fleets - Long Term Implications,* AGARD SMP LS-206.

Sampath S. & Broek D. (1991). Estimation of requirements of inspection intervals for panels susceptible to multiple site damage, In: *Structural Integrity of Aging Airplanes,* Atluri S.N., Sampath, S.G. & Tong, P., Editors, , pp. 339-389, Springer-Verlag, Berlin.

Schijve J., (2001). *Fatigue of Structures and Materials,* Kluwer Academic Publishers, ISBN 0-7923-7013-9, Dordrecht, NL.

Schutz W. (1980). Treatment of scatter of fracture toughness data for design purpose, In: *Practical Applications of fracture Mechanics*, AGARD-AG-257, Liebowitz H. (ed).

Spindel J.E. & Haibach E. (1979). The method of maximum likelihood applied to the statistical analysis of fatigue data. *International Journal of Fatigue*, vol. I, no. 2, (April 1979), pp. 81-88.

Tong Y. C. (2001). Literature Review on Aircraft Structural Risk and Reliability Analysis, Department of Defence DSTO, Melbourne. Available from http://dspace.dsto.defence.gov.au/dspace/bitstream/1947/4289/1/DSTO-TR-1110%20PR.pdf

US Department of Defence (1998). Joint Service Specification Guide - Aircraft Structures, JSSG-2006, Available from
http://www.everyspec.com/USAF/USAF+(General)/JSSG-2006_10206/.

Wang G. S. (1988). An Elastic-Plastic Solution for a Normally Loaded Center Hole in a finite Circular Body, *Int. J. Press-Ves & Piping*, vol. 33, pp. 269-284.

# Potential of MoSi$_2$ and MoSi$_2$-Si$_3$N$_4$ Composites for Aircraft Gas Turbine Engines

Melih Cemal Kushan[1], Yagiz Uzunonat[2],
Sinem Cevik Uzgur[3] and Fehmi Diltemiz[4]
*[1]Eskisehir Osmangazi University*
*[2] Anadolu University*
*[3]Ondokuz Mayis University*
*[4]1st Air Supply and Maintenance Base*
*Turkey*

## 1. Introduction

It has been expected that gas turbine engines in high temperature environments where aggressive mechanical stresses may occur and a good surface stability is needed should operate more efficiently. So the investigations about the materials which will be able to carry the aviation technology to the next level are beginning to accelerate in this direction. And also it expected that those new materials using in gas turbine engines as a high temperature structural material will exceed the superalloys' mechanical and physical limits. The intended development can only be achieved by providing the improvement of the essential properties of the structural materials such as thermal fatigue, oxidation resistance, strength/weigth ratio and fracture toughness. There are two different type of materials which are candidate to resist the operating conditions about 1200°C; first one is structural ceramics such as SiC, Si$_3$N$_4$ and the second one is structural silicides such as MoSi$_2$.

After the propulsion systems with high strength/weight ratio, it observed that development of new materials with high strength and low density was necessary, thus the studies about the intermetallics began. The most important ones of these intermetallic compounds are silicides and aluminides. By the oxide layers in Al$_2$O$_3$, it can be used as a protective material in high temperature applications. Moreover, aluminides such as FeAl, TiAl, Ni$_3$Al, can be suitable for some special applications in low and medium temperatures. In spite of these advantages, they remain inadequate above the temperatures 1200°C for their melting points with 1400-1600°C. Their low strength and creep resistance is not suitable for the temperatures above 1000°C. For this reason, it seems that silicides and aluminides are the proper materials for high service applications (Vaseduvan & Petrovic, 1992).

## 2. Superalloys and their limitations at elevated temperatures

In aviation applications, advanced gas turbine elements are exposed to several mechanic, thermal and corrosive environments and intensive studies for the developing of these parts are still continuing. However, these alloys are needed to be cooled during the operation of the turbine engine and the practical temperature limits for metallic alloys remain below 1100°C. But in this situation, the elevation of turbine inlet temperature will be quite difficult and expansive. Because of these given limitations, there is not any important improvements on nickel based superalloys since 1985 (Soetching, 1995).

The basic facts that can directly effect the performance of superalloys in high temperatures are oxidation, hot corrosion and thermal fatigue. These effects cause the superalloy elements' surfaces may react with hot gases easier, and then their surface stability decreases (Bradley, 1988). Furthermore, during operation and stand-by period of turbine, there occurs a oscillation motion in the hot section elements respectively. This causes thermal fatigues on the superalloy parts.

## 2.1 Oxidation

Oxidation is one of the most serious factors acting on the gas turbine's service life and can be determined as the reaction of materials with oxygen in 2-4 atm. partial pressure (Tein & Caulfield, 1989). Mostly the uniform oxidation is not accepted as a considerable problem in relatively low temperatures (870°C and below). But in temperatures about 1100°C, the aluminum content in the form of $Al_2O_3$ as a protective oxide can not provide the expected protection in long term periods. For this reason, it is necessary to use the silicide based structural composite materials or to make protective coatings with respect to the segment's location in gas turbine.

## 2.2 Hot corrosion

The process of hot corrosion contains a structural element and the reactions occurring in its surroundings. In operating conditions at high temperatures there is a possible accelerated oxidation for superalloys. Another name for this reaction is hot corrosion and it consists of two different mechanisms as low temperature (680-750°C) and high temperature (900-1050°C) hot corrosion (Akkuş, 1999).

The basic principle to avoid from the hot corrosion in superalloys is using of the high content of chrome (≥ %20) during the manufacturing of material. But only a few types of nickel based superalloys have this rate for their high proportion of $\gamma'$ and $\gamma''$ structure.

## 2.3 Thermal fatigue

Heating with non-uniform distribution make interior stresses in the zones hotter than the average temperature of the turbine, and tension stresses in the colder zones. Superalloy turbine vanes are the good examples of elements exposed to thermal fatigue in aeroplane jet engines. During the acceleration, inlet and outlet edges of the turbine vane can heat and expand easier than the medium part under cooling. But in deceleration, inlet and outlet parts can quietly cool off than the medium parts. This case results as fatigue crack at the edges.

## 3. Physical and mechanical properties of MoSi₂

$MoSi_2$ is a potential material for high temperature structural applications primarily due to its high melting point (2020°C), lower density (6.3 $g/cm^3$) compared with superalloys, excellent oxidation resistance, high thermal conductivity, and thermodynamic compatibility with many ceramic reinforcements. However, low fracture toughness at near-ambient temperatures, low strength at elevated temperatures in the monolithic form and tendency to pest degradation at ~500°C have seriously limited the development of MoSi$_2$-based structural materials. Several recent studies have attempted to address these issues and have shown promising results. For example, pest resistant MoSi$_2$-based materials have been developed using silicon nitride reinforcement or alloying with Al.

For general polycrystalline ductility five independent deformation modes are necessary. Changing the critical resolved shear stress of the slip systems through alloying may be a way to activate all three slip vectors, and obtain polycrystalline ductility. In fact, solid solution softening has been observed at room tempera ture in MoSi$_2$ alloyed with Al and transition metals such as Nb, V and Ta. The mechanism of softening is not clearly understood, although first principles calculations indicate that solutes such as Al, Mg, V and Nb may change the Peierls stress so as to enhance relative to cleavage. Clearly, more work is needed to understand how alloying may influence the mechanical behavior of MoSi$_2$.

With regard to elevated temperature strengthening of MoSi$_2$, both alloying with W to form C11b (Mo, W)Si$_2$ alloys and composites with ceramic re inforcements such as SiC have been tried. A (Mo, W)Si$_2$./20 vol.% SiC composite was shown to have significantly higher strength than Mar-M247 superalloy at temperatures above 1000°C. However, the strength of the (Mo, W)Si$_2$./20 vol.% SiC composite dropped by almost an order of mag -nitude from 1200 to 1500°C; the yield strength at 1500°C was only ~75 MPa. A simpler and more e€ective way of strengthening MoSi$_2$ at elevated temperatures is needed where the strength can be better retained with increasing temperature above 1200°C. Our preliminary studies using hot hardness experiments have shown that Re addition to MoSi$_2$ caused signifcant hardening up to 1300°C. Further, it has been reported that alloying with Re, perhaps in synergism with carbon, increased the pesting resistance in the temperature range of 500 ± 800°C. In another preliminary study, polycrystalline (Mo, Re)Si$_2$ alloys exhibited a minimum creep rate of ~5 x 10$^{-6}$/s at 100 MPa applied stress at 1400° C as compared with the ~1 x 10$^{-4}$/s creep rate exhibited by MoSi$_2$. No detailed mechanistic study has been performed to understand the effects of Re alloying on the elevated temperature mechanical behavior of MoSi$_2$. In the present investigation, we have evaluated the mechanical properties, in compression, of arc-melted polycrystalline MoSi$_2$ and (Mo, Re) Si$_2$ alloys. We find that significant strengthening is achieved up to 1600°C by only small additions of Re. The mechanisms of elevated temperature solid solution strengthening are elucidated by considering the generation of constitutional Si vacancies that may pair with Re substitutionals to form tetragonally distorted point defect complexes. Characteristics of MoSi$_2$ make it an interesting material as high temperature structural silicide. Not only it has a low density and a high melting point but also it can excellently resist the free oxygen of air in high temperature environments for a long time period. On the other hand, researchers noticed its potential as a structural material due to its electrical resistance increasing after every use and high modulus of elasticity at high temperatures. This makes MoSi$_2$ a candidate material for structural high temperature applications particularly in gas turbine engines. MoSi$_2$ and its composites offer a higher rate of resistance to oxidizing and aggressive environments during the combustion processes with their high melting points.

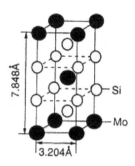

Fig. 1. Unit cell of the body-centered tetragonal C11b structure of MoSi$_2$. (Misra et al., 1999).

Fracture toughness of the material shows similarities with the other silicon based ceramics and yet it receives a brittle fracture resulted with low toughness. Table 1. shows the considerable characteristics of MoSi$_2$.

|  | Metric | English |
|---|---|---|
| Density | 6.23 g/cm$^3$ | 0.225 lb/in$^3$ |
| Molecular Weigth | 152.11 g/mol | 152.11 g/mol |
| Electrical Resistance (20°C) | 3.5x10$^{-7}$ ohm-cm | 3.5x10$^{-7}$ ohm-cm |
| Electrical Resistance(1700°C) | 4.0x10$^{-6}$ ohm-cm | 4.0x10$^{-6}$ ohm-cm |
| Thermal Capacity | 0.437 J/g-°C | 0.104 BTU/lb-°F |
| Thermal Conductivity | 66.2 W/m-K | 459 BTU-in/hr-ft$^2$-°F |
| Melting Point | 2030°C | 4046°F |
| Maximum Service Temp. | 1600°C | 2912°F |
| Crystal Structure | Tetragonal | Tetragonal |

Table 1. Basic characteristics of MoSi$_2$.

The figure below shows the tetragonal lattice structure directions, red and blue points inidicate silicon and molybdenum atoms respectively.

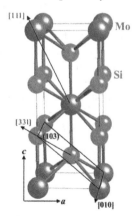

Fig. 2. Tetragonal MoSi$_2$ lattice structure.

One of the most considerable limitations in $MoSi_2$ applications is the structural disintegration during the low temperature oxidation which is known as pesting oxidation (Meschter, 1992). Previously, we noted that $MoSi_2$ has an excellent oxidation resistance above the 1000°C, but at the temperatures about 500°C as it is presented Figure 3., the oxidation mechanism accelerates because of the volume expansion, $MoO_3$ crystals, amorph-shaped $SiO_2$ bulks and $MoSi_2$ particles residual from the reaction. If the material is porous and the surface accuracy is low, this state can be observed along the cracks or grain boundaries, and granular oxide particles occur as a result.

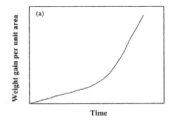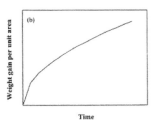

Fig. 3. Isothermal oxidation curves (a) at room temperature, (b) at the temperatures above 1000° C (Liu et al., 2001).

This fact was discovered in 1955 and predicted as the grain boundary fracture due to solution oxygen at the grain boundaries after the short-term cyclic diffusion, even though its complete nature is still a phenomena (Chou & Nieh, 1992, 1993). Methods for preventing from the pest effect are continuing. These methods are; making a protective $SiO_2$ coating on the material and increasing the relative density of $MoSi_2$ in the structure (Wang et al., 2003)

773K 600s        773K 3600s        773K 7200s

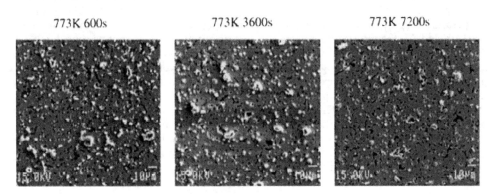

Fig. 4. Surfaces oxidized at 773K for 600-7200s in $O_2$ (Chen et al., 1999).

During the oxidation reactions above 600°C, no pesting effect can be observed. $MoSi_2$ based composites have considerably higher isothermal oxidation resistance than any other titanium, niobium or tantalum based composites, intermetallic compounds and nickel based superalloys, $MoSi_2$ perfectly keeps this condition to 1600°C (Vaseduvan & Petrovic, 1992).

673K 7200s                    773K 7200s                    873K 7200s

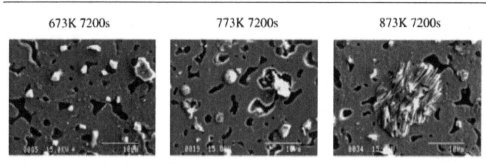

Fig. 5. Surfaces oxidized at 673-873$K$ for 7200s in $O_2$ (Chen et al., 1999).

Despite excellent oxidation resistance, high melting point, and low density, the potentials of molybdenum disilicide as a high temperature structural material have not been utilized due to its brittleness at low temperatures and low strength at high temperatures . For example, below 900° C, the fracture toughness of $MoSi_2$ is in the range of 2-4 $MPam^{1/2}$ , and the 0.2% offset yield strength of $MoSi_2$ at 1600° C is about 20 MPa. Alloying or reinforcing with a second phase may lower the brittle to ductile transition temperature (BDTT) of $MoSi_2$. However, ductile-phase toughening with metallic phases has limited applicability in $MoSi_2$ due to the chemical reaction with silicon to form silicides, and reinforcing with ceramic second phases such as SiC and $ZrO_2$ has only a modest effect on enhancing plastic flow and increasing toughness.

First principles calculations indicate that alloying of $MoSi_2$ while maintaining its body-centered tetragonal (C11b) structure may result in improved mechanical properties. For example, Al and Nb may enhance ductility and Re may increase strength. Improvements in both low and high temperature mechanical properties of $MoSi_2$ have been reported by alloying $MoSi_2$ with small amounts of Al, Nb, and Re (<2 at.%). During alloying, below the solubility limits of alloying elements in the C11b structure of $MoSi_2$, Al substitutes for Si, whereas Re and Nb substitute for Mo. The solubility limits of Re, Nb and Al in $MoSi_2$ have been reported as ~2.5, 1.3 and 2.7 at.%, respectively. Although improvements in the ambient temperature toughness have been reported by alloying of $MoSi_2$ beyond the solubility limits with Nb and Al, the rates of improvement per fraction of solute are not as considerable as those observed in single phase alloys. Furthermore, the presence of secondary phases with a lower high temperature strength than the matrix alloy would degrade the mechanical properties at high temperature (>1500°C) for which applications $MoSi_2$ is an excellent candidate. The aim of this investigation was to explore the possibility of obtaining concurrently enhanced room temperature ductility and high temperature strength in single-phase $MoSi_2$ by combining the high temperature hardening and the low temperature softening effects of Re, Al, and Nb. Hardness testing at room temperature and compression testing at 1600° C are conducted on unalloyed and alloyed $MoSi_2$ samples in order to study both low and high temperature effects of each alloying composition on the mechanical properties of $MoSi_2$. (Sharif et al., 2001).

## 4. Effects of alloying

### 4.1 Hardness

Microhardness testing performed on stoichiometric samples obtained from melting (Mo, Re or Nb)(Si, Al)$_{2.01}$ samples indicated that unalloyed $MoSi_2$ had an average Vickers hardness value of 89968 Hv. Samples containing 2 at.% Al or 1 at.% Nb had average Vickers hardness values of 72928 Hv and 72950 Hv, respectively. 2.5 at.% Re containing samples had the highest hardness value of 103971 Hv. Samples containing 1 at.% Re+2 at.% Al had an average hardness value of 74230 Hv, slightly higher than Al containing samples but significantly lower than both MoSi₂ and (Mo, 1 at.% Re)Si₂ samples. Slip lines were observed around indentations in all samples except the unalloyed MoSi₂ and (Mo, 2.5 at.% Re)Si₂ samples. Samples containing 1 at.% Nb+2 at.% Al did not exhibit any improvements in the mechanical properties and were excluded from further considerations.

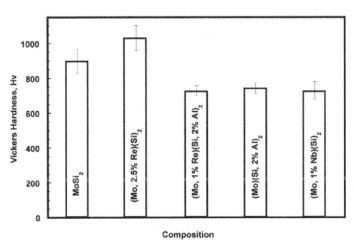

Fig. 6. Vickers hardness values for polycrystalline samples of all composition under investigation. Hardness of polycrystalline Al containing samples is also compared to the values obtained on monocrystalline Al containing samples on (100) and (001) surfaces of the crystal.

## 4.2 Yield strength

Compression testing at room temperature and at 1600°C was used to determine the 0.2% offset yield strength for all polycrystalline samples. Unalloyed MoSi₂ and 2.5 at.% Re containing samples could not be deformed plastically below 900 and 1200° C, respectively. Below these temperatures, the aforementioned samples would undergo brittle fracture during compression testing. The addition of 2.5 at.% Re increased the BDTT, in compression, of MoSi₂ by about 300° C while increasing its yield strength from 14 MPa to 170 at 1600° C. Among alloying elements investigated here, 2.5 at.% Re was most effective in increasing strength at 1600° C. The addition of 2 at.% Al was effective in both increasing the high temperature strength to 55 MPa and lowering the BDTT to 425° C. Mo(Si, 2 at.% Al)₂ samples exhibited the lowest room temperature yield strength of 415 MPa. Addition of 1 at.% Re+2 at.% Al combined the beneficial effects of both alloying elements and resulted in enhanced ambient temperature compressive plasticity and high temperature strength compared to the unalloyed samples.

However, the improvements in room temperature plasticity was less than that of samples alloyed with 2 at.% Al alone as evident from the value of the room temperature yield strength of (Mo, 1 at.% Re)(Si, 2 at.% Al)$_2$ alloy, 670 MPa. Similarly, the enhancement in high temperature strength was less than that of only Re containing samples but greater than that of only Al containing samples. The effects of 1 at.% Nb as an alloying element by itself in lowering BDTT and enhancing high temperature strength of MoSi$_2$ was more pronounced than the combined effects of 1 at.% Re and 2 at.% Al. The room temperature yield strength of (Mo, 1 at.% Nb)Si$_2$, 500 MPa, was higher than that of 2 at.% Al containing samples and lower than that of (Mo, 1 at.% Re)(Si, 2 at.% Al)$_2$ samples. The 0.2% offset strength of (Mo, 1 at.% Nb)Si$_2$ samples at 1600°C, 143 MPa, was an order of magnitude greater than that of unalloyed MoSi$_2$ (Sharif et al., 2001).

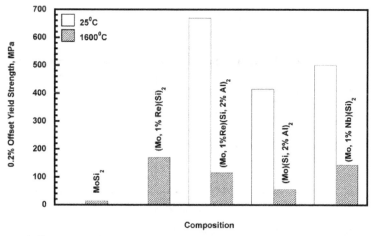

Fig. 7. Effects of alloying on the room temperature and high temperature (1600°C) strength of MoSi$_2$.

## 4.3 Ductility

Molybdenum disilicide crystallizes in an ordered body centered tetragonal structure with a=0.320 nm and c=0.785 nm, formed by alternate stacking of single Mo and double Si (001) layers. With its high temperature ductility and exceptional resistance to corrosion and fatigue crack growth, MoSi$_2$ combines the toughness of a metal with the strength of a ceramic and is a promising candidate to replace nickel alloys in the next generation of high-temperature gas turbines. Unfortunately, it undergoes a ductile–brittle transition (DBT) at 1200°C, with the fracture toughness dropping to 2–3 MPa m$^{1/2}$, well below the minimum of 20 MPa m$^{1/2}$ required for engine applications. This brittleness at low temperature means that MoSi$_2$ must be formed by costly electro-discharge machining and places a severe limitation on its potential technological utility. However, there is a reasonable chance that the DBT in MoSi$_2$ may be manipulated or even eliminated. Many of the slip systems in MoSi$_2$ are ductile and it is only for a stress axis near [001] that a DBT is observed.

It is desirable therefore, to alter the properties of MoSi$_2$ in very specific ways. This can be, and has in the past been, attempted by heuristically changing the composition or structure of the material and studying experimentally the effect of these changes. It will be argued in this paper that advances in the theory of bonding in solids, based on quantum mechanical density functional calculations, offer an alternative route which can be used as a cost-effective precursor to experiment. The need, in the case of $MoSi_2$, is for an element or elements which can be introduced at microalloy levels (less than 5%) and which will perturb the brittle–ductile behavior in favor of ductility without adversely affecting the advantageous physical properties. While the method of choice would normally be an atomistic calculation, bonding in $MoSi_2$ is known to have hybrid metallic and covalent character. Determination of the effects of alloying on such bonding requires accurate quantum mechanical treatment of the electrons, and generation of reliable interatomic potentials, which are an essential prerequisite to atomistic methods, is impractical1. Instead, use is made of recent advances in the theory of dislocation nucleation and mobility which provide approximate links between these properties and the generalized stacking fault energy surface, which can be calculated accurately using first principles quantum mechanical techniques. A similar approach has been used successfully by two of the present authors to investigate the DBT in silicon. Even with these gross approximations, the numerical work is intensive. The calculations are restricted to small supercells, with correspondingly large alloy content, and the effects of true microalloying must be estimated by interpolation. An overview of the experimental background will be presented in the next section.

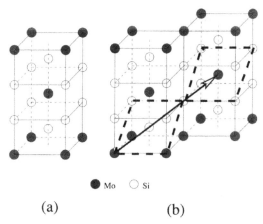

● Mo   ○ Si

(a)                          (b)

Fig. 8. Crystal structure of $MoSi_2$. (a) Unit cell for the body centered C11b structure; solid circles represent Mo atoms and open circles represent Si atoms. (b) (013) plane and the Burgers vector for {013}⟨331⟩ slip systems (Waghmare et al., 1999).

## 4.4 Creep

The creep behavior of $MoSi_2$-based materials has been extensively studied. It has been observed that the grain size has a large effect on creep resistance of monolithic $MoSi_2$. Reinforcing with SiC also refined grain size that enhanced creep rates overshadowing any beneficial effects of reinforcement Increased creep resistance has been noted only when volume fractions of SiC are above 20%. Another important factor strongly affecting the creep strain rate of $MoSi_2$ is the presence of silica particles ($SiO_2$). During high temperature deformation, the $SiO_2$ particles at the grain boundaries flow to form intergranular film, which slides or cracks. A high volume fraction of $SiO_2$ and reduction in grain size, both enhance creep rates; but it is of interest to examine how the two are interrelated. Alloying of polycrystalline $MoSi_2$ with Al and C converts $SiO_2$ to $Al_2O_3$ and SiC, respectively, which leads to the enhancement in creep resistance. When C is added, oxygen is got rid off in the form of CO or $CO_2$, which may leave behind fine pores, which are difficult to close. When Al is added, the reaction between Al and $SiO_2$ forms $Al_2O_3$ and Si. The Si may remain in elemental form or react with $Mo_5Si_3$ particles that are present in small volume fractions. These particles form due to partial oxidation of $MoSi_2$ during hot pressing, particularly when vacuum is low. The probability of the reaction between free Si and residual $Mo_5Si_3$ in the present study is high, as free Si has been observed in the microstructure only very rarely. The figure below presents the effect of alloying of single and polycrystalline $MoSi_2$ with Al, on the creep rates at 1300°C. The single crystals of $Mo(Si_{0.97} Al_{0.03})_2$, with hexagonal C40 or hP9 (Pearson's symbol) structure, have shown higher creep rates, compared to those of single crystals of $MoSi_2$ along the [0 15 1] orientation of stress axis. However, the trend reverses with change of stress-axis to [001] direction. On the other hand, polycrystalline $MoSi_2$-5.5Al alloy has shown improvement in creep resistance, compared to polycrystalline $MoSi_2$ at 1300°C. Unlike $Mo(Si_{0.97} Al_{0.03})_2$, the matrix phase of $MoSi_2$-5.5Al has tetragonal, C11b structure.

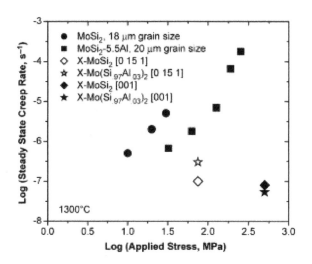

Fig. 9. Comparison of steady state creep rates, measured at 1300°C on $MoSi_2$ and $MoSi_2$-5.5Al alloy, as well as single crystals of $MoSi_2$ and $Mo(Si_{0.97} Al_{0.03})_2$ tested with [0 15 1] and [001] orientations. Single crystals are marked as X (Mitra et al., 2004).

As expected, the creep rates of polycrystalline $MoSi_2$ and $MoSi_2$-5.5Al alloy are higher compared to those of single crystals at 1300°C, because of the role of grain boundaries at 0.68 Tm: In the present investigation, samples of $MoSi_2$ with varying grain sizes and $SiO_2$ contents, as well as those of $MoSi_2$-20 vol% SiC composite and $MoSi_2$-Al alloys have been creep tested at 1200°C and their behaviors analyzed. The values of activation volume and threshold stress have been calculated. These provide an insight into the ratecontrolling and strengthening mechanisms. The creep behavior of the above materials has also been compared with deformation behavior under constant strain rate tests.

## 4.5 Plastic deformation

Monolithic $MoSi_2$ exhibits only a modest value of fracture toughness at low temperatures and inadequate strength at high temperatures. Thus, many of recent studies on the development of $MoSi_2$-based alloys have focused on improving these poor mechanical properties through forming composites with ceramics and with other silicides. These properties have recently been reported to be significantly improved in composites formed with $Si_3N_4$ and SiC. However, the volume fraction of $Si_3N_4$ and SiC ceramic reinforcements in these $MoSi_2$-composites generally exceeds 50%. Further improvements in mechanical properties of these composites will be achieved if those of the $MoSi_2$ matrix phase are improved. The present study was undertaken to achieve this by alloying additions to $MoSi_2$. Transition-metal atoms that form disilicides with tetragonal C11b, hexagonal C40 and orthorhombic C54 structures are considered as alloying elements to $MoSi_2$. These three structures commonly possess (pseudo-) hexagonally arranged $TMSi_2$ layers and differ from each other only in the stacking sequence of these $TMSi_2$ layers; the C11b, C40 and C54 structures are based on the AB, ABC and ADBC stacking of these layers, respectively. W and Re have been known to form a C11b disilicide with Si and they are believed to form a complete C11b solid-solution with $MoSi_2$, although recent studies have indicated that the disilicide formed with Re is an off-stoichiometric (defective) one for-mulated to be $ReSi_{1.75}$ having a monoclinic crystal structure. The details of our crystal structure assessment for $ReSi_{1.75}$ as well as phase equilibria in the $MoSi_2 \pm ReSi_{1.75}$ pseudobinary system will be published elsewhere. Large amounts of alloying additions are possible for these alloying elements, and high temperature strength is expected to be improved through a solid solution hardening mechanism since the hardness of both $WSi_2$ and $ReSi_{1.75}$ is reported to be larger than that of $MoSi_2$. The yield strength of $MoSi_2$ powder compacts is greatly increased when $WSi_2$ is alloyed with $MoSi_2$ by more than 50 vol%. In addition, our previous study on single crystals of $MoSi_2 \pm WSi_2$ solid solutions has indicated that the compression yield stres above 1200° C greatly increases when the $WSi_2$ content in the solutions exceeds 50 vol.%. However, low temperature deformability may be declined upon alloying with these elements because of the increased strength. Indeed, the room temperature hardness of both $MoSi_2 \pm WSi_2$ and $MoSi_2 \pm ReSi_{1.75}$ solid solutions is reported to monotonically increase with the increase in either $WSi_2$ or $ReSi_{1.75}$ content.

V, Cr, Nb and Ta have been known to form a C40 disilicide with Si. Al is also known to transform $MoSi_2$ from the C11b to the C40 structures by substituting it for Si. Of the five slip systems identified to be operative in $MoSi_2$, slip on {110}<111> is operative from 500°C. 1/2<111> dislocations of this slip system are reported to dissociate into two identical 1/4<111> partials separated by a stacking fault. The stacking across the fault is ABC and resembles the stacking of (0001) in the C40 structure. Hence, the addition of elements that form a C40 disilicide may cause the energy difference between C11b and C40 structures to decrease so that the energy of the stacking fault would also be decreased, although the solid solubility of these alloying elements in $MoSi_2$ has been reported to be rather limited to the level of a few atomic %. From this point of view, we may expect that the deformability of $MoSi_2$ at low temperatures increases upon alloying with elements that form a C40 disilicide. This is consistent that V and Nb may enhance the ductility of $MoSi_2$. Indeed, room temperature hardness of $MoSi_2$ polycrystals decreases upon alloying with Cr, Nb, Ta and Al and similar observations were made for Al-bearing $MoSi_2$ polycrystals. Compression deformation experiments made so far on ternary $MoSi_2$ single crystals containing these elements have focused attention to the high temperature deformation behavior. The yield stress of MoSi₂ increases upon allying with Cr above 1100°C. A similar observation was made for Nb-bearing MoSi₂ at 1400°C. However, since these compression experiments were made only at high temperatures above 1100°C, almost nothing is know about the low temperature strength and deformability of these ternary MoSi₂ single crystals. V, Cr, Nb and Al that form a disilicide with the C40 structure and W and Re that form a disilicide with the C11b structure as alloying elements to MoSi₂, and investigated the deformation behavior of single crystals of MoSi₂ containing these elements in a wide temperature range from room temperature to 1500°C. The crystal orientations investigated were the [0 15 1] orientation, in which slip on {110}<111> is operative, and the [001] orientation, in which the highest strength is obtained at high temperatures for binary MoSi₂.

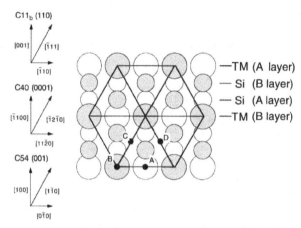

Fig. 10. Atomic arrangement on TMSi₂ layers corresponding to {110}, (0001) and (001) planes in the C11b, C40 and C54 structures, respectively. The stacking positions of A±D and crystallographic directions with respect to these three structures are indicated (Inui et al., 2000)

## 5. Development of MoSi₂ – Si₃N₄ composites

Interest in fiber reinforced ceramic matrix composites (FRCMCs) has increased steadily over the past 15 years, and several refined silicon-base composite systems are now being produced commercially. These composites offer very good structural stiffness, high specific strength to weight, and good high temperature environmental resistance. Industrial applications include, hot gas filters, shrouds, and combuster liners. In addition, silicon-base ceramic composites are being considered for gas turbine hot gas flow path components, e.g. combusters transition pieces, and nozzles. The manufacturers of liquid rocket engines are also looking to ceramic composites in hopes of obtaining better efficiency in the next generation of designs. Applications include inlet nozzles, fuel turbopump rotors, injectors, combustion chambers, nozzle throats, and nozzle extensions. In order to maximize properties, materials developers have now begun to pay more attention to engineered interfaces between the matrix and the fiber reinforcement. If the interfacial debonding energy and sliding resistance is low, the fibers can pull away or out of the matrix and form bridges behind the advancing crack front which renders these otherwise brittle materials acceptably compliant. Unfortunately, most of the incremental toughening and attended fiber pullout occurs at engineering strains that exceed the strain which occurs at the ultimate strength of the material, and the toughening benefit is, therefore, not useful for design purposes. Thus, the degree of incremental toughening that occurs during the inelastic portion of the stress-strain curve up to the ultimate strength will have to be improved before increased use of FRCMCs can be realized.

It has also been shown that $MoSi_2$ offers the potential of combining the effects of second phase reinforcements with metallurgical alloying to improve mechanical properties without degrading oxidation resistance. For example, high temperature (1200°C) creep was reduced by a factor of 10 by alloying with $WSi_2$ and by another factor of 10–15 with the addition of SiC whiskers. Additions of carbon and zirconia have also proven to be beneficial. Carbon reacts with the oxygen impurities in $MoSi_2$ to improve the toughness at high temperatures by removing $SiO_2$, and leaves behind a compatible SiC phase. Zirconia, which is thermochemically stable with $MoSi_2$, can also be used to increase fracture toughness. a 20 v/o loading of particulate $ZrO_2$ increased the low temperature fracture toughness of $MoSi_2$ by a factor of four. The toughening transformation occurs above the ductile–brittle transition and therefore enhances the low temperature properties. In an attempt to reinforce $MoSi_2$ with SCS-6 silicon fibers, it was discovered that the large thermal expansion difference between the matrix and fiber introduced matrix cracking upon cooling from the densification temperature. This problem was solved by adding $Si_3N_4$ to form a two phase composite matrix with a coefficient of expansion that more closely matches the fiber. There is no reaction between $Si_3N_4$ and $MoSi_2$, even at fabrication temperatures as high as 1750°C. No gross cracking occurs on cool down, although some microcracking has been observed and is a function of grain size. The critical particle size below which microcracking will not occur was calculated to be 3 mm. Coarse phase $MoSi_2$– $Si_3N_4$ composites also exhibit higher room temperature toughness than fine phase material, reaching values of 8 $MPam^{1/2}$ . Fracture toughness also increases with temperature and the trend is quite significant above 800°C with toughness values exceeding 10 $MPam^{1/2}$.

However, the fine phase materials are stronger than the coarse phase materials with bend strengths reaching 1000 MPa. The $MoSi_2$– $Si_3N_4$ composites have also been shown to exhibit R-curve behavior, and crack deflection and particle pullout have been observed. Molybdenum disilicide does not have good creep resistance at high temperatures above its brittle-to-ductile transition. When high volume fractions of $Si_3N_4$ are added, creep is improved significantly and the activation energy is comparable to monolithic silicon nitride.

Additions of carbon can also improve creep resistance as well as toughness. In situ processing with carbon additions have produced material with creep resistance comparable to Ni-base superalloys. The silicon-base composite systems of current interest typically utilize carbon or silicon carbide fibers and silicon nitride or silicon carbide matrices. A popular designation is to display the fiber first followed by the matrix phase, e.g. C/SIC, SiC/SiC, and $SiC/Si_3N_4$. Mixtures composed of $MoSi_2$ and $Si_3N_4$ form two phase composites that are also candidates as matrices in C or SiC fiber reinforced composite systems. The combination of a fiber reinforced composite with a composite matrix becomes a little confusing, but can be represented by $SiC/ MoSi_2$– $Si_3N_4$.

The matrix properties for several silicon-base composite systems and their fiber properties are presented below. The high coefficient of thermal expansion of SCS-6, i.e. $4.8 \times 1^{-6}$ °C$^{-1}$, indicates that this fiber will have a larger expansion coefficient than the matrix phase for SiC/SiC and SiC/ Si$_3$N$_4$ fiber reinforced composite systems. It is generally more difficult to weave and fabricate structural components from large diameter fibers. The size can also influence properties. For example, toughness scales directly with fiber radius while the matrix cracking strength is inversely proportional to the radius. The properties for MoSi$_2$-Si$_3$N$_4$ compare quite favorably with both SiC and Si$_3$N$_4$. The fracture toughness is slightly above the middle range for SiC and comparable to Si$_3$N$_4$. The highest matrix toughness values are on the order of 10 MPam$^{1/2}$ for in situ toughened silicon nitride. The toughness for all the candidate matrix materials will depend upon processing conditions and microstructure. It is anticipated that in situ toughening of the silicon nitride phase in the two phase MoSi$_2$-Si$_3$N$_4$ composites should yield further improvements for this matrix candidate.

The onset of nonlinear behavior s often found in tension tests marked by a distinct load drop, indicating the initiation of matrix cracking, whereas in flexure, this important feature may go undetected. It is the region between matrix cracking and fiber bundle failure at maximum load where matrix enhancement can make the greatest contribution. Many FRCMC composites exhibit much of their toughness beyond this point, because as the fibers pull away from the matrix, they bridge cracks and impose traction forces that retard crack growth. Pullout toughening extends life after failure, and adds a margin of safety from the catastrophic nature of failure often found in brittle materials, but this phenomenon is not useful as a design property. Three of the composites exhibit matrix cracking stresses in the range of 150–175 MPa. Of the four systems, the SiC/Si$_3$N$_4$ has the highest matrix cracking stress, which is near 350 MPa. This composite also has the largest coefficient of thermal expansion mismatch, CTE, with the fiber having the larger value.

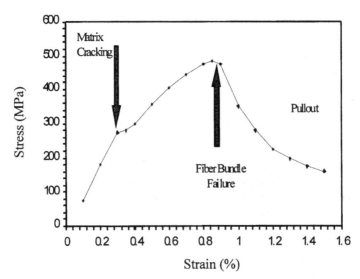

Fig. 11. Stress–strain behaviour for ceramic fiber reinforced: ceramic matrix composites.

Upon cooling from the consolidation temperature, the fiber can theoretically contract and debond from the matrix unless there is enough surface roughness for asperity contact. Upon reloading, the matrix will not efficiently transfer stress to the fiber unless it is in intimate contact. However, the high elastic modulus, as indicated by the stress– strain curve, suggests that good load transfer is occurring in this system. Inelastic behavior starts at about 350 MPa, but the ultimate is reached at about 450 MPa which is well below the expected fiber bundle failure. The ultimate strength for the Silcomp matrix composite is on the order of 650 MPa and is in reasonable agreement with bundle fiber failure, as is the MoSi₂– 0.5Si₃N₄ matrix composite.

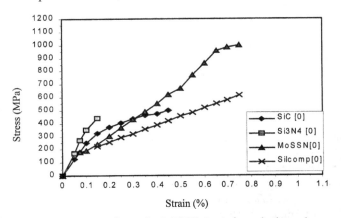

Fig. 12. Tensile stress–strain curves for uniaxial SCS-6 reinforced silicon-base composite systems (Courtright, 1999).

A number of composite approaches have been developed to toughen brittle high temperature structural ceramic materials. Many of these approaches have also been applied to high temperature structural silicides.

The MoSi₂–Si₃N₄ composite system is an interesting and important one. Si₃N₄ is considered to be the most important structural ceramic, due to its high strength, good thermal shock resistance, and relatively high (for a structural ceramic) room temperature fracture toughness. Si₃N₄ and MoSi₂ are thermodynamically stable species at elevated temperatures.

| Property | MoSi₂ | Si₃N₄ |
|---|---|---|
| Density(g/cm³) | 6.2 | 3.2 |
| Thermal expansion coefficent (10⁻⁶/°C) | 7.2 | 3.8 |
| Thermal conductivity (W/mK) | 65 | 37 |
| Melting point(°C) | 2030 | 2100 |
| Creep resistance (°C) | 1200 | 1400 |
| Toughness | Low | Low |
| Oxidation resistance | Good | Excellent |
| Structural stability | Good | Good |
| Intricate machinability | Good | Difficult |
| Cost | Low | High |

Table 2. Some physical and thermal properties of MoSi₂ and Si₃N₄ (Nathesan & Devi, 2000).

When composites were synthesized with elongated $Si_3N_4$ grains toughness can reach to 15 MPa m$^{1/2}$ (Nathal & Hebsur, 1997).

| Designation | Microstructure |
|---|---|
| MS-60 | Fully dense β-$Si_3N_4$, with long whisker-type morphology |
| MS-70 | Fully dense β- $Si_3N_4$, with long whisker-type morphology |
| MS-80 | Not fully dense β- $Si_3N_4$, with blocky morphology |
| MS-50 | Fully dense α- $Si_3N_4$, with blocky morphology |
| MS-40 | Not fully dense α- $Si_3N_4$, fine grained $MoSi_2$ and blocky α- $Si_3N_4$ |

Fig. 3. Microstructures of different $MoSi_2$– $Si_3N_4$ composites.

However, a drawback of transformation toughening is that toughness decreases with increasing temperature, due to the thermodynamics of the phase transformation. Discontinuously reinforced ceramic composites have typically employed ceramic whiskers or particles as the reinforcing phases. An example is SiC whisker reinforced $Si_3N_4$. Toughening mechanisms here are crack deflection and crack bridging. Discontinuous ceramic composites can reach toughness levels of 10 MPam$^{1/2}$. One important variant of this approach is the in-situ toughening of $Si_3N_4$ due to the presence of elongated $Si_3N_4$ grains. By way of comparison to structural ceramics, the room temperature fracture toughness of polycrystalline $MoSi_2$ is approximately 3 MPam$^{1/2}$, while the room temperature fracture toughness of equiaxed polycrystalline $Si_3N_4$ which is densified without densification aids is also 3 MPam$^{1/2}$ (Petrovic, 2000). For comparison, two monolithic ceramics SiC and $Si_3N_4$ are also included in the figure. Further improvement in room temperature fracture can be achieved by microalloying $MoSi_2$ with elements like Nb, Al an Mg or by randomly oriented long whisker type β-$Si_3N_4$ grains (Hebsur, 1999).

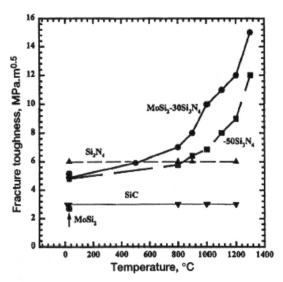

Fig. 13. Temperature dependence of fracture toughness of $MoSi_2$-based materials compared with ceramic matrices (Hebsur, 1999).

MoSi$_2$–Si$_3$N$_4$–SiC hybrid discontinuous particle-continuous fiber composites have been developed with excellent room temperature fracture toughness, thermal shock resistance, and thermo-mechanical impact behavior. These hybrid composites consist of MoSi$_2$–Si$_3$N$_4$ particulate composites which form the matrix for SiC continuous fibers. The MoSi$_2$–Si$_3$N$_4$ portion of the hybrid composites has two functions. First, additions of 30-50.% Si$_3$N$_4$ to the MoSi$_2$ completely eliminates the oxidation pest behavior at the intermediate 500°C temperature. Second, the Si$_3$N$_4$ addition aids to match the thermal expansion coefficient of the matrix to that of the SiC fibers. This prevents thermal expansion coefficient mismatch cracking in the hybrid composite matrix.

Figure 14.(a) shows a SEM back scattered image of a fully dense MoSi$_2$-βSi$_3$N$_4$ composite (MS-70). During processing, the original α-Si$_3$N$_4$ powder particles are transformed into randomly oriented whiskers of β-Si$_3$N$_4$. These long whiskers are well dispersed throughout the material and appear to be quite stable, with very little or no reaction with the MoSi$_2$, even at 1900°C. In some isolated areas, the Mo$_5$Si$_3$ phase is visible. Figure 14.(b) shows a back scattered image of MoSi$_2$-βSi$_3$N$_4$ (MS-80) with the β-Si$_3$N$_4$ exhibiting a blocky aggregate-type morphology.

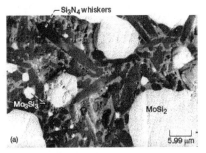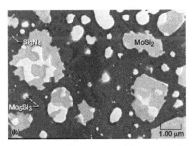

Fig. 14.(a) randomly oriented in-situ grown long whiskers of β-Si$_3$N$_4$ and large MoSi$_2$ particle size, (b) Si$_3$N$_4$ has a blocky particulate structure (Hebsur et al., 2001).

Density of (MS-70) is 4.57±0.01g/cm$^3$ and Vickers microhardness is 10.7±0.6GPa. Figure 15. shows the coefficient of thermal expansion as a function of temperature for (MS-70). From this data the average coefficient for expansion of this composite material is about 4.0ppm/°C.

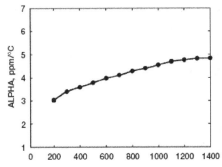

Fig. 15. The coefficient of linear expansion for MoSi$_2$-βSi$_3$N$_4$ (Hebsur et al., 2001).

The oxidation behaviour of a MoSiB alloy is also included for comparison. 500°C is the temperature for maximum accelerated oxidation and pest for $MoSi_2$-base alloys. There is interest in this alloy, over $MoSi_2$, for structural aerospace applications due to its attractive high temperature oxidation resistance (Bose, 1992; Berczik, 1997).

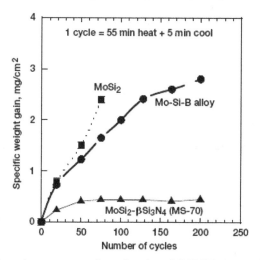

Fig. 16. Specific weight gain versus number of cycles of (MS-70) at 500°C (Hebsur, 2001).

However, the (MS-70) shows very little weight gain compared to binary $MoSi_2$ and the MoSiB alloy, indicating the absence of accelerated oxidation. In contrast the binary $MoSi_2$ and MoSiB alloys exhibits accelerated oxidation followed by pesting.

## 6. Conclusion

Based on the cyclic oxidation properties at 900°C, the family of $MoSi_2$-$Si_3N_4$ composites show promise for aircraft applications. The composites do not exhibit the phenomena of pesting, and the weight gain after 500h is negligible and superior to base line hybrid composites.

A wide spectrum of mechanical and environmental properties have been measured in order to establish feasibility of an $MoSi_2$ composite with $Si_3N_4$ particulate. The high impact resistance of the composite is of particular note, as it was a key property of interest for engine applications. Processing issues have also been addressed in order to lower cost and improve shape making capability. These results indicate that this composite system remains competitive with other ceramics as potential replacement for superalloys.

## 7. References

Vaseduvan, A.K. & Petrovic, J.J. (1992). A Comparative Overview of Molybdenum Disilicide Composites. *Materials Science and Engineering*, A155, pp. 1-17

Tein, J. K. & Caulfield, T. (1989). *Superalloys, Supercomposites, and Superceramics*, Boston Academic Press, New York

Soetching, F.O. (1995). A Design Perspective on Thermal Barrier Coatings. *Proceedings of a Conference at NASA Lewis Research Center*, September 1994

Bradley, E.F. (1988). *Superalloys a Technical Guide*, Metals Park, Ohio

Misra, A.; Sharif, A.A. & Petrovic, J.J. (2000). Rapid Solution Hardening at Elevated Temperatures by Substitional Re Alloying in MoSi$_2$, *Acta Mater*, Vol.48, pp. 925-932

Akkus, I. (1999). The Aluminide Coating of Superalloys with Pack Cementation Method. *Journal of Institution of Science Osmangazi University*, Vol. 18, pp. 27-28

Meschter, P.J. (1992). Low Temperature Oxidation of Molybdenum Disilicide, *Metallurg. Trans. A*, Vol. 23A, pp. 1763-1772

Liu, Q.; Shao G. & Tsakiropoulos, P. (2001). On the Oxidation Behaviour of MoSi$_2$. *Intermetallics*, Vol. 8, pp. 1147-1158

Chou, T.C. & Nieh, T.G. (1992). New Observation of MoSi$_2$ Pest at 500°C. *Script. Metallurg. Mater.*, Vol. 26, pp. 1637-1642

Chou, T.C. & Nieh, T.G. (1993). Pesting of the High Temperature Intermetallic MoSi$_2$. *Journal of Materials*, Vol. 30, pp. 15-22

Wang, G.; Jiang, W. & Bai, G. (2003). Effect of Addition of Oxides on Low-Temperature Oxidation of Molybdenum Disilicide. *Journal of American Ceramic Society*, Vol. 86, pp. 731-734

Chen, J.; Li, C.; Fu, Z.; Tu, X.; Sundberg, M. & Pompe, R. (1999). Low Temperature Oxidation Behaviour of MoSi$_2$Bbased Material. *Materials Science and Engineering*, Vol. A26, pp. 239-244

Sharif, A.A.; Misra, A. & Petrovic, J.J. (2001). Rapid Solution Hardening at Elevated Temperatures by Substitional Re Alloying in MoSi$_2$, *Acta Mater*, Vol. 48, pp. 925-932

Waghmare, U.V.; Bulatov, V. & Kasiras, E. (1999). Microalloying for Ductility in Molybdenum Disilicide, *Materials Science and Engineering*, Vol. A261, pp. 147-157

Mitra, R.; Sadananda, K. & Feng, C.R. (2004). Effect of microstructural parameters and Al Alloying on Creep Behavior, Threshold Stress and Activation Volumes of Molybdenum Disilicides. *Intermetallics*, Vol. 12, pp. 827-836

Inui, H.; Ishikawa, K. & Yamaguchi, M. (2000). Effects of Alloying Elements on Plastic Deformation of Single Crystals of MoSi$_2$. *Intermetallics*, Vol. 8, pp. 1131-1145

Courtright, E.R. (1999). A Comparison of MoSi$_2$ Matrix Composites with Other Silicon-Base Composite Systems. *Materials Science and Engineering*, Vol. A261, pp. 53-63

Natesan, K. & Deevi, S.C. (2000). Oxidation Behaviour of Molybdenum Disilicides and Their Composites. *Intermetallics*, Vol. 8, pp. 1147-1158

Nathal, M.V. & Hebsur, M.G. (1997). Strong, Tough, and Pest-Resistant MoSi$_2$-Base Hybrid Composite for Structural Applications. *Structural Intermetallics*, Warrendale (USA): TMS, pp. 949-953

Petrovic, J.J. (2000). Toughening Strategies for MoSi$_2$-Based High Temperature Structural Silicides, *Intermetallics*, Vol. 8, pp. 1175-1182

Hebsur, M.G. (1999). Development and Characterization of SiC$_{(f)}$/MoSi$_2$-Si$_3$N$_{4(p)}$ Hybrid Composites. *Material Science and Engineering*, Vol. A261, pp. 24-37

Hebsur, M.G.; Choi, S.R.; Whittenberger, J.D.; Salem, J.A. & Noebe, R.D. (2001). Development of Tough, Strong, and Pest-Resistant MoSi$_2$-$\beta$Si$_3$N$_4$ Composites for

High Temperature Structural Applications, *International Symposium on Structural Intermetallics*, NASA, 2001

Bose, S. (1992). *High Temperature Silicides*, North-Holland, NY

Berczik, D.M. (1997). Oxidation Resistant Molybdenum Alloy, U.S. Patent, No. 5, 696, 150

# 5

# ALLVAC 718 Plus™ Superalloy for Aircraft Engine Applications

Melih Cemal Kushan[1], Sinem Cevik Uzgur[2],
Yagiz Uzunonat[3] and Fehmi Diltemiz[4]
*[1]Eskisehir Osmangazi University*
*[2]Ondokuz Mayis University*
*[3]Anadolu University*
*[4]Air Supply and Maintenance Base*
*Turkey*

## 1. Introduction

Innovations on the aerospace and aircraft industry have been throwing light upon building to future's engineering architecture at the today's globalization world where technology is the indispensable part of life. On the basis of aviation sector, the improvements of materials used in aircraft gas turbine engines which constitute 50 % of total aircraft weight must protect its actuality continuously. On the other hand utilization of super alloys in aerospace and defense industries can not be ignored because of excellent corrosion and oxidation resistance, high strength and long creep life at elevated temperatures.

Materials that can be used at the homologous temperature of 0.6 Tm and still remain stable to withstand severe mechanical stresses and strains in oxidizing environments are so-called superalloys, usually based on Ni, Fe or Co (Sims et al, 1987). Nickel-based superalloys are the exceptional group of superalloys with superior materials properties. Their excellent properties range from high temperature mechanical strength, toughness to resistance to degradation in oxidizing and corrosive environment. Therefore they are not only used in aerospace and aircraft industry, but also in ship, locomotive, petro- chemistry and nuclear reactor industries.

Inconel 718 is Ni-based, precipitation- hardening superalloy with Nb as a major hardening element, used for high temperature aerospace applications very widely in recent years (Yaman & Kushan, 1998). However, the metastability of the primary strengthening ($\gamma''$, gamma double prime) phase is typically unacceptable for applications above about 650°C. As a result, other more costly and difficult to process alloys, like Waspalloy, are used in such applications. Although Waspalloy is strengthened primarily by $\gamma'$, it is still more susceptible to weld-related cracking than Inconel 718 (Otti et al, 2005). In these circumstances ALLVAC 718 Plus™ come to stage, which is strengthened with uniform cubic FCC inter metallic $\gamma'$ phase, innovated by ATI ALLVAC Company very recently. In recent years it has been becoming widespread dramatically for using of disc material in aerospace gas turbine engine parts. The most important reason of this is the high yield and ultimate tensile strength and very good corrosion and oxidation resistance of material together with excellent creep resistance at elevated temperatures. Fig. 1 shows that where wrought alloy 718Plus can be used as a disc material for high pressure (HP) compressor as well as for high pressure (HP) turbine discs (Bond & Kennedy, 2005).

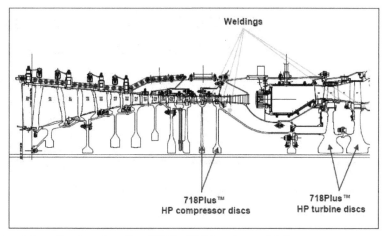

Fig. 1. Potential applications for alloy 718PlusTM in a future high pressure core section (Bond & Kennedy, 2005).

The newly innovated ALLVAC 718 Plus superalloy which is the last version of Inconel 718 has been proceeding in the way to become a material that aerospace and defense industries never replace of any other material with combining its good mechanical properties, easy machinability and low cost.

## 2. Gas turbine engines

Gas turbine engines, also known as jet engines, power most modern civilian and military aircraft. Fig. 2 shows some sections of this kind of an engine. The inlet (intake) directs outside air into the engine. The compressor (shown in a (a) part of Fig. 2) is situated at the exit of the inlet. In order to produce thrust, it is essential to compress the air before fuel is added. In an axial-flow compressor, the air flows in the direction of the shaft axis through alternate rows of stationary and rotating blades, called stators and rotors, respectively. Modern axial-flow compressors can increase the pressure 24 times in 15 stages, with each set of stators and rotors making up a stage. The compressors in most modern engines are divided into low-pressure and high-pressure sections which run off two different shafts. In the combustor, or burner (shown in a (b) part of Fig. 2), the compressed air is mixed with fuel and burned. Fuel is introduced through an array of spray nozzles that atomize it. An electric igniter is used to begin combustion. The combustor adds heat energy to the air stream and increases its temperature (up to about 1930°C), a process which is accompanied by a slight decrease in pressure (~ 1-2%). For best performances, the combustion temperature should be the maximum obtainable from the complete combustion of the oxygen and the fuel. However, turbine inlet temperatures currently cannot exceed about 1100°C because of material limits. Hence, only part of the compressed air is burned in the combustor; the remainder is used to cool the turbine.

Fig. 2. Some basic sections of a gas turbine engine (Eliaz et al, 2002).

Leaving the combustor, the hot exhaust is passed through the turbine (shown in a (c) part of Fig. 2), in which the gases are partially expanded through alternate stator and rotor rows. Depending on the engine type, the turbine may consist of one or several stages. Like the compressor, the turbine is divided into low-pressure and high-pressure sections (shown in Fig. 3), the latter being closer to the combustor.

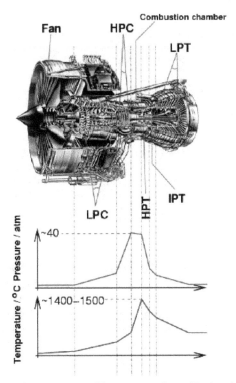

Fig. 3. The temperature and pressure profile in gas turbine (Carlos & Estrada, 2007).

The turbine provides the power to turn the compressor, to which it is connected via a central shaft, as well as the power for the fuel pump, generator, and other accessories. From thermodynamics, the turbine work per mass airflow is equal to the change in the specific enthalpy of the flow from the entrance to the exit of the turbine. This change is related to the temperatures at these points. The temperature at the entrance to the turbine can be as high as 1650°C, considerably above the melting point of the material from which the blades are made.

The gases, leaving the turbine at an intermediate pressure, are finally accelerated through a nozzle to reach the desired high jet-exit velocity. Because the exit velocity is greater than the free stream velocity, thrust is produced. The amount of thrust generated depends on the rate of mass flow through the engine and the leaving jet velocity, according to Newton's Second Law. Thus, the gas is accelerated to the rear, and the engine (as well as the aircraft) is accelerated in the opposite direction according to Newton's Third Law (Eliaz et al, 2002).

Modern gas turbines have the most advanced and sophisticated technology in all aspects; construction materials are not the exception due their extreme operating conditions. The most difficult and challenging point is the one located at the turbine inlet, because, several difficulties associated to it; like extreme temperature, high pressure, high rotational speed, vibration, small circulation area, and so on. These rush characteristics produce effects on the gas turbine components that are shown on the Table 1 (Carlos & Estrada, 2007).

|  | Oxidation | Hot corrosion | Interdiffusion | Thermal Fatigue |
|---|---|---|---|---|
| Aircraft | Severe | Moderate | Severe | Severe |
| Land-based Power Generator | Moderate | Severe | Moderate | Light |
| Marine Engines | Moderate | Severe | Light | Moderate |

Table 1. Severity of the different surface related problems for gas turbine applications (Carlos & Estrada, 2007).

In order to overcome those barriers, gas turbine components are made using advanced materials and modern alloys (superalloys) that contains up to ten significant alloying elements.

## 3. Superalloys

These alloys have been developed for high-temperature service and include iron-, cobalt- and nickel based materials, although nowadays they are principally nickel based. These materials are widely used in aircraft and power-generation turbines, rocket engines, and other challenging environments, including nuclear power and chemical processing plants. The aero gas turbine was the impetus for the development of superalloys in the early 1940s, when conventional materials available at that time were insufficient for the demanding environment of the turbine. Therefore it can be said that "The development of superalloys made the modern gas turbine possible".

A major application of superalloys is in turbine materials, jet engines, both disc and blades. Initial disc alloys were *Inco 718* and *Inco 901* produced by conventional casting ingot, forged billet and forged disc route. These alloys were developed from austenitic steels, which are still used in industrial turbines, but were later replaced by *Waspaloy* and *Astroloy* as stress and temperature requirements increased. These alloys were turbine blade alloys with a suitably modified heat-treatment for discs. However, blade material is designed for creep, whereas disc material requires tensile strength coupled with low cycle fatigue life to cope with the stress changes in the flight cycle. To meet these requirements *Waspaloy* was thermomechanically processed (TMP) to give a fine-grain size and a 40% increase in tensile strength over the corresponding blade material, but at the expense of creep life. Similar improvements for discs have been produced in *Inco 901* by TMP. More highly-alloyed nickel-based discs suffer from excessive ingot segregation which makes grain size control difficult. Further development led to alloys produced by powder processing by gas atomization of a molten stream of metal in an inert argon atmosphere and consolidating the resultant powder by HIPing to near-net shape. Such products are limited in stress application because of inclusions in the powder and, hence, to realize the maximum advantage of this process it is necessary to produce 'superclean' material by electron beam or plasma melting.

Improvements in turbine materials were initially developed by increasing the volume fraction of $\gamma'$ in changing *Nimonic 80A* up to *Nimonic 115*. Unfortunately increasing the (Ti +Al) content lowers the melting point, thereby narrowing the forging range makes processing more difficult. Improved high-temperature oxidation and hot corrosion performance has led to the introduction of aluminide and overlay coatings and subsequently the development of *IN 738* and *IN 939* with much improved hot-corrosion resistance.

Further improvements in superalloys have depended on alternative manufacturing routes, particularly using modern casting technology like Vacuum casting (Smallman & Bishop, 1999).

In these alloys $\gamma'$ ($Ni_3Al$) and $\gamma''$ ($Ni_3Nb$) are the principal strengtheners by chemical and coherency strain hardening. The ordered $\gamma'$-$Ni_3Al$ phase is an equilibrium second phase in both the binary Ni–Al and Ni–Cr–Al systems and a metastable phase in the Ni–Ti and Ni–Cr–Ti systems, with close matching of the $\gamma'$ and the FCC matrix. The two phases have very similar lattice parameters and the coherency confers a very low coarsening rate on the precipitate so that the alloy overages extremely slowly even at 0.7Tm. In alloys containing Nb, a metastable $Ni_3Nb$ phase occurs but, although ordered and coherent, it is less stable than $\gamma'$ at high temperatures (Smallman & Ngan, 2007).

In high-temperature service, the properties of the grain boundaries are as important as the strengthening by $\gamma'$ within the grains. Grain boundary strengthening is produced mainly by precipitation of chromium and refractory metal carbides; small additions of Zr and B improve the morphology and stability of these carbides. Optimum properties are developed by multistage heat treatment; the intermediate stages produce the desired grain boundary microstructure of carbide particles enveloped in a film of $\gamma'$ and the other stages produce two size ranges of $\gamma'$ for the best combination of strength at both intermediate and high temperatures (Smallman & Ngan, 2007). Table 2 indicates the effect of the different alloying elements and Table 3 indicates the common ranges of main alloying additions and their effects on superalloy properties.

| Influence | Cr | Al | Ti | Co | Mo | W | B | Zr | C | Nb | Hf | Ta |
|---|---|---|---|---|---|---|---|---|---|---|---|---|
| Matrix strengthening | ✓ | | | ✓ | ✓ | ✓ | | | | | | |
| γ' formers | | ✓ | ✓ | | | | | | | ✓ | | ✓ |
| Carbide formers | ✓ | | ✓ | | ✓ | ✓ | | | | ✓ | ✓ | ✓ |
| Grain boundary strengthening | | | | | | | ✓ | ✓ | ✓ | ✓ | | |
| Oxide scale formers | ✓ | ✓ | | | | | | | | | | |

Table 2. The effect of the different alloying elements (Smallman & Ngan, 2007).

| Element | Range, wt.% | Effect |
|---|---|---|
| Cr | 5-25 | Oxidation and hot corrosion resistance; carbides; solution hardening |
| Mo, W | 0-12 | Carbides; solution hardening |
| Al | 0-6 | Precipitation hardening; oxidation resistance; γ' former |
| Ti | 0-6 | Precipitation hardening; carbides; γ' former |
| Co | 0-20 | Affects amount of precipitate |
| Ni | Balance | Stabilizes γ phase; forms hardening precipitates |
| Nb | 0-5 | Carbides; solution hardening; precipitation hardening |
| Ta | 0-12 | Carbides; solution hardening; oxidation resistance; γ' former |

Table 3. Common Ranges of Main Alloying Additions and Their Effects on Superalloys.

The wide range of applications for superalloys has expanded many other areas since they were developed and now includes aircraft and land-based gas turbines, rocket engines, chemical, and petroleum plants. The performance of an industrial gas turbine engines depends strongly on service conditions and the environment in which it operates.

### 3.1 Iron-nickel-based superalloys

Iron-nickel base superalloys evolved from austenitic stainless steels and are based on the principle of combining both solid-solution hardening and precipitate forming elements. As a class, the iron nickel superalloys have useful strengths to approximately 650°C (1200°C). The austenitic matrix based on nickel and iron, with at least 25 wt % Ni needed to stabilize the FCC phase. Other alloying elements, such as Chromium partition primarily to austenite to provide solid-solution hardening. Most alloys contain 25 to 45 wt % Nickel. Chromium in the range of 15 to 28 wt% is added for oxidation resistance at elevated temperature, while 1 to 6 wt% Mo provides solid solution strengthening. The main elements that facilitate precipitation hardening are titanium, aluminum and niobium.

The strengthening precipitates are primarily γ' ($Ni_3Al$), η ($Ni_3Ti$), and γ" ($Ni_3Nb$). Elements that partition to grain boundaries, such as Boron and Zirconium, suppress grain boundary creep, resulting in significant increases in rupture life. Boron in quantities of 0.003 to 0.03 wt% and, less frequently, small additions of zirconium are added to improve stress-rupture properties and hot workability. Zirconium also forms the MC carbide ZrC. Another MC carbide (NbC) is found in alloys that contain niobium such as Inconel 706 and Inconel 718. Vanadium is also added in small quantities to iron-nickel superalloys to improve both notch ductility at service temperatures and hot workability. Based on their composition and strengthening mechanisms, there are several groupings of iron-nickel superalloys (Campbell, 2008).

The most common precipitate is γ', typified by A-286, V-57 or Incoloy 901. Some alloys, typified by Inconel (IN)- 718, which precipitate γ", were formerly classed as iron-nickel-base superalloys but now are considered to be nickel-base.

The most common type of iron-nickel-base superalloys is INCONEL 718 which is a precipitation- hardening alloy used for high-temperature applications. In particular, the reputation of wrought Inconel 718 for being relatively easy to weld is generally attributed to the sluggish precipitation kinetics of the tetragonal γ" strengthening phase. Inconel 718 is a relatively recent alloy as its industrial use started in 1965. It is a precipitation hardenable alloy, containing significant amounts of Fe, Nb and Mo. Minor contents of Al and Ti are also present. Inconel 718 combines good corrosion and high mechanical properties with and excellent weldability. It is employed in gas turbines, rocket engines, turbine blades, and in extrusion dies and containers.

Ni and Cr contribute to the corrosion resistance of this material. They crystallize as a γ phase (face centred cubic). Nb is added to form hardening precipitates γ" (a metastable inter metallic compound $Ni_3Nb$, centred tetragonal crystal). Ti and Al are added to precipitate in the form of intermetallic γ' ($Ni_3$(Ti,Al), simple cubic crystal). They have a lower hardening effect than particles. C is also added to precipitate in the form of MC carbides (M = Ti or Nb). In this case the C content must be low enough to allow Nb and Ti precipitation in the form of γ' and γ" particles. Mo is also frequent in Inconel 718 in order to increase the mechanical resistance by solid solution hardening. Finally, a β phase (intermetallic $Ni_3Nb$), (sometimes called δ phase) can also appear. It is an equilibrium particle with orthorhombic structure. All theses particles can precipitate along the grain boundaries of the γ matrix increasing the intergranular flow resistance of the present alloy. A typical precipitation time temperature (PTT) diagram for this alloy is shown in Fig. 4 (Thomas et al, 2006).

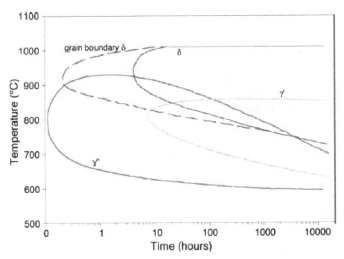

Fig. 4. PTT diagram of different phases in Inconel 718 (Thomas et al, 2006).

## 3.2 Nickel-based Superalloys

Nickel-based superalloys are an unusual class of metallic materials with an exceptional combination of high temperature strength, toughness, and resistance to degradation in corrosive or oxidizing environments. The nickel-based alloys show a wider range of application than any other class of alloys.

The austenitic stainless steels were developed and utilized early in the 1900s, whereas the development of the nickel-based alloys did not begin until about 1930. In aerospace applications nickel-based superalloys are used widely as components of jet engine turbines. Therefore important position of super-alloys in this area is manifested by the fact that they represent at present more than 50 % of mass of advanced aircraft engines. Extensive use of super-alloys in turbines, supported by the fact that thermo-dynamic efficiency of turbines increases with increasing temperatures at the turbine inlet, became partial reason of the effort aimed at increasing of the maximum service temperature of high-alloyed alloys (Jonsta et al, 2007). Therefore in gas turbine applications alloys with good stability and very low crack-growth rates that are readily inspectable by nondestructive means are desired. Fuel efficiency and emissions are also key commercial and environmental drivers impacting turbine-engine materials. To meet these demands, modern nickel-based alloys offer an efficient compromise between performance and economics. The chemistries of several common and advanced nickel-based superalloys are listed in Table 4 and the parts of gas turbine engine in which Nickel-based superalloys (marked red) commonly used are shown in Fig. 5.

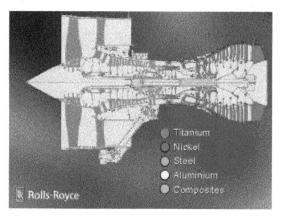

Fig. 5. Commonly used materials in gas turbine engine components.

In the environmental series nickel is nobler than iron but more active than copper. Reducing environments, such as dilute sulfuric acid, find nickel more corrosion resistant than iron but not as resistant as copper or nickel-copper alloys. The nickel molybdenum alloys are more corrosion resistant to reducing environment than nickel or nickel- copper alloys (Philip & Schweitzer, 2003). Nickel-based superalloys are extremely prone to weld cracking.

High-temperature strength of Ni-base superalloys depends mainly, on the volume fraction and morphology of $\gamma'$ precipitates. Several basic factors contribute to the magnitude of hardening of the alloy (Sajjadi & Zebarjad, 2006).

| Alloy | Cr | Ni | Co | Mo | W | Nb | Ti | Al | Fe | C | B | Other |
|---|---|---|---|---|---|---|---|---|---|---|---|---|
| A286 | 15 | 26 | — | 1.25 | — | — | 2 | 0.2 | 55.2 | 0.04 | 0.005 | 0.3 V |
| AF115 | 10.7 | 56 | 15 | 2.8 | 5.9 | 1.7 | 3.9 | 3.8 | — | 0.05 | 0.02 | 0.75 Hf; 0.05 Zr |
| AF2-1DA | 12 | 59 | 10 | 3 | 6 | — | 3 | 4.6 | <0.5 | 0.35 | 0.015 | 1.5 Ta, 0.1 Zr |
| AF2-1DA6 | 12 | 59.5 | 10 | 2.75 | 6.5 | — | 2.8 | 4.6 | <0.5 | 0.04 | 0.015 | 1.5 Ta, 0.1 Zr |
| Alloy 706 | 16 | 41.5 | — | — | — | — | 1.75 | 0.2 | 37.5 | 0.03 | — | 2.9 (Nb+Ta), 0.15 Cu |
| Alloy 718 | 19 | 52.5 | — | 3 | — | 5.1 | 0.9 | 0.5 | 18.5 | 0.08 | — | 0.15 Cu |
| APK12 | 18 | 55 | 15 | 3 | 1.25 | — | 5 | 2.5 | — | 0.03 | 0.035 | 0.035 Zr |
| Astroloy | 15 | 56.5 | 15 | 5.25 | — | — | 3.5 | 4.4 | <0.3 | 0.06 | 0.03 | 0.06 Zr |
| Discaloy | 14 | 26 | — | 3 | — | — | 1.7 | 0.25 | 55 | 0.06 | — | — |
| IN100 | 10 | 60 | 15 | 3 | — | — | 4.7 | 5.5 | <0.6 | 0.15 | 0.015 | 0.06 Zr, 1.0 V |
| KM-4 | 12 | 56 | 18 | 4 | — | 2 | 4 | 4 | — | 0.03 | 0.03 | 0.03 Zr |
| MERL-76 | 12.4 | 54.4 | 18.6 | 3.3 | — | 1.4 | 4.3 | 5.1 | — | 0.02 | 0.03 | 0.35 Hf; 0.06 Zr |
| N18 | 11.5 | 57 | 15.7 | 6.5 | — | — | 4.35 | 4.35 | — | 0.015 | 0.015 | 0.45 Hf; 0.03 Zr |
| PA101 | 12.5 | 59 | 9 | 2 | 4 | — | 4 | 3.5 | — | 0.15 | 0.015 | 4.0 Ta; 1.0 Hf; 0.1 Zr |
| René 41 | 19 | 55 | 11 | 10 | — | — | 3.1 | 1.5 | <0.3 | 0.09 | 0.01 | |
| René 88 | 16 | 56.4 | 13.0 | 4 | 4 | 0.7 | 3.7 | 2.1 | — | 0.03 | 0.015 | 0.03 Zr |
| René 95 | 14 | 61 | 8 | 3.5 | 3.5 | 3.5 | 2.5 | 3.5 | <0.3 | 0.16 | 0.01 | 0.05 Zr |
| Udimet 500 | 19 | 52 | 19 | 4 | — | — | 3 | 3 | <4.0 | 0.08 | 0.005 | |
| Udimet 520 | 19 | 57 | 12 | 6 | 1 | — | 3 | 2 | — | 0.08 | 0.005 | |
| Udimet 700 | 15 | 55 | 17 | 5 | — | — | 3.5 | 4 | <1.0 | 0.07 | 0.02 | 0.02 Zr |
| Udimet 710 | 18 | 55 | 14.8 | 3 | 1.5 | — | 5 | 2.5 | — | 0.07 | 0.01 | |
| Udimet 720 | 18 | 55 | 14.8 | 3 | 1.25 | — | 5 | 2.5 | — | 0.035 | 0.033 | 0.03 Zr |
| Udimet 720LI | 16 | 57 | 15.0 | 3 | 1.25 | — | 5 | 2.5 | — | 0.025 | 0.018 | 0.03 Zr |
| V57 | 14.8 | 27 | — | 1.25 | — | — | 3 | 0.25 | 48.6 | 0.08 | 0.01 | 0.5 V |
| Waspaloy | 19.5 | 57 | 13.5 | 4.3 | — | — | 3 | 1.4 | <2.0 | 0.07 | 0.006 | 0.09 Zr |

Table 4. The chemical compositions of several superalloys (wt.%) (Furrer & Fecht, 1999).

## 3.3 Cobalt-based Superalloys

The cobalt-based superalloys (Table 5) are not as strong as nickel-based superalloys, but they retain their strength up to higher temperatures. They derive their strength largely from a distribution of refractory metal carbides (combinations of carbon and metals such as Mo and W), which tend to collect at grain boundaries (Fig. 6). This network of carbides strengthens grain boundaries and alloy becomes stable nearly up to the melting point. In addition to refractory metals and metal carbides, cobalt superalloys generally contain high levels of Cr to make them more resistant to corrosion that normally takes place in the presence of hot exhaust gases. The Cr atoms react with oxygen atoms to form a protective layer of $Cr_2O_3$ which protects the alloy from corrosive gases. Being not as hard as nickel-based superalloys, cobalt superalloys are not so sensitive to cracking under thermal shocks as other superalloys. Co-based superalloys are therefore more suitable for parts that need to be worked or welded, such as those in the intricate structures of the combustion chamber (Jovanović et al).

| Alloy | C | Mn | Si | Cr | Ni | Mo | W | Fe | Co |
|---|---|---|---|---|---|---|---|---|---|
| X-45 | 0.25 | .5 | 0.9 | 25 | 10 | - | 7.5 | <2 | Bal. |
| X-40 | 0.5 | .5 | 0.9 | 25 | 10 | - | 7.5 | <2 | Bal. |
| FSX-414 | 0.35 | .5 | 0.9 | 29.5 | 10 | - | 7.5 | <2 | Bal. |
| WI-52 | 0.45 | .4 | 0.4 | 21 | - | - | 11 | 2 | Bal. |
| Haynes -25 | 0.1 | 1.2 | 0.8 | 20 | 10 | - | 15 | <3 | Bal. |
| F-75 | 0.25 | .5 | 0.8 | 28 | <1 | 6 | <.2 | <0.75 | Bal. |
| Haynes Ultimet | 0.06 | .8 | 0.3 | 25 | 9 | 5 | 2 | 3 | Bal. |
| Co 6 | 1.1 | | 0.8 | 29 | <3 | <1.5 | 5.5 | <3 | Bal. |

Table 5. The chemical composition of some Cobalt-based superalloys (Jovanović et al).

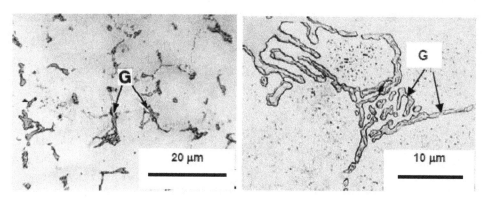

Fig. 6. Optical micrograph of Haynes-25. G mainly $M_6C$ carbides (Jovanović et al).

## 4. ALLVAC 718 plus™

Inconel 718 is a nickel base superalloy that is used extensively in aerospace applications because of its unique high temperature mechanical properties. Since it was invented by Eiselstein, it has been used as a material of construction for aero-engine and land based turbine components. The reasons for alloy 718's popularity include excellent strength, good hot and cold workability, the best weldability of any of the superalloys and last, but not least, moderate cost. However, the application of the alloy has been limited to a temperature below 650 ∘C, as its properties deteriorate rapidly on exposure above this temperature due to the instability of the main strengthening phase of the alloy, γ″ (Idowu & Ojo, 2007). With prolonged exposure at this temperature or higher, γ″ rapidly overages and transforms to the equilibrium δ phase with an accompanying loss of strength and especially creep life (Kennedy, 2005).

Other wrought, commercial superalloys exist which have significantly greater temperature capability such as Waspaloy and René 41. These alloys are typically γ′ hardened and are significantly more difficult to fabricate and weld. Because of this and because of their intrinsic raw material content, these alloys are significantly more expensive than alloy 718. There have been numerous attempts to develop an affordable, workable 718-type alloy with increased temperature capability. After a number of years of systematic work, including both computer modeling and experimental melting trials, ATI Allvac has developed a new alloy, Allvac® 718Plus™, which offers a full 55°C temperature advantage over alloy 718. The alloy maintains many of the desirable features of alloy 718, including good workability, weldability and moderate cost (Kennedy, 2005).

ATI Allvac has extensively investigated the 718Plus alloy billet properties, both as an internal program and as part of the Metals Affordability Initiative program entitled "Low-Cost, High Temperature Structural Material" for turbine engine ring-rolling applications. The objective of all these programs is to develop an alloy with the following characteristics:

- 55°C temperature advantage based on the Larson-Miller, time-temperature parameter
- Improved thermal stability; equal to Waspaloy at 704°C
- Good weldability; at least intermediate to 718 and Waspaloy alloys

- Minimal cost increase; intermediate to 718 and Waspaloy alloys
- Good workability; better than Waspaloy alloy

The use of 718Plus alloy in elevated temperature applications is of interest for military systems. In particular, the manufacturing difficulties associated with alloys such as Waspaloy provide a need for a material with similar component capabilities, but with better producibility. Initial characterization shows that the alloy exhibits many similarities to Alloy 718, including good workability, weldability and intermediate temperature strength capability (Bergstrom & Bayhan, 2005).

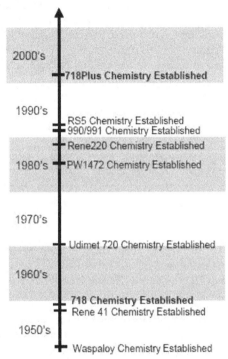

Fig. 7. Developments leading up to alloy 718 and subsequent efforts to improve capability over 718 (Otti et al, 2005).

Since the advent of the first superalloys over 60 years ago, alloy developers have worked to promote strength and high temperature stability while balancing processability. Processing constraints for many alloy systems preclude their general use for cast and wrought forging applications. Instead these compositions are used in the cast form, or are producible only using powder metallurgy. The development and introduction of alloy 718 in the late 1950's offered a significant breakthrough in malleability and weldability relative to other high strength alloys available at that time including Waspaloy and René 41 which are primarily gamma prime strengthened. Since the introduction of alloy 718 a significant number of alloys have been examined, including cast as well as wrought alloys, with the primary intent to maintain or improve properties and provide increased thermal stability while maintaining favorable processability. Some of the alloys developed subsequently are shown along with 718, Waspaloy, and René 41 in the development timeline of Fig. 7. A key requirement beyond strength, toughness, fatigue, creep, crack growth resistance, and processability which has also driven composition development is weldability (Otti et al, 2005).

## 4.1 Chemistry

There are lots of wrought alloys in use for gas turbine engine parts, such as Waspalloy, which have high temperature capability. But they are typically much more difficult to manufacture and fabricate into finished parts and also significantly more expensive than alloy 718 (Bond & Kennedy, 2005). Therefore when ALLVAC 718 plus is compared to Inconel 718 this newly modified super alloy has the higher content of Al+ Ti, the higher ratio of Al/ Ti and the addition of W and Co instead of Fe. As a result it provides increased temperature capacity up to 55°C and impressive thermal stability. Therefore it closes the gap between Inconel 718 and Waspalloy, as combining the good processability and weldability of Inconel 718 with the temperature capability of Waspalloy. (Schreiber et al, 2006). The chemical compositions of the ALLVAC 718 plus with Inconel 718 and Waspalloy are given in Table 6.

| Alloy | Chemistry, wt% | | | | | | | | | | |
|---|---|---|---|---|---|---|---|---|---|---|---|
|  | C | Cr | Mo | W | Co | Fe | Nb | Ti | Al | P | B |
| 718Plus | 0.025 | 18.0 | 2.70 | 1.0 | 9.0 | 10.0 | 5.40 | 0.70 | 1.45 | 0.007 | 0.004 |
| 718 | 0.025 | 18.1 | 2.90 | – | – | 18.0 | 5.40 | 1.00 | 0.45 | 0.007 | 0.004 |
| Waspaloy | 0.035 | 19.4 | 4.25 | – | 13.25 | – | – | 3.00 | 1.30 | 0.006 | 0.006 |

Table 6. Nominal chemistry comparison of the ALLVAC 718 plus, Inconel 718 and Waspalloy (Cao, 2005).

Alloy 718Plus has a much larger content of γ′ and γ″ than alloy 718 and a smaller amount of δ phase. Solvus temperatures for γ′ and γ″ are also higher in alloy 718Plus. All of these points likely contribute to improved high temperature properties. One of the major differences between alloy 718 and Waspaloy is the speed of the precipitation reaction. The γ″ precipitation in alloy 718 is very sluggish and accounts in part for the good weldability and processing characteristics of the alloy.

In 718- type alloys primarily Fe, Co, Mo and W are the matrix elements. The effects of alloying elements on microstructure, mechanical properties, thermal stability and processing characteristics of alloy are important factors. Niobium is one of the major hardening elements and the other two is Al and Ti. The change in Al/ Ti ratio and the increase in Al+ Ti content converts the alloy into a predominantly γ′ strengthening alloy and it gives the alloy an improved thermal stability. Furthermore the modification on the content of Al and Ti develop the optimum mechanical properties of the alloy. Another factor on the improvement of the mechanical properties and thermal stability is the addition of Co up to about 9 wt%. Still further improvement occurs with Fe content of 10 wt%, 2.8 wt% Mo and 1 wt% W (Cao & Kennedy, 2004). Very small additions of P and B further increases stress rupture and creep resistance.

## 4.2 Strengthening mechanisms

As mentioned before, the primary strengthening phase is γ' with a volume fraction ranging from 19.7-23.2 %, depending on the quantity of δ phase. Gamma prime strengthened alloys like Waspaloy and René 41 have much greater stability at higher temperature than γ" strengthened alloys like 718 since γ" grows rapidly and partially decomposes to equilibrium δ phase at temperatures in the 650–760°C range. Studies of the γ' phase in 718Plus alloy show it to be high in Nb and Al, which is very different from the γ' present in Waspaloy and René 41 and may account for its unique precipitation behavior and strengthening effects.

Like most superalloys there is a strong relationship between processing, structure and properties for alloy 718Plus. Optimum mechanical properties are achieved with a microstructure which has a small amount of rod shaped δ particles on the grain boundaries like that shown in Fig. 8 (a). Excessively high forging temperatures or high solution heat treating temperatures will result in structures with little or no δ phase precipitates that are prone to notch stress rupture failure. It is reported that no notch problems have been experienced using the 954°C solution temperature, probably because some δ phase can be precipitated at this temperature. However, excessively long heating times and possibly large amounts of stored, strain energy can result in large amounts of δ phase appearing on grain boundaries, twin lines and intragranularly, Fig. 8 (b). Such structures can lead to lower than expected tensile and rupture strength (Kennedy, 2005).

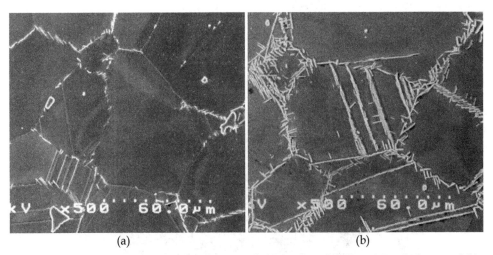

(a)                                                             (b)

Fig. 8. SEM Micrographs of Alloy 718Plus™ with (a) Preferred δ Phase Morphology and (b) Excessive δ Phase (Kennedy, 2005).

Alloy 718Plus does contain δ phase which is beneficial for conferring stress rupture notch ductility and controlling microstructure during thermo-mechanical processing. However, the volume fraction of the delta phase is considerably less than is found in alloy 718 and tends to be more stable with a much slower growth rate at elevated temperatures. Some γ" may also be present in 718Plus alloy but in a much lower quantity, less than 7% (Jeniski & Kennedy, 2006).

When Inconel 718 is compared with the ALLVAC 718 plus it is reported that the size of strengthening phases increases in both alloys after long time thermal exposure (Fig. 9), but more significantly in alloy 718. In alloy 718, the average size of γ″+γ′ grows from about 15 nm at as heat-treated condition to almost 100 nm after 500 hrs long time aging at 760°C as indicated in Fig. 10 and the main strengthening phase γ″ grows to about 200nm in estimation (see Fig. 11). However, in alloy 718Plus, the main strengthening phase γ′ coarsens slowly and the average size of γ′ is still about 70 nm as indicated in Figure 9. These important quantitative phase analyses results convince us that alloy 718Plus has a superior stable microstructure in comparison with alloy 718 (Xie et al, 2005).

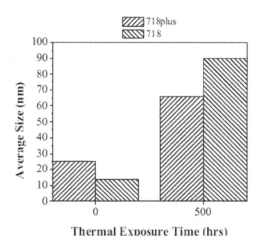

Fig. 9. The size of strengthening phases in Alloy 718 and Alloy 718 plus (Xie et al, 2005).

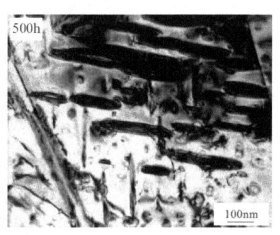

Fig. 10. The coarsening of strengthening phases γ″ and γ′ in alloy 718 after 760°C (Xie et al, 2005).

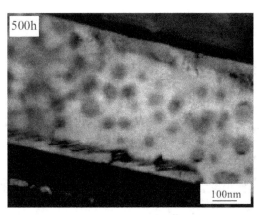

Fig. 11. The coarsening of strengthening phase γ´ in alloy 718 plus after 760°C Aging (Xie et al, 2005).

## 4.3 Microstructure

The microstructure of 718 plus in as received hot-rolled condition consists of FCC austenitic matrix with an average grain size of 50 µm. Fig. 12 shows the optical micrograph of the alloy. It can be seen that precipitates with round-to-blocky morphology are randomly dispersed within the microstructure (Vishwakarma et al, 2007).

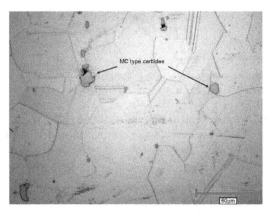

Fig. 12. Optical microstructure of as received 718 plus alloy (Vishwakarma et al, 2007).

SEM/EDS analysis of the precipitates shows them to be mainly Nb-rich MC type carbides containing Ti and C. As laves phase can be eliminated by high temperature homogenization and thermo–mechanical processing of wrought Inconel 718 and 718 Plus type of alloys, it is therefore not observed in the microstructure of the as received material. The delta phase, which is commonly observed in Inconel 718, is not observed in the as received microstructure of 718 plus alloy (Vishwakarma et al, 2007).

Heat treated at 950°C for 1 hour microstructure of 718 plus alloy, grain size of 54 µm, which has normal B and P concentrations, has shown in Fig. 13. It can be seen that needle like δ phase is observed on the grain boundaries and occasionally intra-granularly on the twin boundaries. And also seen in the microstructure, round and blocky shaped MC type carbide particles are randomly distributed. Ti-rich carbo-nitride particles can be also observed (Vishwakarma & Chaturvedi, 2008). Intermetallic phases like FCC γ′ and BCT γ″ are expected to form in 718 plus alloys but γ′ is the main strengthening phase in these superalloys (Cao & Kennedy, 2004).

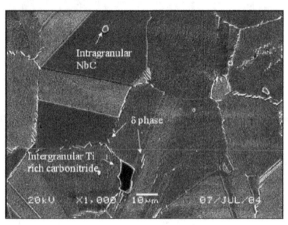

Fig. 13. Microstructure of 718 plus alloy heat treated at 950°C for 1 h (Vishwakarma & Chaturvedi, 2008).

## 4.4 Mechanical properties

Alloy 718Plus™ has a significant strength advantage over alloy 718 at temperatures above 650°C and over the entire temperature range compared to Waspaloy and A286. Elongation for alloy 718Plus™ over the entire temperature range remained high at 18% minimum. These data are consistent with comparisons of alloy 718, Waspaloy, A286 and alloy 718Plus™ in other product forms, including billet, rolled rings, forgings and sheet (Bond & Kennedy, 2005). In Fig. 14 shows the effect of temperature on room temperature ultimate tensile strength for several alloys. The tensile strength of ALLVAC 718 plus and Waspaloy is shown in Fig. 15.

It is reported by ATI ALLVAC Cooperation that extensive studies demonstrated that this alloy has shown superior tensile and stress rupture properties to alloy 718 and comparable properties to Waspaloy at the temperature up to 704°C. However, relatively speaking, the data on fatigue crack propagation (FCP) resistance of this alloy are still insufficient. Alloys 718Plus, 718 and Waspaloy have similar fatigue crack growth rates under 3 seconds triangle loading at 650°C with 718Plus being slightly better. Waspalloy shows the best resistance to fatigue crack growth under hold time fatigue condition while the resistance of 718Plus is better than that of Alloy 718 (Liu et al, 2005).

Examination of the fatigue fracture surfaces by scanning electronic microscope (SEM) revealed transgranular crack propagation with striations for 718Plus at room temperature. The fracture mode at 650°C is the mixture of intergranular and transgranular modes (Liu et al, 2004).

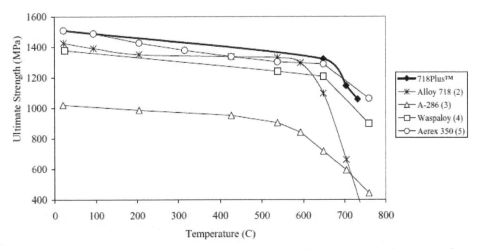

Fig. 14. Effect of test temperature on room temperature tensile ultimate tensile strength for several alloys (Bond & Kennedy, 2005).

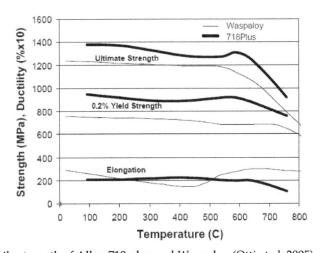

Fig. 15. The tensile strength of Alloy 718 plus and Waspaloy (Otti et al, 2005).

Direct aging can be effectively applied to alloy 718Plus to improve its mechanical properties, including strength and stress rupture life of alloy 718Plus. Considering the fine grain size and high strength resulting from direct aging, the low cycle fatigue resistance of this alloy should also be significantly improved although further experimental verification is necessary. DA processing of this alloy is also different from Waspaloy in that hot working at temperatures above the γ' solvus can achieve a good, direct age response (Cao & Kennedy, 2005).

## 4.5 Weldability

There are numerous types of superalloys with a difference in weldability among the types. The solid solution alloys are the easiest to weld because they don't undergo drastic metallurgical changes when heated and cooled. Because of their limited strength, however, they are only used in certain areas of a gas turbine, such as the combustor.

The precipitation-strengthened alloys are more demanding during welding and post welding because of the precipitation of the hardening phase that usually contains aluminum, titanium, or niobium. These elements oxidize very easily and, therefore, alloys that contain them need better gas protection during welding. A third type of superalloy is the mechanically alloyed materials that cannot be welded without suffering a drastic drop in strength. These alloys are usually joined by mechanical means or diffusion bonding. In addition to those elements that enable a superalloy to undergo precipitation hardening, such as aluminum, titanium, and niobium, other elements are added to enhance mechanical properties or corrosion resistance. These include boron and zirconium, which are often intentionally added to some alloys to improve high temperature performance but at a cost to weldability. There are numerous other elements that are not intentionally added but can be present in very small quantities that are harmful, such as lead and zinc. These are practically insoluble in superalloys and can cause hot cracking during solidification of the welds. Small quantities of these elements on the surface of a metal can cause localized weld cracking. Sulfur is considered detrimental if present in too large a quantity, but can cause low weld penetration problems if present in very low amounts (Donald & Tillack, 2007).

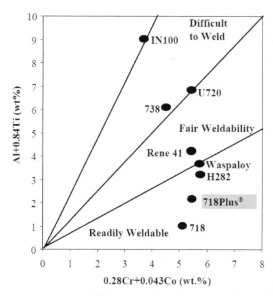

Fig. 16. Effect of chemistry on post-weld heat cracking (Jeniski & Kennedy, 2006).

It is reported that limited weldability testing has been conducted on 718Plus alloy but results have been encouraging. Weldability of alloy 718Plus is believed to be quite good, at least intermediate to alloys 718 and Waspaloy (Kennedy, 2005). Improved weldability over Waspaloy is one of the primary drivers for 718Plus alloy in engine applications. Figure 16 shows the weld cracking tendency for a number of well known commercial alloys and illustrates the good welding characteristics expected with 718Plus alloy based on its chemistry (Jeniski & Kennedy, 2006). Some micrographs of the Electron Beam Welded ALLVAC 718 plus, Inconel 718 and Waspaloy rings are shown in Fig. 17.

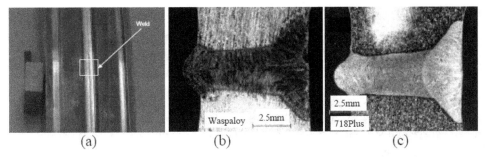

Fig. 17. EB welding of 890 mm diameter rolled rings, (a) typical weld location for 718, Waspaloy, and 718Plus welds and typical welds for (b) Waspaloy, and (c) 718Plus (Otti et al, 2005).

## 4.6 Cost and Applications

The first commitment to a production use of alloy 718Plus has been made for a high temperature tooling application, replacing Waspaloy as a hot shear knife. Other applications include aero and land-base turbine disks, forged compressor blades, fasteners, engine shafts and fabricated sheet/plate components. Product forms include rolled or flash butt welded rings, closed die forgings, bar, rod, wire, sheet, plate and castings.

The alloy also can be used for flash-butt welded ring applications. Sheet form of the alloy is being considered for fabricated engine parts such as turbine exhaust cases and engine seals. Fasteners remain another potential application for 718Plus alloy. The property advantages for 718Plus alloy have also led to its being considered for rotating parts. Cao and Kennedy have shown that 718Plus alloy is capable of direct aging (DA), low temperature working followed by aging with no prior solution heat treatment. DA processing resulted in the production of very fine grain material with yield strength improvement at 704°C of 70-100 MPa. The alloy is also being considered for blading applications in areas where alloy 718 is limited due to elevated operating temperatures.

The alloy has also other applications outside of the jet and power turbine engines. Any application that currently uses alloys 718, Waspaloy, René 41 or other nickel-based superalloys can consider 718Plus alloy as a substitute for reasons of cost savings or increased temperature capability. Other markets where 718Plus alloy has potential are automotive turbo-chargers or industrial markets like chemical process or oil and gas where alloy 718 is used (Jeniski & Kennedy, 2006).

The cost of finished components of alloy 718 Plus is expected to be intermediate to alloys 718 and Waspaloy (Kennedy, 2005).

## 5. Acknowledgment

The author would like to thank Eskisehir Osmangazi University for supporting the research study with Eskisehir Osmangazi University Research Funding Project, Project No: 2011/15020.

## 6. Conclusion

A superalloy is a metallic alloy which is developed to resist most of all high temperatures, usually in cases until 70 % of the absolute melting temperature. All of these alloys have an excellent creep, corrosion and oxidation resistance as well as a good surface stability and fatigue life.

The main alloying elements are nickel, cobalt or nickel – iron, which can be found in the VIII group of the periodic system of the elements. Fields of application are found particularly in the aerospace industry and in the nuclear industries, e.g. for engines and turbines.

The development of these advanced alloys allows a better exploitation of engines, which work at high temperatures, because the Turbine Inlet Temperature ( TIT ) depends on the temperature capability of the material which forms the turbine blades. Nickel-based superalloys can be strengthened through solid-solution and precipitation hardening.

Nickel-based superalloys can be used for a higher fraction of melting temperature and are therefore more favourable than cobal-based and iron-nickel-based superalloys at operating temperatures close to the melting temperature of the materials.

The newly innovated nickel based ALLVAC 718 Plus superalloy which is the last version of Inconel 718 has been proceeding in the way to become a material that aerospace and defense industries never replace of any other material with combining its good mechanical properties, easy machinability and low cost.

## 7. References

Bergstrom D. S., and Bayhan. T. D., (2005) Properties and Microstructure Of ALLVAC 718 Plus Alloy Rolled Sheet, *Superalloys 718, 625, 706 and Derivatives Edited by E.A. Loria TMS (The Minerals, Metals & Materials Society)*.

Bond. B.J. and Kennedy. R.L., (2005) Evaluation of ALLVAC 718 plus Alloy In the Cold Worked and Heat Treated Condition, *Superalloys 718, 625, 706 and Derivatives Edited by E.A. Loria TMS (The Minerals, Metals & Materials Society)*.

Campbell F. C. (2008). *Elements Of Metallurgy And Engineering Alloys*, ASTM International., ISBN: 978-0-87170-867-0, USA.

Cao. W., (2005) Solidification and Solid State Phase Transformation of ALLVAC 718 Plus Alloy, *Superalloys 718, 625, 706 and Derivatives 2005, TMS*.

Cao. W. and Kennedy. R., (2004) Role of Chemistry in 718-Type Alloys- Alloy 718 plus Development, *Edited by K.A. Green, T.M. Pollock, H. Harada T.E. Howson, R.C. Reed, J.J. Schirra, and S, Walston, Superalloys, TMS (The Minerals, Metals & Materials Society)*

Cao. W., and Kennedy. R. L., (2005) Application Of Direct Aging To ALLVAC 718 Plus Alloy For Improved Performance, *Superalloys 718, 625, 706 and Derivatives Edited by E.A. Loria TMS (The Minerals, Metals & Materials Society)*.

Carlos A. E. M., (2007) New Technology Used In Gas Turbine Blade Materials, *Scientia et Technica Ano XIII, No:36, ISSN: 0122-1701*

Donald. T. J., (2007) Welding Superalloys For Aerospace Applications, *Welding Journal*. pp 28-32 January.

Eliaz. N., Shemesh. G., Latanision. R.M., (2002) Hot Corrosion in Gas Turbine Components, *Engineering Failure Analysis* 9 31–43

Furrer. D., Fecht. H., (1999) Ni-Based Superalloys For Turbine Discs, *JOM*

Idowu, O.A. , Ojo, O.A., Chaturvedi, M.C. (2007) Effect of heat input on heat affected zone cracking in laser welded ATI Allvac 718Plus superalloy. *Materials Science and Engineering*. Vol. A No. 454–455 pp.389–397.

Jeniski. R. A., Jr. and Kennedy. R. L., (2006) Development of ATI Allvac 718Plus Alloy and Applications, *II. Symposium on Recent Advantages of Nb-Containing Materials in Europe*

Jonšta Z., Jonšta P., Vodárek V., Mazanec K. (2007) Physical-Metallurgical Characteristics Of Nickel Super-Alloys Of Inconel Type. *Acta Metallurgica Slovaca*, 13, 4 (546 - 553).

Jovanović T. M., Lukic. B., Miskovic. Z., Bobic. I., Ivana, Cvijovic. B. D., Processing And Some Applications Of Nickel, Cobalt And Titanium-Based Alloys, *Association of Metallurgical Engineers of Serbia Review paper*, MJoM *Metalurgija Journal Of Metallurgy*

Kennedy. R. L., (2005) Allvac® 718plus™ Superalloy For The Next Forty Years, *Superalloys 718, 625, 706 and Derivatives 2005, TMS*.

Liu. X., Xu. J., Deem. N., Chang. K., Barbero. E., Cao. W., Kennedy. R. L., Carneiro. T., (2005) Effect Of Thermal-Mechanical Treatment On The Fatigue Crack Propagation Behavior Of Newly Developed Allvac 718plus Alloy, *Superalloys 718, 625, 706 and Derivatives 2005, TMS*.

Liu. X., Rangararan. S., Barbero. E., Chang. K., Cao. W., Kennedy. R.L., and Carneiro. T., (2004) Fatigue Crack Propagation Behaviors Of New Developed Allvac 718plus Superalloy, *Superalloys 2004, TMS*.

Otti. E.A., Grohi. J. And Sizek. H., (2005) Metals Affordability Initiative: Application of Allvac Alloy 718Plus for Aircraft Engine Static Structural Components, *Superalloys 718, 625, 706 and Derivatives Edited by E.A. Loria TMS (The Minerals, Metals & Materials Society)*.

Philip A. Schweitzer, P. E., (2003). *Metallic Materials: Physical, Mechanical and Corrosion Properties*, Marcel Dekker, Inc., ISBN: 0-8247-0878-4, USA.

Sajjadi, S.A., Zebarjad, S.M. (2006) Study of fracture mechanisms of a Ni-Base superalloy at different temperatures. *Journal of Achievements in Materials and Manufacturing Engineering*. Vol. 18 Issue 1-2.

Schreiber. K., Loehnert. K., Singer. R.F., (2006) Opportunities and Challenges for the New Nickel-Based Alloy 718 Plus, *II. Symposium on Recent Advantages of Nb-Containing Materials in Europe*.

Sims. C.T., Stoloff. N.S. and Hagel. W.C., (1987) Superalloys II- High Temperature Materials for Aerospace and Industrial Power, *John Wiley & Sons Inc.*, USA.

Smallman R. E. Ngan, A. W. H., (2007). *Physical Metallurgy and Advanced Materials*, Elsevier Ltd., ISBN: 978 0 7506 6906 1, UK.

Smallman R. E. Bishop, R. J., (1999). *Modern Physical Metallurgy and Materials Engineering*, Reed Educational and Professional Publishing Ltd., ISBN: 0 7506 4564 4, UK.

Thomas. A., El-Wahabi. M., Cabrera. J.M., Prado. J. M., (2006) High Temperature Deformation of Inconel 718, *Journal of Materials Processing Technology* 177 469–472.

Vishwakarma. K.R., Richards. N.L., Chaturvedi. M.C., (2007) Microstructural Analysis of Fusion and Heat Affected Zones in Electron Beam Welded ALLVAC® 718PLUSTM superalloy, *Materials Science and Engineering* A 480 517–528

Vishwakarma. K.R. and Chaturvedi. M.C., (2007) A Study Of Haz Microfissuring In A Newly Developed Allvac® 718 Plus Tm Superalloy.

Yaman. Y.M., Kushan. M.C., (1998) Hot Cracking Susceptibilities In the Heat Affected Zone of Electron Beam Welded Inconel 718, *Journal of Materials Science Letters* 17, 1231–1234.

Xie. X., Wang. G., Dong. J., Xu. C., Cao. W., Kennedy. R. L., (2005) Structure Stability Study On A Newly Developed Nickel-Base Superalloy- ALLVAC 718 Plus, *Superalloys 718, 625, 706 and Derivatives Edited by E.A. Loria TMS (The Minerals, Metals & Materials Society)*.

# Part 2

# Aircraft Control Systems

# 6

# An Algorithm for Parameters Identification of an Aircraft's Dynamics*

I. A. Boguslavsky

*State Institute of Aviation Systems, Moskow Physical Technical Institute*
*Russia*

## 1. Introduction

Development of efficient parameter identification methods for the model of a dynamic system based on real-time measurements of some components of its state vector should be taken as one of the most important problems of applied statistics and computational mathematics. Calculating the motion of the system given the initial conditions and its mathematical model is conventionally called the direct problem of dynamics. Then, the inverse problem of dynamics would be the problem of identifying the system model parameters based on measurements of certain components of the state vector provided that the general structural scheme of the model is known from physical considerations. Such an inverse problem corresponds to identification problem for the dynamic system representing an aircraft. In this case, the general structural scheme of the model (motion equations) follows from the fundamental laws of aerodynamics.

In many cases, modern computational methods and wind tunnel experiments can provide sufficient data on nominal parameters of the mathematical model - nominal aerodynamic characteristics of the aircraft. Nevertheless, there exist problems [1] that require correcting nominal parameters based on measurements taken in real flights. These imply

(1) verifying and interpreting theoretical predictions and results of wind tunnel experiments (flight data can also be used to improve ground prediction methods),

(2) obtaining more exact and complete mathematical models of the aircraft dynamics to be applied in designing stability enhancement methods and flight control systems,

(3) designing flight simulators that require more accurate dynamic aircraft profile in all flight modes (many motions of aircrafts and flight conditions can be neither reconstructed in the wind tunnel nor calculated analytically up to sufficient accuracy or efficiency),

(4) extending the range of flight modes for new aircrafts, which can include quantitative determination of stability and impact of control when the configuration is changed or when special flight conditions are realized,

(5) testing whether the aircraft specification is compliant.

*This work has been supported by the Russian Foundation for Basic Research.

Furthermore, dimensionless numbers at the nodes of one-or two-dimensional tables found in wind tunnel experiments serve as nominal values in the aerodynamic parameter identification problem of the aircraft. This causes the vector that corrects these parameters determined by the algorithm processing digital data flows received from the aircraft sensors to have a significant dimension of the order about several tens or hundreds.

It is worth noting that the USA (NASA) is doing extensive work on theoretical and practical aircraft identification by test flights. In 2006 alone, in addition to many journal publications, American Institute of Aeronatics and Astronatics (AIAA) published three fundamental monographs [1-3] on the subject. An implementation of multiple NASA recommended algorithms for identification problems, SIDPAS (Systems Identification Programs for Aircraft) software package written in MATLAB M-files language is available on the Internet as an appendix to [1]. Various existing identification methods published in monographs on statistics and computational mathematics are widely reviewed in [1].

For the most general identification method, one should take the known nonlinear least squares method [4] that forms the sum of errors squared - differences between the real measurements and their calculated analogues obtained by numerical integration of motion equations of the system for some realization of the vector of unknown parameters.

Successful identification yields the vector of parameters that delivers the global minimum to the above mentioned sum of errors squared. Still, this criterion is statistically valid only for linear identification problems, in which measurements are linear with respect to the unknown vector of parameters.

Implementing the nonlinear least squares method to correct nominal parameters of the aircraft based on its test flight data involves computational challenges. These arise when the dimension of the correction vector is big and the sum of errors squared as the function of the correction vector has multiple relative minimums or when variations of the Newton's method are applied, with the sequence of local linearizations performed to find stationary points of this function. In [1], the regression method supported by *lesq.m, smoo.m, derive.m, and xstep.m* files in SIDPAS is recommended for practical applications.

Suppose the motion equations of the system and the sequence of measurements have the form

$$dx/dt = f(x, \vartheta + \eta, u), ...(0.1)$$

$$y_k = H_k(x(t_k)) + \xi_k, ...(0.2)$$

where $x(t_k)$ is the $n \times 1$-dimensional vector of the system states at the current instant $t$ and at the given instants $t_k, k = 1, ..., N, \vartheta$ is the $r \times 1$-vector of nominal (known) parameters of the system, $\eta$ is the vector of unknown parameters that serves as the correction vector for the nominal vector $\vartheta$ after the results of measurements are stochastically processed, $u$ is the control vector of the system, $f(...)$ is the given vector-function, $y_k$ is the sequence of vectors-results of measurements, $H_k(...)$ is the given vector function, and $\xi_k, k = 1, ..., N$ is the sequence of random vectors-errors of measurements with the given random generator for the mathematical simulation.

We can state the identification problem for the vector $\eta$ as follows. Find the estimate as the function of the vector $Y_N$ formed of the results of all measurements $y_1, ... y_N$.

The regression method given in [1] solves this problem under the following limitations

(1) all components of the state vector can be measured : $y_k = x(t_k) + \xi_k$,

(2) at the measurement instants $t_k$, the algorithm constructs the estimate of the vector of derivatives $dx/dt$,

(3) the vector function $f(x, \vartheta + \eta, u)$ linearly depends on the vector $\eta$.

These fundamental limitations of the regression method duplicate features of the identification algorithm from [5]. The substantial drawback of the algorithm [5] and the algorithm of the regression method is that they do not allow using the mathematical model to analyze theoretically (without applying the Monte-Carlo method) observability conditions of components of the vector of parameters to be identified for the preliminary given control law for the test flight of the aircraft and information on random errors of its sensors. Note that this is the drawback of all known numerical methods that solve nonlinear identification problems.

Relations (0.1) and (0.2) show that when conditions (1)-(3) are met and $N$ is sufficiently big, the estimation vector satisfies the overdetermind system of linear algebraic equations, with methods to solve it being well known. The given conditions seem to be rather rigid and may be hard-to-implement. For instance, it is arguable whether one can construct the vector of derivatives dx/dt sufficiently accurately given the real turbulent atmosphere conditions, which imply that the outputs of the angle of attack and sideslip sensors inevitably include random and unpredictable frequency components.

All this justifies the development of new identification algorithms that can be applied to dynamic systems of a rather general class and do not possess drawbacks of NASA algorithms. The proposed multipolynomial approximation algorithm (MPA algorithm) serves as such a new identification algorithm.

## 2. Statement of the problem and basic scheme of the proposed identification algorithm

The general scheme for identifying aerodynamic characteristics of the aircraft by the test flight data is as follows [1]. Motion equations of the aircraft (0.1) and system (0.2) of measurements of motion characteristics of the aircraft are given. The vector $\vartheta$ is the vector of nominal aerodynamic parameters determined in the wind tunnel experiment. Calculated by the results of real (test) flight, the vector $\eta$ is used to correct the vector $\vartheta$.

When the aircraft flies, its computer fixes the digital array of initial conditions and time functions, viz. current control surface angles and measurements of some motion parameters of the aircraft (some components of the vector x(t) of the state of the aircraft) received from its sensors. Note that selecting the criterion for optimal or, at least, rational mode to control the test flight is a separate problem and lies beyond our further consideration. The current motion characteristics measured as the time function such as angles of attack and sideslip and components of the vector of angular velocity and g-load obtained by the inertial system of the aircraft are registered for real (not known for sure) aerodynamic parameters of the aircraft (parameters $\vartheta + \eta$) and can be called measured characteristics of the perturbed motion.

Once the flight under the mentioned (given) initial conditions and time functions (control surface angles) is completed, nominal motion equations (equations of form (1) for $\eta = 0$) are integrated numerically for the nominal aerodynamic parameters of the aircraft. For the calculated characteristics of the nominal motion of the aircraft one should take the obtained data - components of the state vector of the aircraft as the function of discrete time. Differences between measurable characteristics of the perturbed motion and calculated characteristics of the nominal motion serve as carriers of data on the unknown vector $\eta$ that shows the difference between real and nominal aerodynamic parameters.

The input of the MPA identification algorithm receives the vector of initial conditions and control surface angles as functions of time and arrays of characteristics of nominal and perturbed motions.

The output of the algorithm is $\hat{\eta}(Y_N)$, which is the correction vector for nominal aerodynamic parameters.

The identification algorithm is efficient if the motion equations integrated numerically with the corrected aerodynamic parameters yield such motion characteristics $\vartheta + \hat{\eta}(Y_N)$ (*corrected characteristics*, in what follows) that are close to real (measurable) characteristics.

In this work, we consider the technology of applying the Bayes MPA algorithm [6, 7] to solve identification problems on the example of the aircraft, for which nominal aerodynamic parameters of the pitching motion are the nominal parameters of one of an "pseudo" F-16 aircraft.

We replace real flights by mathematical simulation, with characteristics of the perturbed motion obtained by integrating the motion equations of the aircraft numerically. In these equations, nominal aerodynamic parameters at the nodes of the corresponding tables are changed to random values that do not exceed in modulus the given $25 \div 50$ percents of nominal values at these nodes.

Fundamentally, the MPA algorithm assumes that the vector of unknown parameters $\eta$ is random on the set of possible flights. We assume that the a priori statistical-generator for computer generated random vectors $\eta$ and $\xi_k$ is given. This generator makes the algorithm estimating components of the vector $\eta$ (the identification algorithm) Bayesian. Further, for particular calculations, we assume that random components of the mentioned vectors are distributed uniformly and can be called by the standard Random program in Turbo Pascal.

The MPA algorithm provides the approximation method we implement with the multidimensional power series of the vector $E(\eta|Y_N)$ of the conditional mathematical expectation of the vector $\eta$ if the vector of measurements $Y_N$ is fixed and a priori statistical data on random vectors $\eta$ and $\xi_k$ are given.

The vector $E(\eta|Y_N)$ is known to be optimal, in root-mean-square sense, estimate of the random vector $\eta$.

We describe the steps of operation of the MPA algorithm when it identifies the vector $\eta$[6, 7].

Step 1. Suppose $d$ is a given positive integer number and the set of integer numbers $a_1, ..., a_N$ consists of all nonnegative solutions of the integer inequality $a_1 + ... + a_N \leq d$, the number of which we denote by $m(d, N)$. The value $m(d, N)$ is given by the recurrent formula proved by induction.

$$m(d, N) = m(d - 1, N) + (N + d - 1) \cdots N/d!, m(1, N) = N.$$

We obtain the vector $W_N(d)$ of dimension $m(d, N) \times 1$, the components $w_1, ..., w_{m(d,N)}$ of which are all possible values $y_1^{a_1}...y_N^{a_n}$ of the form that represent the powers of measurable values.

Then, we construct the base vector $V(d, N)$ of dimension $(r + m(d, N)) \times 1, V(d, N) = \|\eta W_N(d)\|$.

Step 2. We use a known statistical generator of random vectors $\eta$ and $\xi_k$ to solve repeatedly the Cauchy problem for Eq.(1) for given initial conditions $x(0)$, a control law $u(t)$ and various realizations of random vectors $\eta$ and $xi_k$.

We apply the Monte-Carlo method to find the prior first and second statistical moments of the vector $V(d, N)$, i.e., the mathematical expectation $\bar{V}(d, N)$, and the covariance matrix $C_V(d, N) = E((V(d, N) - \bar{V}(d, N))(V(d, N) - \bar{V}(d, N))^T)$.

Implementation of step 2 is a learning process for the algorithm, adjusting it to solve the particular problem described by Eqs. (1) and (2).

Step 3. For given $d$ and $N$ and a fixed vector $Y_N$, we assign the vector $\hat{\eta}(W_N(d))$ to be the solution to the estimation problem. This vector gives an approximate estimate of the vector $E(\eta|Y_N)$ that is optimal in the root-mean-square sense on the set of vector linear combinations of components of the vector $W_{N_1}(d)$

$$\hat{\eta}(W_N(d)) = \sum_{a_1 + ... + a_N \leq d} \lambda(a_1, ..., a_N) y_1^{a_1} \cdots y_N^{a_N}. \quad (1.1)$$

The vector $\bar{V}(d, N)$ and the matrix $C_V(d, N)$ are the initial conditions for the process of recurrent calculations that realizes the principle of observation decomposition [6] and consists of $m(d, N)$ steps. Once the final step is performed, we obtain vector coefficients $\lambda(a_1, ..., a_N)$ for (1.1). Moreover, we determine the matrix $C(d, N)$, which is the covariance matrix of the estimation errors for the vector $E(\eta_N|Y_N)$ of conditional mathematical expectation estimated by the vector $\hat{\eta}(W_N(d))$.

Calculating the elements of the matrix $C(d, N)$, we have the method of preliminary (prior to the actual flight) analysis of observability of identified parameters for the given control law, structure of measurements and their expected random errors. Recurrent calculations do not require matrix inversion and indicate the situations when the next component of the vector $W_N(d)$ is close to linear combination of its previous components. To implement the recursion, we process the components of the vector $W_N(d)$ one after another. However, the adjustment of the algorithm performed by applying the Monte-Carlo method to find the vector $\bar{V}(d, N)$ and the matrix $C_V(d, N)$ takes into account a priori ideas on stochastic structure of components of the whole set of possible vectors $W_N(d)$ that can appear in any realizations of the random vectors $\eta$ and $\xi_k$ allowed by the a priori conditions.

This adjustment is the price we have to pay if we want the MPA algorithm to solve nonlinear identification problems efficiently. This is what makes the MPA algorithm differ fundamentally from, for instance, the standard Kalman filter designed to solve linear identification problems only or from multiple variations of algorithms resulted from attempts to extend the Kalman filter to nonlinear filtration problems.

In [6], a multidimensional analogue of the K. Weierstrass theorem (the corollary of the M. Stone theorem [9]) is used to prove that when the integer $d$ is increases then the error estimates of the vector $E(\eta|Y_N)$ the vector $|\hat{\eta}(W_N(d)) - E(\eta|Y_N)|$ tend to zero uniformly on some region. Formulas of the recurrent algorithm are given and justified in [6, 7] and in the Appendix.

This scheme for the MPA algorithm operation shows that it can be applied to identify parameters of almost any dynamic system provided that the structures of the motion equations and measurements of form (0.1) and (0.2) and prior statistical generators of random unknown parameters and errors of measurements are given. The MPA algorithm is devoid of the above listed limitations and drawbacks, which gives it substantial advantages over NASA identification algorithms. Apart from errors of computations, the algorithm does not add any other errors (such as errors due to linearization of nonlinear functions) into the identified parameters. Therefore, one should expect that the priori spread of identifiable parameters to be always greater than the posterior spread. This is why we can use iterations.

Let us compare the sequential steps of the standard discrete Kalman filter and the MPA algorithm.

(1) The Kalman filter identifies the vector $\eta$, which can be represented by part of components of the state vector of the linear dynamic system for the observations that linearly depend on state vectors. The a priori data are the first and second moments of components of random initial state vectors, uncorrelated random vectors of perturbations and observation errors. We need these data for sequential (recurrent) construction of the estimation vector that is root-mean-square optimal. Usually assigned, a priori data can be also determined by the Monte-Carlo method if the complex mechanism of their appearance is given.

(2) To find an asymptotic solution to the nonlinear identification problem, the MPA algorithm, unlike the Kalman filter, requires a priori statistical data on both the initial and all hypothesized future state vectors of the dynamic system and observations. These a priori data are represented by the first and second statistical moments for the random vector V(d, N): the vector $\overline{V}(d, N))$ and the matrix $C_V(d, N)$. These moments are calculated using the Monte-Carlo method. However, there are cases when they can be obtained by numerical multidimensional region integration.

(1.1) Once conditions from (1) are met, the Kalman filter constructs the recurrent process, at every step of which the current estimation vector optimal in the root-mean-square sense and the covariance matrix of errors of the estimate are calculated.

(2.1) Based on (2), the MPA algorithm implements the recurrent computational process that do not require matrix inversion. At each step of the process, we construct

i. the current estimation vector $\hat{\eta}(W_N(d))$ linear with respect to components of the vector $W_N(d)$ and optimal in the root-mean-square sense on the set of linear combinations of components of this vector; moreover, the uniform convergence $\hat{\eta}(W_N(d)) \to E(\eta|Y_N), d \to \infty$. is attained on some region,

ii. the current covariance matrix of estimation errors (we emphasize that known numerical methods of constructing approximations of the vector of nonlinear estimates cannot calculate current covariance matrices of estimation errors).

Implementation of items 2 and 2.1 makes the MPA algorithm more efficient than any known linear identification algorithm since it

i. does not involve linearization,

ii. does not apply variants of the Newton method to solve systems of nonlinear algebraic equations,

iii. forms the estimation vector that tends uniformly to the vector of conditional mathematical expectation for the growing integer $d$,

iiii. obtains the covariance matrix of estimation errors.

It is worth emphasizing that in this work we just develop the fundamental ground of computational technique for solving the complex problem of aircraft parameter identification.

## 3. Testing the MPA algorithm: Problem reconstruction (identification of parameters) for the attractor from units of an electrical chain

We consider the boundary inverse problem for the attractor whose equations are presented in [8 ]. The three parameters are the initial conditions: $X_1[0] = \eta_1, X_2[0] = \eta_2, X_3[0] = \eta_3$. The six parameters $\eta_{3+i}, i = 1, ..., 6$ correspond to combinations of the inductance, the resistances and the two capacitances of a circuit.

The equations of the mathematical model of the circuit take the following form [8]:

$$X_1[k-1] < -\eta_{3+6} : f = \eta_{3+5};$$

$$-\eta_{3+6} < X_1[k-1] < \eta_{3+6} : f = X_1[k-1](1 - X_1[k-1]^2);$$

$$X_1[k-1] > \eta_{3+6} : f = -\eta_{3+5};$$

$$X_1[k] = X_1[k-1] + \delta X_2[k-1];$$

$$X_2[k] = X_2[k-1] + \delta(-X_1[k-1] - \eta_{3+1}X_2[k-1] + X_3[k-1]);$$

$$X_3[k] = X_3[k-1] + \delta(\theta_{3+2}(\eta_{3+3}f - X_3[k-1]) - \eta_{3+4}X_2[k-1]);$$

where $X_1[k]$ corresponds to a voltage, $X_2[k]$ to a current and $X_3[k]$ to another voltage.

We suppose that by $i = 1, 2, 3$

$$\eta_i \in 1 + (\varepsilon_i - 0.5).$$

We also suppose [ 8] that

$$\eta_{3+1} \in 0.5(1 + (\varepsilon_1 - 0.5)); \eta_{3+2} = 0.3(1 + (\varepsilon_2 - 0.5)); \eta_{3+3} = 15(1 + (\varepsilon_3 - 0.5));$$

$$\eta_{3+4} \in 1.5(1 + (\varepsilon_4 - 0.5)); \eta_{3+5} = 0.5(1 + (\varepsilon_5 - 0.5)); \eta_{3+6} = 1.2(1 + (\varepsilon_6 - 0.5));$$

$$y_k = X_1(t_k)) + \zeta_k$$

$$\delta = 0.01, k = 1, ..., N = 1200, z_1 = \sum_{k=1}^{k=N/T} y_k, z_2 = \sum_{k=1+N/T}^{k=2 \times N/T} y_k, ...$$

The algorithm uses approximations of parameters by means of linear combinations of the constructed values $z_i(d = 1)$. Values $z_1, z_2$ - are the sums of values of flowing observations - serve as inputs of MPA algorithm

The relative errors of the boundary problem are

$$
\begin{array}{llll}
i & 1 & 2 & 3 \\
T = 24 \\
\quad \Delta_i & 0.025 & 0.264 & 0.272 \\
T = 48 \\
\quad \Delta_i & 0.0007 & -0.003 & 0.046 \; . \\
T = 120 \\
\quad \Delta_i & 0.00005 & -0.00264 & 0.01687 \\
T = 240 \\
\quad \Delta_i & 0.00001 & -0.00049 & 0.02686
\end{array}
$$

The relative errors of the inverse problem are

$$
\begin{array}{lllllll}
i & 1 & 2 & 3 & 4 & 5 & 6 \\
T = 24 \\
\quad \Delta_{3+i} & -0.347 & 0.198 & -0.250 & 0.097 & 0.095 & 0.136 \\
T = 48 \\
\quad \Delta_{3+i} & -0.140 & 0.234 & -0.222 & 0.104 & 0.143 & 0.133 \; . \\
T = 120 \\
\quad \Delta_{3+i} & -0.169 & 0.205 & -0.167 & 0.094 & 0.179 & 0.097 \\
T = 240 \\
\quad \Delta_{3+i} & -0.042 & 0.129 & -0.031 & -0.0001 & 0.151 & 0.146
\end{array}
$$

The resulted tables show, that corresponding adjustment the MPA algorithm - a corresponding selection of value $T$ allows to make small relative errors of an estimation of parameters of the non-linear dynamic system.

## 4. Identification of aerodynamic coefficients of the pitching motion for an pseudo f-16 aircraft

We illustrate efficiency of offered MPA algorithm on an example of identification of 48 dimensionless aerodynamic coefficients for the aircraft of near F-16. The aircraft we shall conditionally name " pseudo F-16 ". The term "near" is justified by that, what is the coefficients are taken from SIDPAS [1], but are perturbed by addition of some random numbers.

The tables resulted below, show, that errors of identification are small also modules of their relative values do not surpass several hundredth. The considered problem corresponds to minimization of object function of 48 variables, which it is made of the sum of squares of differences of actual and computational angles of attack, g-load, pitch angles, observable with frequency 10 hertz during 25 sec. flight of the aircraft maneuvering in a vertical plane.

| number $\alpha_i$ | $C_{Z_0}(\alpha_i)$ | $C_{m_0}(\alpha_i)$ | $C_{Z_q}(\alpha_i)$ | $C_{m_q}(\alpha_i)$ |
|---|---|---|---|---|
| 1 | 0.7700 | −0.1740 | −8.8000 | −7.2100 |
| 2 | 0.2410 | −0.1450 | −25.8000 | −5.4000 |
| 3 | −0.1000 | −0.1210 | −28.9000 | −5.2300 |
| 4 | −0.4160 | −0.1270 | −31.4000 | −5.2600 |
| 5 | −0.7310 | −0.1290 | −31.2000 | −6.1100 |
| 6 | −1.0530 | −0.1020 | −30.7000 | −6.6400 |
| 7 | −1.3660 | −0.0970 | −27.7000 | −5.6900 |
| 8 | −1.6460 | −0.1130 | −28.2000 | −6.0000 |
| 9 | −1.9170 | −0.0870 | −29.0000 | −6.2000 |
| 10 | −2.1200 | −0.0840 | −29.8000 | −6.4000 |
| 11 | −2.2480 | −0.0690 | −38.3000 | −6.6000 |
| 12 | −2.2290 | −0.0060 | −35.3000 | −6.0000 |

Table 1. Nominal values of the functions $C_{Z_0}(\alpha), C_{m_0}(\alpha), C_{Z_q}(\alpha), C_{m_q}(\alpha)$

## 4.1 Pitching motion equations

We use the rectangular coordinate system XYZ adopted in NASA. Then for the unperturbed atmosphere and conditions $V = const$, pitching motion equations have the form [1]:

$$da/dt = \omega_Y + (g/V)(N_Z + \cos(\theta - \alpha)),$$
$$d\omega_Y/dt = M_Y/J_Y,$$
$$d\theta/dt = \omega_Y,$$
$$N_Z = C_Z(\alpha, \delta_s)qS/G,$$
$$M_Y = C_m(\alpha, \delta_s)qSb,$$

where $V$=300 ft/sec,$H$=20000 ft, $\alpha$ is the angle of attack, $N_Z$ is the g-load, which is the vector of aerodynamic forces projected onto the axis $Z$ and divided by the weight of the aircraft, $M_Y$ is the vector of the moment of aerodynamic forces projected onto the axis $Y$, $\omega$ is the vector of the angular velocity of the aircraft projected onto the axis $Y$,$\theta$ is the angle between the the axis $X$ and the horizontal plane, $q$ is the value of the dynamic pressure, $G$ is the weight, $J_Y$ is the moment of inertia with respect to the axis $Y$, $S$ is the area of the surface generating aerodynamic forces, $b$ is the mean aerodynamic of the wing, $C_Z(\alpha, \delta)$ and $C_m(\alpha, \delta)$ are dimensionless coefficients of the aerodynamic force and moment,$\delta_s$ is the angle of the stabilator devlection measured in degrees.

Functions $C_Z(\alpha, \delta_s)$ and $C_m(\alpha, \delta_s)$ are given by the relations [1],:

$$C_Z(\alpha, \delta_s) = C_{Z_0}(\alpha) - 0.19(\delta_s/25) + C_{Z_q}(\alpha)(b/(2V))\omega,$$
$$C_m(\alpha, \delta_s) = C_{m_0}(\alpha)\delta_s + C_{m_q}(\alpha)(b/(2V))\omega + 0.1C_Z.$$

## 4.2 Parametric model aerodynamic forces and moments

Nominal values of 4 functions of the angle of attack $C_{Z_0}(\alpha), C_{m_0}(\alpha), C_{Z_q}(\alpha), C_{m_q}(\alpha)$ are given with the argument step $(55 - 1)/12$ degree at 12 nodes (Table 1) in range $-10° \leq \alpha \leq 45°$ .

To determine values of functions between the nodes, we use linear interpolation. Having analyzed Table 1, we can see that functions $C_{Z_0}(\alpha_i), C_{m_0}(\alpha_i), C_{Z_q}(\alpha_i), C_{m_q}(\alpha_i)$ are essentially

| number $\alpha_i$ | $\Delta(C_{Z_0}(\alpha_i))$ | $\Delta(C_{m_0}(\alpha_i))$ | $\Delta(C_{Z_q}(\alpha_i))$ | $\Delta(C_{m_q}(\alpha_i))$ |
|---|---|---|---|---|
| 1 | 0.7700 | −0.1740 | −8.8000 | −7.2100 |
| 2 | −0.5290 | 0.0290 | −17.0000 | 1.8100 |
| 3 | −0.3410 | 0.0240 | −3.1000 | 0.1700 |
| 4 | −0.3160 | −0.0060 | −2.5000 | −0.0300 |
| 5 | −0.3150 | −0.0020 | 0.2000 | −0.8500 |
| 6 | −0.3220 | 0.0270 | 0.5000 | −0.5300 |
| 7 | −0.3130 | 0.0050 | 3.0000 | 0.9500 |
| 8 | −0.2800 | −0.0160 | −0.5000 | −0.3100 |
| 9 | −0.2710 | 0.0260 | −0.8000 | −0.2000 |
| 10 | −0.2030 | 0.0030 | −0.8000 | −0.2000 |
| 11 | −0.1280 | 0.0150 | −8.5000 | −0.2000 |
| 12 | 0.0190 | 0.0630 | 3.0000 | 0.6000 |

Table 2. Nominal values of increment $\Delta(C_{Z_0}(\alpha_i)), \Delta(C_{m_0}(\alpha_i)), \Delta(C_{Z_q}(\alpha_i)), \Delta(C_{m_q}(\alpha_i))$

nonlinear. Table 2 confirms this visual impression. In it increments are presented 4 functions on each step of Table 1. Apparently, increments noticeably vary.

We study the identification problem for the perturbed analogues of the functions $C_{Z_0}(\alpha), C_{m_0}(\alpha), C_{Z_q}(\alpha), C_{m_q}(\alpha)$. The number of nominal coefficients that determine these functions is 12+12+12+12 = 48. Let us single out the problem which is the most complex for the MPA algorithm, when the actual coefficients differs from the nominal coefficients by the unknown bounded by the prior limits value $\eta_i$ at each point of the table. Then, for accumulated results of measurements of parameters of the perturbed motion, the MPA algorithm is to estimate 48 components of the vector of random estimates, - the vector of differences between actual and nominal coefficients.

Suppose $\vartheta_i$ and $B_i$ are the i-th components of the nominal and actual (perturbed) vectors of aerodynamic coefficients, $i = 1, ..., 48$, i.e. the number of actual coefficients to be identified is 48 in this case. We assume that the parametric model

$$B_i = \vartheta_i + \eta_i.$$

holds. The vector $\eta$ serves as the vector of perturbations of nominal data errors of aerodynamic parameters, and identification yields the estimates of its components. We give the structure of these components by the formula $\eta_i = \vartheta_i \rho_i \varepsilon_i, 0 < \rho_i < 1, -1 < \varepsilon_i < 1$. The positive number $\rho_i$ gives the maximum value that, by identification conditions, can be attained by the ratio of the absolute values of the random value of perturbations $\eta_i$ and nominal coefficients $\vartheta_i$ .

## 4.3 Transient processes of characteristics of nominal motions

We wish to identify-estimate - during one test flight the 48 unknown aerodynamic coefficients for 12 nodes-12 the set angles of attack $\alpha_i, i = 1..., 12$. For a testing maneuver the characteristics $\alpha(t), N_Z(t), \theta(t)$ of Transient Processes are carrier of information of the the the identified coefficients. Therefore during flight the aircraft should "visit" vicinities of angles of attack $-10° \le \alpha \le 45°$

| number. obs. $k$ | $\delta_s(k)$ | $\alpha(k)$ | $N_Z(k)$ | $\theta(k)$ |
|---|---|---|---|---|
| 1 | −0.0200 | 3.6820 | 0.1021 | 0.0132 |
| 3 | −0.0600 | 5.1462 | −0.3525 | 0.1388 |
| 5 | −0.1000 | 6.0119 | −0.2956 | 0.1689 |
| 7 | −0.1400 | 6.7707 | −0.2493 | 0.0300 |
| 9 | −0.1800 | 7.5061 | −0.2085 | −0.1851 |
| 11 | −0.2200 | 8.2964 | −0.1685 | −0.3945 |
| 13 | −0.2600 | 9.2186 | −0.1253 | −0.5227 |
| 15 | −0.3000 | 10.2083 | −0.6016 | −0.5119 |
| 17 | −0.3400 | 10.6145 | −0.5691 | −0.6187 |
| 19 | −0.3800 | 10.8889 | −0.5477 | −0.8891 |
| 21 | −0.4200 | 11.0977 | −0.5334 | −1.2461 |
| 23 | −0.4600 | 11.2993 | −0.5223 | −1.6252 |
| 25 | −0.5000 | 11.5494 | −0.5114 | −1.9688 |
| 27 | −0.5400 | 11.9047 | −0.4974 | −2.2222 |
| 29 | −0.5800 | 12.4277 | −0.4774 | −2.3286 |
| 31 | −0.6200 | 13.1919 | −0.4477 | −2.2247 |
| 33 | −0.6600 | 14.2870 | −0.4043 | −1.8352 |
| 35 | −0.7000 | 15.4810 | −0.8822 | −1.1481 |
| 37 | −0.7400 | 16.1493 | −0.8343 | −0.6582 |
| 39 | −0.7800 | 16.6530 | −0.7993 | −0.3828 |
| 41 | −0.8200 | 17.0629 | −0.7728 | −0.2382 |
| 43 | −0.8600 | 17.4401 | −0.7511 | −0.1552 |
| 45 | −0.9000 | 17.8400 | −0.7310 | −0.0747 |
| 47 | −0.9400 | 18.3168 | −0.7095 | 0.0571 |
| 49 | −0.9800 | 18.9260 | −0.6838 | 0.2926 |
| 51 | −1.0200 | 19.7287 | −0.6512 | 0.6855 |
| 53 | −1.0600 | 20.1833 | −1.1389 | 1.1343 |
| 55 | −1.1000 | 20.1954 | −1.1266 | 1.3075 |
| 57 | −1.1400 | 19.9812 | −1.1273 | 1.2486 |
| 59 | −1.1800 | 19.9888 | −1.1293 | 1.1572 |
| 61 | −1.2200 | 19.9789 | −1.1308 | 1.1146 |
| 63 | −1.2600 | 20.0083 | −1.1316 | 1.1129 |
| 65 | −1.3000 | 20.0049 | −1.1325 | 1.1437 |
| 67 | −1.3400 | 20.0371 | −1.1330 | 1.2113 |
| 69 | −1.3800 | 20.0328 | −1.1338 | 1.3073 |
| 71 | −1.4200 | 20.0598 | −1.1344 | 1.4359 |
| 73 | −1.4600 | 20.0636 | −1.1346 | 1.6053 |
| 75 | −1.5000 | 20.0993 | −1.1344 | 1.8069 |
| 77 | −1.5400 | 20.1945 | −1.1326 | 2.0671 |
| 79 | −1.5800 | 20.3760 | −1.1278 | 2.4096 |
| 81 | −1.6200 | 20.6696 | −1.1186 | 2.8558 |
| 83 | −1.6600 | 21.1005 | −1.1037 | 3.4247 |
| 85 | −1.7000 | 21.6926 | −1.0820 | 4.1331 |
| 87 | −1.7400 | 22.4690 | −1.0523 | 4.9948 |
| 89 | −1.7800 | 23.4509 | −1.0133 | 6.0212 |

```
 91  -1.8200 24.6576 -0.9641  7.2201
 93  -1.8600 25.7086 -1.3912  8.6103
 95  -1.9000 26.9187 -1.3506 10.2913
 97  -1.9400 28.6178 -1.2942 12.4085
 99  -1.9800 30.7696 -1.6768 15.0754
101  -2.0200 32.4354 -1.6031 17.5706
103  -2.0600 33.8743 -1.5431 19.7731
105  -2.1000 35.1432 -1.8357 21.7958
107  -2.1400 35.7317 -1.8014 23.4808
109  -2.1800 35.9655 -1.7815 24.7918
111  -2.1800 35.9159 -1.7715 25.8128
113  -2.1400 35.6010 -1.7680 26.5691
115  -2.1000 34.9990 -1.7695 27.0436
117  -2.0600 34.7166 -1.4423 27.4338
119  -2.0200 34.4743 -1.4525 27.8824
121  -1.9800 34.2309 -1.4610 28.3457
123  -1.9400 33.9488 -1.4692 28.7851
125  -1.9000 33.5921 -1.4785 29.1649
127  -1.8600 33.1250 -1.4900 29.4506
129  -1.8200 32.5103 -1.5050 29.6075
131  -1.7800 31.7080 -1.5248 29.5992
133  -1.7400 30.6740 -1.5505 29.3864
135  -1.7000 29.6897 -1.1403 29.0369
137  -1.6600 29.5906 -1.1751 29.2924
139  -1.6200 30.0407 -1.6327 30.1792
141  -1.5800 30.0324 -1.1643 30.9954
143  -1.5400 30.0007 -1.1623 31.6784
145  -1.5000 29.9971 -1.1623 32.3405
147  -1.4600 29.9834 -1.6165 32.9999
149  -1.4200 29.9916 -1.1620 33.6324
151  -1.3800 29.9805 -1.1621 34.2532
153  -1.3400 30.0190 -1.1626 34.8756
155  -1.3000 29.9687 -1.6164 35.4719
157  -1.2600 30.0181 -1.1623 36.0635
159  -1.2200 29.9808 -1.1614 36.6311
161  -1.1800 29.9772 -1.1614 37.1835
163  -1.1400 29.9490 -1.6153 37.7184
165  -1.1000 29.9574 -1.1606 38.2407
167  -1.0600 29.9417 -1.6141 38.7453
169  -1.0200 29.9178 -1.1614 39.1922
171  -0.9800 29.8635 -1.1625 39.6161
173  -0.9400 29.7346 -1.1650 39.9729
175  -0.9000 29.4842 -1.1711 40.2178
177  -0.8600 29.0596 -1.1829 40.3020
179  -0.8200 28.3990 -1.2028 40.1699
```

| | | | |
|---|---|---|---|
| 181 | −0.7800 | 27.4286 | −1.2336 | 39.7559 |
| 183 | −0.7400 | 26.0585 | −1.2787 | 38.9816 |
| 185 | −0.7000 | 24.4358 | −0.8717 | 37.7788 |
| 187 | −0.6600 | 22.7903 | −0.9362 | 36.2669 |
| 189 | −0.6200 | 20.7622 | −1.0163 | 34.4409 |
| 191 | −0.5800 | 18.7935 | −0.5743 | 32.3636 |
| 193 | −0.5400 | 17.0475 | −0.6702 | 30.3402 |
| 195 | −0.5000 | 15.1676 | −0.7674 | 28.2787 |
| 197 | −0.4600 | 13.9500 | −0.3141 | 26.3775 |
| 199 | −0.4200 | 13.1154 | −0.3750 | 24.8646 |
| 201 | −0.3800 | 12.4703 | −0.4193 | 23.5913 |
| 203 | −0.3400 | 11.9129 | −0.4537 | 22.4424 |
| 205 | −0.3000 | 11.3566 | −0.4837 | 21.3242 |
| 207 | −0.2600 | 10.7223 | −0.5143 | 20.1561 |
| 209 | −0.2200 | 9.9324 | −0.5497 | 18.8636 |
| 211 | −0.1800 | 9.8365 | −0.0427 | 17.6449 |
| 213 | −0.1400 | 9.9588 | −0.0463 | 16.6494 |
| 215 | −0.1000 | 9.9853 | −0.0452 | 15.7456 |
| 217 | −0.0600 | 9.9853 | −0.0437 | 14.8087 |
| 219 | −0.0200 | 10.0132 | −0.0415 | 13.8278 |
| 221 | 0.0200 | 9.9999 | −0.0398 | 12.7981 |
| 223 | 0.0600 | 9.9409 | −0.5639 | 11.7156 |
| 225 | 0.1000 | 9.9557 | −0.0371 | 10.5715 |
| 227 | 0.1400 | 9.9311 | −0.0359 | 9.3815 |
| 229 | 0.1800 | 9.8312 | −0.0369 | 8.1116 |
| 231 | 0.2200 | 9.6174 | −0.0423 | 6.7279 |
| 233 | 0.2600 | 9.2466 | −0.0540 | 5.1941 |
| 235 | 0.3000 | 8.6685 | −0.0746 | 3.4700 |
| 237 | 0.3400 | 7.8225 | −0.1068 | 1.5084 |
| 239 | 0.3800 | 6.6340 | −0.1539 | −0.7475 |
| 241 | 0.4200 | 5.0095 | −0.2203 | −3.3684 |
| 243 | 0.4600 | 3.6383 | 0.2178 | −6.2070 |
| 245 | 0.5000 | 2.2053 | 0.1484 | −9.0796 |
| 247 | 0.5400 | 0.5330 | 0.0717 | −12.0920 |
| 249 | 0.5800 | −0.9453 | 0.5373 | −15.2658 |

Table 3. The characteristics $\alpha(t)$, $N_Z(t)$, $\theta(t)$ of the nominal motions for the chosen control law $\delta_s(t)$.

### 4.4 Estimating identification accuracy of 48 errors of aerodynamic parameters of the aircraft

Primary task of MPA algorithm consists in identification - estimation-48 increments of 4 functions. If entry conditions and increments are determined, values of the unknown coefficients follow from obvious recurrent formulas.

To estimate the accuracy, we assume that the current values of $\alpha, N_Y, \theta$ are measured every 0.1 sec. during 25 seconds .We assume that random errors of measurement represent the discrete white noise bounded by the true measurable value multiplied by the given value $\epsilon$. An amount of the primary observations equal 3*250=750.

| number $\alpha_i$ | nom.koef. $C_{Z_0}(\alpha_i)$ | perturb.koef. $C_{Z_0}(\alpha_i)$ | $\delta(C_{Z_0}(\alpha_i))$ |
|---|---|---|---|
| 1 | 0.6512 | 0.6326 | 0.02854 |
| 2 | 0.0205 | 0.0260 | −0.26410 |
| 3 | −0.3778 | −0.3646 | 0.03491 |
| 4 | −0.7395 | −0.7213 | 0.02456 |
| 5 | −1.0610 | −1.0657 | −0.00443 |
| 6 | −1.4038 | −1.4016 | 0.00159 |
| 7 | −1.7679 | −1.7424 | 0.01444 |
| 8 | −2.0582 | −2.0453 | 0.00627 |
| 9 | −2.2774 | −2.3388 | −0.02693 |
| 10 | −2.4568 | −2.5459 | −0.03625 |
| 11 | −2.5639 | −2.6698 | −0.04130 |
| 12 | −2.5404 | −2.6505 | −0.04334 |

Table 4. The Relative errors of the identifications of $C_{Z_0}(\alpha_i)$ by $\rho = 0.25$

| number $\alpha_i$ | nom.koef. $C_{m_0}(\alpha_i)$ | perturb.koef. $C_{m_0}(\alpha_i)$ | $\delta(C_{m_0}(\alpha_i))$ |
|---|---|---|---|
| 1 | −0.2130 | −0.2054 | 0.03582 |
| 2 | −0.1816 | −0.1783 | 0.01851 |
| 3 | −0.1567 | −0.1550 | 0.01061 |
| 4 | −0.1618 | −0.1611 | 0.00439 |
| 5 | −0.1634 | −0.1631 | 0.00209 |
| 6 | −0.1427 | −0.1388 | 0.02754 |
| 7 | −0.1372 | −0.1338 | 0.02439 |
| 8 | −0.1495 | −0.1502 | −0.00467 |
| 9 | −0.1175 | −0.1220 | −0.03771 |
| 10 | −0.1139 | −0.1190 | −0.04484 |
| 11 | −0.0957 | −0.1043 | −0.08937 |
| 12 | −0.0399 | −0.0394 | 0.01236 |

Table 5. The Relative errors of the identifications of $C_{m_0}(\alpha_i)$ by $\rho = 0.25$

We compress primary observations for a smoothing the high-frequency errors and reduction of a dimension of matrixes covariance . The file of the primary observations is divided into 12 groups and as an input of the algorithm of the identification the vector of the dimension $12 \times 1$ serves. Components of this vector are the sums of elements of each of 12 groups.

To characterize the accuracy of identification of the random parameter $\eta_i$ the degree of perturbation of the aerodynamic coefficients $\vartheta$ , we determine the relative errors of estimation $(\eta_i - \hat{\eta}_i)/\eta_i$ for every component the identifiable functions . The relative errors designate $\delta(C_{Z_0}(\alpha_i)), \delta(C_{m_0}(\alpha_i)), \delta(C_{Z_q}(\alpha_i)), \delta(C_{m_q}(\alpha_i)), i = 1, ..., 12.$

Apparently, relative errors of identification are small and do not surpass several hundredth at $\rho = 0.25$

| number $\alpha_i$ | nom.koef. $C_{Z_q}(\alpha_i)$ | perturb.koef. $C_{Z_q}(\alpha_i)$ | $\delta(C_{Z_q}(\alpha_i))$ |
|---|---|---|---|
| 1 | −9.9636 | −8.8984 | 0.10691 |
| 2 | −25.2235 | −26.1655 | −0.03735 |
| 3 | −28.4644 | −29.2857 | −0.02885 |
| 4 | −31.4821 | −31.8270 | −0.01096 |
| 5 | −31.3125 | −31.6274 | −0.01006 |
| 6 | −30.8417 | −31.1249 | −0.00918 |
| 7 | −27.5461 | −28.0921 | −0.01982 |
| 8 | −28.1388 | −28.6036 | −0.01652 |
| 9 | −28.9682 | −29.4069 | −0.01515 |
| 10 | −29.7908 | −30.2114 | −0.01412 |
| 11 | −38.6789 | −38.7933 | −0.00296 |
| 12 | −35.7355 | −35.8053 | −0.00195 |

Table 6. The Relative errors of the identifications of $C_{Z_q}(\alpha_i)$ by $\rho = 0.25$

| number $\alpha_i$ | nom.koef. $C_{m_q}(\alpha_i)$ | perturb.koef. $C_{m_q}(\alpha_i)$ | $\delta(C_{m_q}(\alpha_i))$ |
|---|---|---|---|
| 1 | −5.5807 | −6.1771 | −0.10686 |
| 2 | −4.1294 | −4.3066 | −0.04291 |
| 3 | −3.9913 | −4.1368 | −0.03645 |
| 4 | −4.0250 | −4.1662 | −0.03510 |
| 5 | −4.9363 | −5.0012 | −0.01315 |
| 6 | −5.5024 | −5.5314 | −0.00527 |
| 7 | −4.5272 | −4.5870 | −0.01320 |
| 8 | −4.8711 | −4.8936 | −0.00462 |
| 9 | −5.0970 | −5.0915 | 0.00108 |
| 10 | −5.3245 | −5.2912 | 0.00626 |
| 11 | −5.5637 | −5.4908 | 0.01310 |
| 12 | −4.8726 | −4.8937 | −0.00434 |

Table 7. The Relative errors of the identifications of $C_{m_q}(\alpha_i)$ by $\rho = 0.25$

| number $\alpha_i$ | nom.koef. $C_{Z_0}(\alpha_i)$ | perturb.koef. $C_{Z_0}(\alpha_i)$ | $\delta(C_{Z_0}(\alpha_i))$ |
|---|---|---|---|
| 1 | 0.5324 | 0.4255 | 0.20083 |
| 2 | −0.1999 | −0.2092 | −0.04637 |
| 3 | −0.6556 | −0.5969 | 0.08959 |
| 4 | −1.0629 | −0.9303 | 0.12481 |
| 5 | −1.3911 | −1.2772 | 0.08188 |
| 6 | −1.7546 | −1.6697 | 0.04839 |
| 7 | −2.1699 | −2.0331 | 0.06304 |
| 8 | −2.4704 | −2.3417 | 0.05209 |
| 9 | −2.6379 | −2.6342 | 0.00138 |
| 10 | −2.7936 | −2.8450 | −0.01839 |
| 11 | −2.8799 | −2.9723 | −0.03208 |
| 12 | −2.8518 | −2.9530 | −0.03548 |

Table 8. The Relative errors of the identifications of $C_{Z_0}(\alpha_i)$ by $\rho = 0.50$

| number $\alpha_i$ | nom.koef. $C_{m_0}(\alpha_i)$ | perturb.koef. $C_{m_0}(\alpha_i)$ | $\delta(C_{m_0}(\alpha_i))$ |
|---|---|---|---|
| 1 | −0.2520 | −0.2441 | 0.03123 |
| 2 | −0.2183 | −0.2166 | 0.00781 |
| 3 | −0.1924 | −0.1934 | −0.00523 |
| 4 | −0.1966 | −0.1994 | −0.01457 |
| 5 | −0.1979 | −0.2014 | −0.01792 |
| 6 | −0.1834 | −0.1747 | 0.04747 |
| 7 | −0.1773 | −0.1697 | 0.04301 |
| 8 | −0.1860 | −0.1858 | 0.00145 |
| 9 | −0.1481 | −0.1599 | −0.08004 |
| 10 | −0.1438 | −0.1569 | −0.09149 |
| 11 | −0.1225 | −0.1420 | −0.15942 |
| 12 | −0.0738 | −0.0811 | −0.09934 |

Table 9. The Relative errors of the identifications of $C_{m_0}(\alpha_i)$ by $\rho = 0.50$

| number $\alpha_i$ | nom.koef. $C_{Z_q}(\alpha_i)$ | perturb.koef. $C_{Z_q}(\alpha_i)$ | $\delta(C_{Z_q}(\alpha_i))$ |
|---|---|---|---|
| 1 | −11.1272 | −8.6840 | 0.21957 |
| 2 | −24.6470 | −25.6672 | −0.04139 |
| 3 | −28.0288 | −28.8049 | −0.02769 |
| 4 | −31.5642 | −31.3356 | 0.00724 |
| 5 | −31.4249 | −31.1306 | 0.00937 |
| 6 | −30.9833 | −30.6296 | 0.01142 |
| 7 | −27.3921 | −27.6113 | −0.00800 |
| 8 | −28.0776 | −28.1104 | −0.00117 |
| 9 | −28.9364 | −28.9144 | 0.00076 |
| 10 | −29.7817 | −29.7303 | 0.00172 |
| 11 | −39.0577 | −38.2130 | 0.02163 |
| 12 | −36.1709 | −35.1346 | 0.02865 |

Table 10. The Relative errors of the identifications of $C_{Z_q}(\alpha_i)$ by $\rho = 0.25$

| number $\alpha_i$ | nom.koef. $C_{m_q}(\alpha_i)$ | perturb.koef. $C_{m_q}(\alpha_i)$ | $\delta(C_{m_q}(\alpha_i))$ |
|---|---|---|---|
| 1 | −3.9514 | −6.9359 | −0.75528 |
| 2 | −2.8588 | −5.1596 | −0.80480 |
| 3 | −2.7526 | −4.9893 | −0.81258 |
| 4 | −2.7899 | −5.0189 | −0.79894 |
| 5 | −3.7625 | −5.8672 | −0.55939 |
| 6 | −4.3649 | −6.3936 | −0.46477 |
| 7 | −3.3644 | −5.4530 | −0.62079 |
| 8 | −3.7422 | −5.7627 | −0.53993 |
| 9 | −3.9940 | −5.9631 | −0.49299 |
| 10 | −4.2490 | −6.1601 | −0.44976 |
| 11 | −4.5274 | −6.3594 | −0.40464 |
| 12 | −3.7451 | −5.7631 | −0.53882 |

Table 11. The Relative errors of the identifications of $C_{m_q}(\alpha_i)$ by $\rho = 0.50$

## 5. Conclusions

The presented data show that the multipolynomial approximation algorithm can provide a computational ground for developing an efficient parameter identification technique for the nonlinear dynamic system, including identification of aerodynamic parameters of an aircraft. We emphasize that tables characterizing a sufficiently high accuracy of aerodynamic parameter identification are obtained when there are no iterations and d = 1, which corresponds to the case when the estimation vector $(\vartheta \hat{+} \eta)(W_N(d))$ is represented by the vector linear combination of measured data that is optimal on the family of linear operators over the vector of measurements. This is due to good (in terms of the identification problem) properties of the parametric system of equations of the pitching motion of the "pseudo F-16 " aircraft. It can become much more complicated when it comes to the identification problem of the parametric system of equations of complete (spatial) motion of the aircraft. In such case, we may need to use polynomials of the power d > 1 and increase requirements on the computer performance and RAM. This was the case for identification attempts made for some parameters of F-16 complete motion equations. We emphasize that the inputs of the MPA algorithm we considered were not real (were not the results of operation of real sensors of the aircraft during its test flight); they were determined by mathematical simulation - by means the numerically integrations motion equations for perturbed parameters of aerodynamic forces and moments.

## 6. Appendix A: An estimate of the vector of the conditional mathematical expectation that is optimal in the root-mean-square sense

### A.1. An algorithm fundamental (AF)

We consider the algorithm fundamental (AF) for solving the problem of finding the estimate of the vector $E(\eta|Y_N)$ that is optimal in the root-mean-square sense. This vector is known to be the estimate in the root-mean-square sense of the vector $\eta$ once the vector $Y_N$ is fixed. Therefore, it is justified that it is the vector of conditional expectation that AF tends to estimate.

We construct AF that ensures polynomial approximation of the vector $E(\eta|Y_N)$. To do this, we find the approximate estimate of the vector $E(\eta|Y_N)$, which is linear with respect to components of the vector $W_N(d)$ and optimal in the root-mean-square sense. We denote the vector of this estimate by $\hat{\eta}(W_N(d))$ . To obtain the explicit expression for the estimation vector, we calculate elements of the vector $\overline{V}(d,N)$ and the covariance matrix $C_V(d,N)$ that are the first and second (centered) statistical moments for the vector $V(d,N)$. These vector and matrix can be divided into blocks of the following structure

$$E(E(\eta|Y_N)) = E(\eta);$$

$$E(E(W(d,N)|Y_N)) = E(W(d,N));$$
$$= E((E(\eta|Y_N) - E(\eta))(E(\eta|Y_N) - E(\eta))^T) =$$
$$E((\eta - E(\eta))(\eta - E(\eta))^T).$$
$$L_N(d) = E((E(\eta|Y_N) - E(E(\eta|Y_N(d))))(W_N(d) - E(W_N(d)))^T) =$$
$$E(\eta)W_N(d)^T) - E(\eta)E(W_N(d))^T,$$

$$Q_N(d) = E((W_N(d) - E(W_N(d)))(W_N(d) - E(W_N(d)))^T);$$

The right-hand sides of these blocks are the first and second (centered) statistical moments calculated by the Monte-Carlo method. However, their left-hand sides also serve as the first and second (centered) statistical moments of components of the vector of conditional mathematical expectations. Hence, we can use mathematical models of form (0.1) and (0.2) to find these statistical moments experimentally for vectors of conditional expectations as well. This obvious proposition gives us the basis for practical implementation of the computational procedure of estimating the vector of the conditional expectation.

We introduce

$$\hat{\eta}(W_N(d)) = E(\eta) + \Lambda_N(d)(W_N(d) - E(W_N(d))), \quad (A.1)$$

where the matrix $\Lambda_N(d), r \times m(d, N)$ satisfies the equation

$$\Lambda_N(d)Q_N(d) = L_N(d).$$

We also introduce

$$\tilde{\eta}(W_N(d)) = z + \tilde{\Lambda}_N(d)(W_N(d) - E(W_N(d))), \quad (A.2)$$

where $z$ and $\tilde{\Lambda}_N(d)$ are the arbitrary vector and matrix of dimensions $r \times 1$ and $r \times m(d, N)$. Suppose $C(d, N)$ and $\tilde{C}(d, N)$ are the covariance matrices of estimation errors for the vector $E(\eta|Y_N)$ generated by the estimates $\hat{\eta}(W_N(d))$ and $\tilde{\eta}(W_N(d))$.

Lemma. The matrix $\tilde{C}(d, N) - C(d, N)$ is a nonnegative definite matrix : $C(d, N) \le \tilde{C}(d, N)$.

The lemma follows from the identity

$$\tilde{C}(d, N) = C(d, N) + (\Lambda_N(d) - \tilde{\Lambda}_N(d))(\Lambda_N(d) - \tilde{\Lambda}_N(d))^T +$$

$$(\Lambda_N(d)Q_N(d) - L_N(d))(\tilde{\Lambda}_N(d) - \Lambda_N(d))^T +$$

$$((\Lambda_N(d)Q_N(d) - L_N(d))(\tilde{\Lambda}_N(d) - \Lambda_N(d))^T)^T + (z - E(\eta)(z - E\eta)^T. \quad (A.3)$$

Corollary of the lemma. For the vector $E(\eta|Y_N)$, the vector $\hat{\eta}(W_N(d))$ is the estimate optimal in the root-mean-square sense among the set of estimates linear with respect to components of the vector $W_N(d)$. If $Q_N(d) > 0$, the estimation vector is unique and

$$\hat{\eta}(W_N(d)) = E(\eta) + L_N(d)Q_N(d)^{-1}(W_N(d) - E(W_N(d))). \quad (A.4)$$

The covariance matrix $C(d, N)$ of estimation errors of the vector $E(\eta|Y_N)$ is given by the formula

$$C(d, N) = C_\eta - \Lambda_N(d)L_N(d).(A.5)$$

If $Q_N(d) \ge 0$, the vectors that provide linear and optimal in the root-mean-square sense estimate are not unique; however, the variances of components of the difference between these vectors are zeros.

Formula (A.1) gives explicit expressions for the vector coefficients of the form $\lambda(a_1, ..., a_N)$ in (1.1). To find these relations, we open the explicit expressions for components of the vector

$W_N(d)$ and the right-hand side of (A.1) and equate them to the right-hand side of formula (1.1).

We consider asymptotic estimation errors when we use (A.1). Suppose the vector $Y_N$ is fixed. We assume that the vector $E(\eta|Y_N$ is given by the function of $Y_N$ on some a priori region that is a compact; the function is continuous on this region. Then, the following theorem holds.
Theorem.

$$Sup \ _{Y_N \in \Omega_{Y_N}} |\hat{\eta}(W_N(d)) - E(\eta|Y_N)| \Rightarrow 0, d \Rightarrow \infty. \quad (A.6)$$

Proof. The multidimensional analogue of the K. Weierstrass theorem, which is the corollary of the M. Stone theorem [9], states that for any number $\varepsilon > 0$ there exists a multidimensional polynomial $P(W_N(d_\varepsilon))$ such that

$$Sup \ _{Y_N \in \Omega_{Y_N}} |P(W_N(d_\varepsilon)) - E(\eta|Y_N)| < \varepsilon.$$

We can rewrite this relation as

$$Sup \ _{Y_N \in \Omega_{Y_N}} |P(W_N(d)) - E(\eta|Y_N)| \Rightarrow 0, \quad d \Rightarrow \infty. \quad (A.7)$$

We assume that C is the covariance matrix of the random vector $P(W_N(d) - E(\theta|Y_N)$ :

$$C = E((P(W_N(d)) - E(\eta|Y_N))(P(W_N(d)) - E(\eta|Y_N))^T.$$

It follows from (A.7) that
$$C \Rightarrow 0_n, \quad d \Rightarrow \infty \quad (A.8).$$

By construction, the vector $\hat{\eta}(W_N(d))$ provides the estimate of the vector $\eta$ that is linear with respect to components $W_N(d)$ and optimal in the root-mean-square sense. However, it follows from the lemma that for any other non-optimal linear estimate, including estimates of the form $P(W_N(d))$, the relation $C \geq C(d, N)$ holds. Hence, taking into account (A.8), we obtain

$$C(d, N) \Rightarrow 0_n, d \Rightarrow \infty. \quad (A.9)$$

Proposition (A.9) is equivalent to (A.6) if we recall that

$$C(d, N) = \int (Z_{E(\eta|Y_N)}(W_N(d)) - E(\eta|Y_N))(Z_{E(\eta|Y_N)}(W_N(d)) - E(\eta|Y_N))^T$$

$$p(\eta, Y_N)d\eta dY_N,$$

where $p(\eta, Y_N)$ is the joint probability density of the random vectors $\eta$ and $Y_N$. The theorem is proved.

Thus, by (A.1), AF determines the vector series that, with the increasing number $m(d, N)$ of its terms, approximates the vector of conditional mathematical expectation of the vector $\theta$ of the estimated parameters with an arbitrary uniformly small root-mean-square error.

### A.2. Recurrent (Realizable) MPA algorithm

To use formula (A.1), we need to find the matrix inverse to the matrix $Q_N(d)$. When the dimension $m(d, N) \times m(d, N)$ of the matrix $Q_N(d)$ is high and $Q_N(d)$ is close to the singular matrix, it is difficult to calculate elements of the inverse matrix. Below, we give the recurrent computational process based on the principle of decomposing observations, described in [6, 7]. Above, we specified the vector $W_N(d)$ of dimension $m(d, N) \times 1$ with the components $w_1, ..., w_{m(d,N)}$. The computational process consists of $m(d, N)$ successive steps. At each step, we use new updated prior data to find the new estimate of the vector $\theta$ and perform the prediction, which provides estimates for the rest part of the observation vector. At the same time, we determine the covariance matrix of the estimation errors attained at this step. There is no prediction at the last $m(d, N)-$ th step, and the vector $\theta$ is refined for the last time.

Let us construct the recurrent algorithm (the MPA algorithm) that does not calculate inverse matrices and consists of $m(d, N)$ steps of calculating the first and second statistical moments for the sequence of special vectors $V_1, ..., V_i, ..., V_{m(d,N)}$ performed after prior moments $\bar{V}(d, N)$ and $C_V(d, N)$ are found for the basic vector $V(d, N)$. We assume that $V_1$ is composed of $r + m(d, N) - 1$ components of the basic vector $V(d, N)$ left after the component $w_1$ was excluded, $w_1, ...$; $V_i$ composed of components of the vector $V_{i-1}$ left after the component $w_i$ was excluded, etc. The component $w_{m(d,N)}$ is the last component of the vector $W_N(d)$, and once we exclude it, the resulting vector $V_{m(d,N)}$ turns out to equal the estimation vector $\hat{\eta}(W_N(d))$.

At step 1, we use the particular case of formulas of form (A.1) and (A.5) to calculate the vector $\bar{V}_1$ that estimates the vector $V_1$ and is optimal in the root-mean-square sense and linear with respect to $w_1$, and the covariance matrix of the estimation errors $C(V_1)$.

The estimation vector is formed of the estimate of the vector of conditional mathematical expectation $E(\eta|Y_N)$ and the vector of dimension $m(d, N) - 1) \times 1$. Once we fix the value $w_1$, the latter becomes the vector of statistical prediction of the mean values of "future" values $w_2, ..., w_{m(d,N)}$. We emphasize that calculations performed at step 1 are based on the preliminary found prior , $\bar{V}(d, N), C_V(d, N)$.

Suppose steps $1, ..., i$ of the computational process yielded the vector $\bar{V}(d, N)$ and the matrix $C(V_i)$ after the values $w_1, ...w_i$, wi were fixed. At step $i + 1$, we have from the particular case of formulas (A.1) and (A.5) the vector $\bar{V}_{i+1}$ that estimates the vector $V_{i+1}$ and is optimal in the root-mean-square sense and linear with respect to $w_1, ..., w_{i+1}$, and the covariance matrix $C(V_{i+1})$ of estimation errors. The vector $\bar{V}_{i+1}$ is still formed of the estimate of the vector of conditional mathematical expectation $E(\eta|w_1, ..., w_{i+1}$ (first r components of the vector $\bar{V}_{i+1}$) and the vector of statistical prediction of mean values of "future" - h values $w_{i+2}, ..., w_{m(d,N)}$ after $w_1, ..., w_{i+1}$ (the rest $m(d, N) - (i + 1)$ components of the vector $\bar{V}_{i+1}$ ) are fixed. We emphasize that calculations at step $i + 1$ are based on the preliminary found $\bar{V}_i$and $C(V_i)$, which can be naturally called the first and second statistical moments for "future" random values $w_{i+1}, ..., w_{m(d,N)}$. These vectors and matrices represent a priori data on statistical moments of components of the vector $V_{i+1}$ before the algorithm receives the value $w_{i+1}$ at its input.

Recurrent formulas that corresponds exactly to the above given qualitative description of the computational process have the form

$$\bar{V}_{i+1} = \bar{V}_i^1 + q_i^{-1} b_i (w_{i+1} - z_{w_{i+1}}), \quad (A.10)$$

$$C(V_{i+1}) = C(V_i)^1 - q_i^{-1} b_i b_i^T, \quad (A.11)$$

where the scalar $z_{w_{i+1}}$ is the $(r+1)$-th component of the vector $\bar{V}_i$, the scalar is the linear and optimal in the root-mean-square sense estimate of the component after the algorithm has processed the components $w_1, ..., w_i$, $\bar{V}_i^{-1}$ is the vector obtained from the vector $\bar{V}_i$ by eliminating its component $z_{w_{i+1}}$, the scalar $q_i$ is the $(r+1)$-th diagonal element of the matrix $C(V_i)$, which is the variance of the estimation error of the component $w_{i+1}$ after components $w_1, ..., w_i$ were processed, $C(V_i)^1$ is the matrix formed of $C(V_i)$ after the $(r+1)$-th row vector and $(r+1)$-th column vector were excluded, and $b_i$ is the $(r+1)$-th column vector of the matrix $C(V_i)$ with its $(r+1)$-th component deleted.

If the scalar $q_i$ turned out to be close to zero, the component $w_{i+1}$ corresponds to a linear combination of components $w_1, ..., w_i$. Then, $w_{i+1}$ do not give any new information on $\theta$ and should be excluded from the computational process. Note that the sequence of random variables like $(w_{i+1} - z_{w_{i+1}})$ forms an updating sequence. The upper left block of the $(r \times r)$-matrix $C(V_i)$ includes the covariance matrix $C(d, i)$ of estimation errors of the vector $E(\eta | w_i, ..., w_i)$ after the algorithm processed the vector $W_i(d)$.

We assume that $l(i)$ is the vector composed of r first components of the vector $b_i$. The formula representing the evolution of the covariance matrix $C(d, i)$ in the function of the number $i$ of observable components of the vector $W_N(d)$ has the form

$$C(d, i) = C_\eta(0) - q_1^{-1} l(1) l(1)^T - ... - q_i^{-1} l(i) l(i)^T. \quad (A.12)$$

To test this MPA algorithm, we solved numerically several problems of estimating the components of the state vector for essentially nonlinear dynamic systems. The estimated components are unknown random constant parameters $\eta_1, ..., \eta_r$ of the dynamic system.

As for particular applied problems, we considered smoothing problems and the filtration problem.

In the above examples, we applied the Monte-Carlo method for the number of random realizations lying within 5000 - 10000. This number does not affect the estimation errors provided by the MPA algorithm significantly. The estimated random parameters are assumed to be statistically independent and are a priori uniformly distributed. The value of the root-mean- square deviation $\sigma(i, theo)$ is determined theoretically by calculating variances :the diagonal elements of the covariance matrix $C(d, N)$. The value of the root-mean-square deviation $\sigma(i, exp)$ is obtained experimentally by applying the Monte-Carlo method for 5000 realizations. Experimental and theoretical root-mean-square deviations almost coincide, which proves that the above given formulas of the MPA algorithm are correct.

## 7. Acknowledgments

This study was financially supported by the Russian Foundation for Basic Research (grant no. 10-08-00415a).

## 8. References

[1] V. Klein and A. G. Morelli, Aircraft System Identification: Theory and Methods (American Institute of Aeronautics and Astronautics, Reston 2006).

[2] M. B. Tischle and R. K. Remple, Aircraft and Rotorcraft System Identification: Engineering Methods with Flight Test Examples (American Institute of Aeronautics and Astronautics, Reston, 2006).

[3] R. Jategaonkar, Flying Vehicle System Identification: A Time Domain Methodology (American Institute of Aeronautics and Astronautics, Reston, 2006).

[4] L. Ljung, System Identification: Theory for the USER (Prentice Hall, 1987).

[5] B. K. Poplavskii and G. N. Sirotkin, Integrated Approach to Analysis of Processes of Identification of Model Parameters of a Spacecraft, in Transactions No. 429 of p/ya V-8759 [in Russian].

[6] J. A. Boguslavskiy, Bayes Estimators of Nonlinear Regression and Allied Issues, Izv. Ross. Akad. Nauk, Teor. Sist. Upr., No. 4, 14.24 (1996) [Comp. Syst. Sci. 35 (4), 511.521 (1996)].

[7] J. A. Boguslavskiy, A Polynomial Approximation for Nonlinear Problems of Estimation and Control (Fizmatlit,Moscow, 2006) [in Russian].

[8] J. Timmer, H.Rust, W.Horbelt, H.U. Voss, Parametric, nonparametric and parametric modelling of a chaotic circuit time series, Physics Letters A 274 (2000) 123 - 134

[9] A. F. Timan, The Theory of Approximation of Functions of a Real Variable (Nauka, Moscow, 1960) [in Russian].

# 7

# Influence of Forward and Descent Flight on Quadrotor Dynamics

Matko Orsag and Stjepan Bogdan
*LARICS-Laboratory for Robotics and Intelligent Control Systems*
*Department of Control and Computer Engineering,*
*Faculty of Electrical Engineering and Computing, University of Zagreb, Zagreb*
*Croatia*

## 1. Introduction

The focus of this chapter is an aircraft propelled with four rotors, called the quadrotor. Quadrotor was among the first rotorcrafts ever built. The first successful quadrotor flight was recorded in 1921, when De Bothezat Quadrotor remained airborne for two minutes and 45 seconds. Later he perfected his design, which was then powered by 180-horse power engine and was capable of carrying 3 passengers on limited altitudes. Quadrotor rotorcrafts actually preceded the more common helicopters, but were later replaced by them because of very sophisticated control requirements Gessow & Myers (1952). At the moment, quadrotors are mostly designed as small or micro aircrafts capable of carrying only surveillance equipment. In the future, however, some designs, like Bell Boeing Quad TiltRotor, are being planned for heavy lift operations Anderson (1981); Warwick (2007).

In the last couple of years, quadrotor aircrafts have been a subject of extensive research in the field of autonomous control systems. This is mostly because of their small size, which prevents them to carry any passengers. Various control algorithms, both for stabilization and control, have been proposed. The authors in Bouabdallah et al. (2004) synthesized and compared PID and LQ controllers used for stabilization of a similar aircraft. They have concluded that classical PID controllers achieve more robust results. In Adigbli et al. (2007); Bouabdallah & Siegwart (2005) "Backstepping" and "Sliding-mode" control techniques are compared. The research presented in Adigbli et al. (2007) shows how PID controllers cannot be used as effective set point tracking controller. Fuzzy based controller is presented in Varga & Bogdan (2009). This controller exhibits good tracking results for simple, predefined trajectories. Each of these control algorithms proved to be successful and energy efficient for a single flying manoeuvre (hovering, liftoff, horizontal flight, etc.).

This chapter examines the behaviour of a quadrotor propulsion system focusing on its limitations (i.e. saturation and dynamic capabilities) and influence that the forward and descent flights have on this propulsion system. A lot of previous research failed to address this practical problem. However, in case of demanding flight trajectories, such as fast forward and descent flight manoeuvres, as well as in the presence of the In Ground Effect, these aerodynamic phenomena could significantly influence quadrotor's dynamics. Authors in Hoffmann et al. (2007) show how control performance can be diminished if aerodynamic effects are not considered. In these situations control signals could drive the propulsion system well within the region of saturation, thus causing undesired or unstable quadrotor behaviour. This effect is especially important in situations where the aircraft is operating at its limits (i.e. carrying heavy load, single engine breakdown, etc.).

The proposed analysis of propulsion system is based on the thin airfoil (blade element) theory combined with the momentum theory Bramwell et al. (2001). The analysis takes into account the important aerodynamic effects, specific to quadrotor construction. As a result, the chapter presents analytical expressions showing how thrust, produced by a small propeller used in quadrotor propulsion system, can be significantly influenced by airflow induced from certain manoeuvres.

## 2. Basic dynamic model

This section introduces the basic quadrotor dynamic modeling, which includes rigid body dynamics (i.e. Euler equations), kinematics and static nonlinear rotor thrust equation. This model, based on the first order approximation, has been successfully utilized in various quadrotor control designs so far. Nevertheless, recent shift in Unmanned Aerial Vehicle research community towards more payload oriented missions (i.e. pick and place or mobile manipulation missions) emphasized the need for a more complete dynamic model.

### 2.1 Kinematics

Quadrotor kinematics problem is, actually, a rigid-body attitude representation problem. Rigid-body attitude can be accurately described with a set of 3-by-3 orthogonal matrices. Additionally, the determinant of these matrices has to be one Chaturvedi et al. (2011). Since matrix representation cannot give a clear insight into the exact rigid body pose, attitude is often studied using parameterizations Shuster (1993). Regardless of the choice, every parameterization at some point fails to fully represent rigid body pose. Due to the gimbal lock, Euler angles cannot globally represent rigid body pose, whereas quaternions cannot define it uniquely.Chaturvedi et al. (2011)

Although researchers proved the effectiveness of using quaternions in quadrotor control Stingu & Lewis (2009), Euler angles are still the most common way of representing rigid body pose. To uniquely describe quadrotor pose using Euler angles, a composition of 3 elemental rotations is chosen. Following $X - Y - Z$ convention, a world reference coordinate system is first rotated $\Psi$ degrees around $X$ axis. After this, a $\Theta$ degree rotation around an intermediate $Y$ axis is applied. Finally, a $\Phi$ degree rotation around a newly formed $Z$ axis is applied to yield a transformation matrix from the world coordinate system $\mathfrak{W}$ to the body frame $\mathfrak{B}$, as shown in figure 1. Equations 1 and 2 formalize this procedure:

$$Rot\left(\Phi,\Theta,\Psi\right) = Rot\left(z^{w},\Phi\right)Rot\left(y^{w},\Theta\right)Rot\left(x^{w},\Psi\right) \tag{1}$$

$$Rot\left(\Phi,\Theta,\Psi\right) = \begin{bmatrix} c\phi c\theta & c\phi s\theta s\psi - s\phi c\psi & c\phi s\theta c\psi + s\phi s\psi \\ s\phi c\theta & s\phi s\theta s\psi + c\phi c\psi & s\phi s\theta c\psi - c\phi s\psi \\ -s\theta & c\theta s\psi & c\theta c\psi \end{bmatrix} \tag{2}$$

where $c\phi$ and $s\phi$ stand for $cos(\phi)$ and $sin(\phi)$, respectively. The same abbreviations are applied to other angles as well.

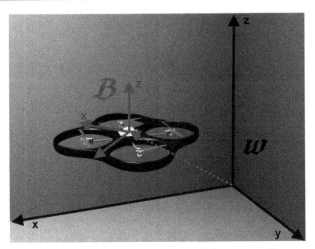

Fig. 1. Transformation from the body frame to the world frame coordinate system

## 2.2 Dynamic motion equations

Forces and torques, produced from the propulsion system and the surroundings, move and turn the quadrotor. In this paragraph, the quadrotor is viewed as a rigid body with linear and circular momentum, $\overrightarrow{L}$ and $\overrightarrow{M}$ respectively. According to the 2nd Newtons law, the force applied to the body equals the change of linear momentum. Using the principal of the change of momentum used in Jazar (2010), the following equation maps the change of quadrotor's position with respect to the applied force:

$$\overrightarrow{\mathbf{F}} = \frac{\partial \overrightarrow{\mathbf{L}}}{\partial t} = \frac{\partial \overrightarrow{\mathbf{v}}}{\partial t} m_q + \frac{\partial m_q}{\partial t} \overrightarrow{\mathbf{v}}$$

$$\overrightarrow{\mathbf{F}} = \frac{\partial \overrightarrow{\mathbf{v}}}{\partial t} m_q = m_q \left[ \frac{\partial^2 x}{\partial t^2} \ \frac{\partial^2 y}{\partial t^2} \ \frac{\partial^2 z}{\partial t^2} \right]^T$$

(3)

where $m_q$ represents quadrotor mass and $\overrightarrow{\mathbf{v}}$ its velocity vector. Due to the fact that most unmanned quadrotors are electrically driven, it is safe to assume that quadrotor mass does not change over time, resulting in a simple equation 3.

Same analysis can be applied to angular momentum, having in mind, the angular momentum is produced from the quadrotor motion as well as from the rotors spinning to produce the desired thrust. There are four important variables concerning angular momentum: quadrotor angular speed vector - $\overrightarrow{\omega}$, rotor angular speed vector - $\overrightarrow{\Omega}$, quadrotor inertia tensor - $\mathbf{I_q}$ and rotor inertia tensor - $\mathbf{I_r}$. Angular motion equations can be derived as follows:

$$\overrightarrow{\mathbf{M}} = \overrightarrow{\omega} + \mathbf{I_r} \overrightarrow{\Omega}$$

$$\overrightarrow{\mathbf{T}} = \frac{\partial \overrightarrow{\mathbf{M}}}{\partial t} + \overrightarrow{\omega} \times \overrightarrow{\mathbf{M}} = \frac{\partial \overrightarrow{\omega}}{\partial t} \mathbf{I_q} + \overrightarrow{\omega} \times \mathbf{I_q} \overrightarrow{\omega} + \overrightarrow{\omega} \times \mathbf{I_r} \overrightarrow{\Omega}$$

(4)

Quadrotors are normally constructed to be completely symmetric. Therefore, their tensor of inertia is a diagonal matrix 5. The same rule applies for rotors as well(otherwise they would be misbalanced and completely useless). Furthermore, rotors spin in one direction only, so the rotor angular speed vector $\vec{\Omega}$ has only one component $\Omega_z$. Evaluating 3 yields a circular motion equation 6.

$$\mathbf{I_q} = \begin{bmatrix} I_{xx} & 0 & 0 \\ 0 & I_{yy} & 0 \\ 0 & 0 & I_{zz} \end{bmatrix} \tag{5}$$

$$M_x = I_{xx}\frac{d\omega_x}{dt} - \left(I_{yy} - I_{zz}\right)\omega_y\omega_z + I_r\omega_y\Omega_z$$

$$M_y = I_{yy}\frac{d\omega_y}{dt} - \left(I_{zz} - I_{xx}\right)\omega_x\omega_z - I_r\omega_x\Omega_z \tag{6}$$

$$M_z = I_{zz}\frac{d\omega_z}{dt} - \left(I_{xx} - I_{yy}\right)\omega_x\omega_y$$

Equation 6 calculates rotation speeds in the body frame coordinate system. To transform these body frame angular velocities into world frame rotations, one needs a transformation matrix 8. This matrix is derived from successive elemental transformations 7 similarly as kinematics equation 2. Infinitesimal changes in Euler angles, affect the rotation vector in a way that the first Euler angle $\Psi$ undergoes two additional rotations, the second angle $\Theta$ only one additional rotation, and the final Euler angle $\Phi$ no additional rotations Jazar (2010):

$$\begin{bmatrix} \omega_x \\ \omega_y \\ \omega_z \end{bmatrix}^{\mathcal{B}} = \begin{bmatrix} 0 \\ 0 \\ \dot{\Phi} \end{bmatrix}^{\mathcal{W}} + Rot\left(\Phi, z^{\mathcal{W}}\right)^T \begin{bmatrix} 0 \\ \dot{\Theta} \\ 0 \end{bmatrix}^{\mathcal{W}} + Rot\left(\Phi, z^{\mathcal{W}}\right)^T Rot\left(\Theta, y^{\mathcal{W}}\right)^T \begin{bmatrix} \dot{\Psi} \\ 0 \\ 0 \end{bmatrix}^{\mathcal{W}} \tag{7}$$

$$\mathbf{J} = \begin{bmatrix} cos(\Psi)/cos(\Theta) & sin(\Psi)/cos(\Theta) & 0 \\ -sin(\Psi) & cos(\Psi) & 0 \\ cos(\Psi)tan(\Theta) & sin(\Psi)tan(\Theta) & 1 \end{bmatrix} \tag{8}$$

## 2.3 Rotor forces and torques

Four quadrotor blades are placed in a square shaped form. Blades that are next to each other spin in opposite directions, thus maintaining inherent stability of the aircraft. The same four blades that make the quadrotor hover enable it to move in the desired direction. Therefore, in order for quadrotor to move, it has to be pitched and rolled in the desired direction. To pitch and roll the quadrotor, some blades need to spin faster, while others spin slower. This produces the desired torques, which in term affect aircraft attitude and position Orsag et al. (2010).

Depending on the orientation of the blades, relative to the body coordinate system, there are two basic types of quadrotor configurations: cross and plus configuration shown in figure 2. In the plus configuration, a pair of blades spinning in the same direction, are placed on $x$ and $y$ coordinates of the body frame coordinate system. With this configuration it is easier to control the aircraft, because each move (i.e. $x$ or $y$ direction) requires a controller to disbalance only the speeds of two blades placed on the desired direction.

The cross configuration, on the other hand, requires that the blades are placed in each quadrant of the body frame coordinate system. In such a configuration each move requires all four blades to vary their rotation speed. Although the control system seems to be more complex, there is one big advantage to the cross construction. Keeping in mind that the amount of torque needed to rotate the aircraft is very similar for both configurations, it takes less change per blade if all four blades change their speeds. Therefore, when the aircraft carries

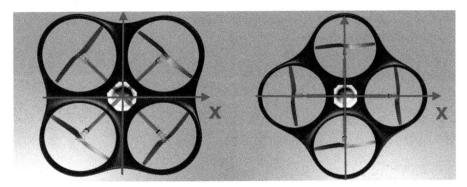

Fig. 2. A side by side image of X and Plus quadrotor configurations

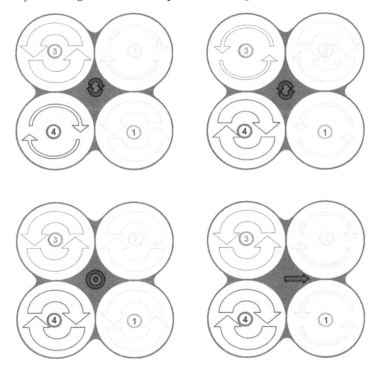

Fig. 3. Plus configuration control inputs for rotation, lift and forward motion. Arrow thickness stands for higher speed.

payload and operates near the point of saturation, it is wiser to use the cross configuration. Changing the speed of each blade for a small amount, as opposed to changing only two blades but doubling the amount of speed change, will keep the engines safe from saturation point. Basic control sequences of cross configuration are shown in figure 3. First approximation of rotor dynamics implies that rotors produce only the vertical thrust force. As the rotors are displaced from the axis of rotation (i.e. $x$ and $y$ axis) they produce corresponding torques,

$\overrightarrow{\mathbf{M_x}} = \overrightarrow{\mathbf{F}} \times \overrightarrow{\mathbf{r_x}}$ and $\overrightarrow{\mathbf{M_y}} = \overrightarrow{\mathbf{F}} \times \overrightarrow{\mathbf{r_y}}$ respectively. Torque $\overrightarrow{\mathbf{M_z}}$ comes from the spinning of each rotor blade $\mathbf{I_r}\overrightarrow{\Omega}$. Adding the corresponding thrust forces and torques yields the following equation:

$$\overrightarrow{\mathbf{F_{tot}}} = \overrightarrow{\mathbf{T_1}} + \overrightarrow{\mathbf{T_2}} + \overrightarrow{\mathbf{T_3}} + \overrightarrow{\mathbf{T_4}}$$
$$M_x^{tot} = M_x^2 + M_x^3 - M_x^1 - M_x^4$$
$$M_y^{tot} = M_y^3 + M_y^4 - M_y^1 - M_y^2$$
$$M_z^{tot} = M_z^2 + M_z^4 - M_z^1 - M_z^3$$

$$(9)$$

## 3. Aerodynamics

As the quadrotor research shifts to new research areas (i.e. Mobile manipulation, Aerobatic moves, etc.) Korpela et al. (2011); Mellinger et al. (2010), the need for an elaborate mathematical model arises. The model needs to incorporate a full spectrum of aerodynamic effects that act on the quadrotor during climb, descent and forward flight. To derive a more complete mathematical model of a quadrotor, one needs to start with basic concepts of momentum theory and blade elemental theory.

### 3.1 Combining momentum and blade elemental theory

The momentum theory of a rotor, also known as classical actuator disk theory, combines rotor thrust, induced velocity (i.e. airspeed produced in rotor) and aircraft speed into a single equation. On the other hand, blade elemental theory is used to calculate forces and torques acting on the rotor by studying a small rotor blade element modeled as an airplane wing so that the airfoil theory can be applied.Bramwell et al. (2001) A combination of these two views, macroscopic and microscopic, yields a base ground for a good approximative mathematical model.

### 3.1.1 Momentum theory

Basic momentum theory offers two solutions, one for each of the two operational states in which the defined rotor slipstream exists. The solutions refer to rotorcraft climb and descent, the so called helicopter and the windmill states. Quadrotor in a combined lateral and vertical move is shown in figure 4. The figure shows the most important airflows viewed in Momentum theory: $V_z$ and $V_{xy}$ that are induced by quadrotor's movement, together with the induced speed $v_i$ that is produced by the rotors.

Unfortunately, classic momentum theory implies no steady state transition between the helicopter and the windmill states. Experimental results, however, show that this transition exists. In order for momentum theory to comply with experimental results, the augmented momentum theory equation 10 is proposed Gessow & Myers (1952),

$$T = 2\rho R^2 \pi v_i \sqrt{(v_i + V_z)^2 + V_{xy}^2 + \frac{V_z^2}{7.67}} \tag{10}$$

where $\frac{V_z^2}{7.67}$ term is introduced to assure that the augmented momentum theory equation complies with experimental results, $R$ stands for rotor radius and $\rho$ is the air density. It is easy to show that in case of autorotation with no forward speed, thrust in equation 10 becomes

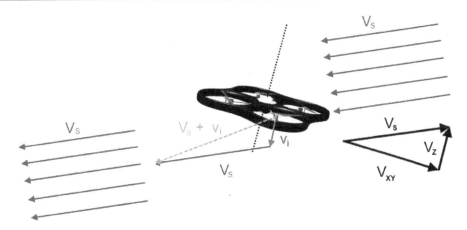

Fig. 4. Momentum theory - horizontal motion, vertical motion and induced speed total airflow vector sum

equal to the drag equation $D = \frac{1}{2}C_D\rho R^2\pi V_z^2$ of a free-falling plate with a drag coefficient $C_D = 1$.

### 3.1.2 Blade element theory

Blade element theory observes a small rotor blade element $\Delta r$ 5. Figure 5 shows this infinitesimal part of quadrotor's blade together with elemental lift and drag forces it produces Bramwell et al. (2001). For better clarity angles are drawn larger than they actually are:

$$\frac{\Delta L}{\Delta R} = \frac{1}{2}\rho V_{str}C_L S$$
$$\frac{\Delta D}{\Delta R} = \frac{1}{2}\rho V_{str}C_D S \tag{11}$$

where $C_L$ and $C_D$ are lift and drag coefficients, $S$ is the surface of the element and $V_{str}$ the airflow around the blade element. The airflow is mostly produced from the rotor spin $\Omega R$ and therefore depends on the distance of each blade element to the center of blade rotation. Adding to this airflow is the total air stream coming from quadrotor's vertical and horizontal movement, $V_S = V_{xy} + V_z$. Finally, blade rotation produces additional induced speed $v_i$. The ideal airfoil lift coefficient $C_L$ can be calculated using equation 12 Gessow & Myers (1952).

$$C_L = a\alpha_{ef} = 2\pi\alpha_{ef} \tag{12}$$

where $a$ is an aerodynamic coefficient, ideally equal to $2\pi$. The effective angle of attack $\alpha_{ef}$, is the angle between the airflow and the blade. Its value changes with the change of airflow direction and due to the blade twist.

Standard rotor blades are twisted because the dominant airflow coming from blade rotation increases linearly towards the end of the blade. According to equation 11 this causes the increase of lift and drag forces. The difference in forces produced near and far from the center of rotation would cause the blade to twist, and ultimately brake. To avoid that, a linear twist,

$$\alpha_m(r) = \Theta_0 - \frac{r}{R}Q_{tw} \tag{13}$$

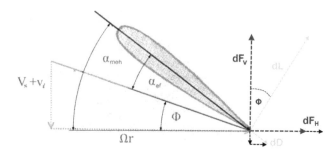

Fig. 5. Infitesimal rotor blade element $\Delta r$ in surrounding airflow Orsag & Bogdan (2009)

is introduced to the blade design.

The effect of varying airflow can be calculated separating the vertical components $V_z + v_i$ and horizontal ones $V_{xy} + \Omega r$. The airflow direction angle $\Phi$ can be easily calculated from the equation

$$\Phi = \arctan\left(\frac{V_z + v_i}{V_{xy} + \Omega r}\right) \approx \arctan\left(\frac{V_z + v_i}{\Omega r}\right) \tag{14}$$

As lift and drag forces are not aligned with body frame of reference, horizontal and vertical projection forces need to be derived. Keeping in mind that $\Omega r \gg \{V_z, v_i, V_{xy}\}$ small angle approximations $\cos(\Phi) \approx 1$ and $\sin(\Phi) \approx \Phi$ can be used. Moreover, in a well balanced rotor blade, drag force should be negligible compared to the lift Gessow & Myers (1952). Applying this considerations to 11 and keeping in mind the relations from figure 5 enables the derivation of horizontal and vertical force equations 15.

$$\frac{dF_V}{dr} = \frac{dL}{dr}\cos(\Phi) + \frac{dD}{dr}\sin(\Phi) \approx \frac{dL}{dr} = \rho V_{tot}^2 c\pi\alpha_{ef}$$
$$\frac{dF_H}{dr} = \frac{dL}{dr}\sin(\Phi) + \frac{dD}{dr}\cos(\Phi) \approx \frac{1}{2}\rho V_{tot}^2 C_D S + \frac{1}{2}\rho V_{tot}^2 C_L S\Phi \tag{15}$$

### 3.1.3 Applying blade element theory to quadrotor construction

This section continues with the observation of a small rotor blade element $\Delta r$ from the previous section, placing it in real surroundings shown in figure 6. Since the blades rotate, the forces produced by blade elements tend to change both in size and direction. This is the reason why an average elemental thrust of all blade elements should be calculated.

Figure 6 shows the relative position of one rotor as it is seen from quadrotor's body frame. This rotor is displaced from the body frame origin and forms an angle of 45° with quadrotor's body frame $x$ axis. Similar relations can be shown for other rotors. Accounting for the number of rotor blades $N$, the following equation for rotor vertical thrust force calculation is proposed Orsag & Bogdan (2009):

$$T = F_V = \frac{1}{2\pi}\int_0^{2\pi}\int_0^R N\frac{\Delta F_V}{\Delta R}drd\psi \tag{16}$$

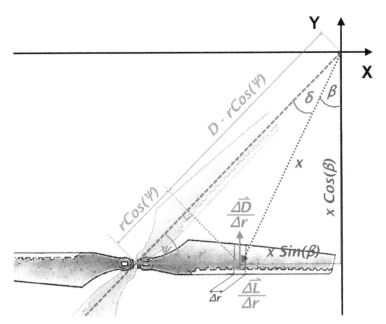

Fig. 6. Blade element in quadrotor coordinate system

where $\psi$ is the blade angle due to rotation, taken at a certain sample time. Solving integral equation 16 yields the expression for rotor thrust (i.e. vertical force) Orsag & Bogdan (2009):

$$F_V = \frac{N\rho a\bar{c}R^3\Omega^2}{4}\left[\left(\frac{2}{3}+\mu^2\right)\Theta_0 - \left(1+\mu^2\right)\frac{\Theta_{tw}}{2} - \lambda_i - \lambda_c\right] \tag{17}$$

The term inside the brackets of equation 17 is known as a thrust coefficient, and is given separately in 18.

$$C_T = \left(\frac{2}{3}+\mu^2\right)\Theta_0 - \left(1+\mu^2\right)\frac{\Theta_{tw}}{2} - \lambda_i - \lambda_c \tag{18}$$

Variables $\mu, \lambda_i$ and $\lambda_c$ are speed coefficients $\frac{V_{xy}}{R\Omega}, \frac{V_z}{R\Omega}$ and $\frac{v_i}{R\Omega}$ respectively. New constant $\bar{c}$ is the average cord length of the blade element shown in figure 5.

The same approach can be applied for the calculation of horizontal forces and torques produced within the quadrotor Orsag & Bogdan (2009). Calculated lateral force has x and y components, coming both from the drag and lift of the rotor, given in 19.

$$C_{Hx} = \cos(\alpha)\,\mu\left[\frac{C_D}{a} + (\lambda_i + \lambda_c)\left(\Theta_0 - \frac{\Theta_{tw}}{2}\right)\right]$$

$$C_{Hy} = \sin(\alpha)\,\mu\left[\frac{C_D}{a} + (\lambda_i + \lambda_c)\left(\Theta_0 - \frac{\Theta_{tw}}{2}\right)\right] \tag{19}$$

In case of torque equations the angles between the forces and directions are easily derived from basic geometric relations shown in figure 6, resulting in the elemental torque equations

Orsag & Bogdan (2009):

$$\frac{\Delta M_z}{\Delta r} = -\frac{\Delta F_H}{\Delta r}\left(D\cos\left(\Psi - \frac{pi}{4}\right) - r\right)$$

$$\frac{\Delta M_{xy}}{\Delta r} = -\frac{\sqrt{2}\Delta F_V}{2\Delta r}\left(D - r\cos\left(\Psi - \frac{\pi}{4}\right) \pm r\sin\left(\Psi - \frac{pi}{4}\right)\right)$$

(20)

Using the same methods which were used for force calculation, the following momentum coefficients were calculated:

$$C_{Mz} = R\left[\frac{1+\mu^2}{2a}C_D - C_T\left(\mu,\lambda,\lambda_i\right)|_{\mu=0}\right] \pm D\mu\cos\left(\frac{\pi}{4} + \phi\frac{C_{Hx}}{\cos(\phi)\mu}\right)$$

$$C_{Mx} = D\frac{\sqrt{2}}{2}C_T \pm R\mu\sin(\phi)\left[\frac{2}{3}\Theta_0 - \frac{1}{2}\left(\Theta_{tw} + \lambda\right)\right]$$  (21)

$$C_{My} = D\frac{\sqrt{2}}{2}C_T \pm R\mu\cos(\phi)\left[\frac{2}{3}\Theta_0 - \frac{1}{2}\left(\Theta_{tw} + \lambda\right)\right]$$

It is important to notice that equations 20 have two solutions, since the rotors spin in different directions, as seen in figure 3. Different rotational directions have the opposite effect on torques. This is why the $\pm$ sign is used in torque equations. These differences, induced from the specific quadrotor construction, along with the augmented momentum equation provide an improved insight to quadrotor aerodynamics. Regardless of the flying state of the quadrotor, by using these equations one can effectively model its behavior.

### 3.2 Building a more realistic rotor model

Building a more realistic rotor model begins with redefining its widely accepted static thrust equation 22 with real experimental results. No matter how precise, static equation is valid only when quadrotor remains stationary (i.e. hover mode). In order for the equation to be valid during quadrotor maneuvers, aerodynamic effects from 3.1 need to be incorporated into the equation.

$$T \sim k_T\Omega^2$$  (22)

### 3.2.1 Experimental results

This section presents the experimental results of a static thrust equation for an example quadrotor. Most of researched quadrotors use DC motors to drive the rotors. Although new designs use brushless DC motors (BLDC), brushed motors are still used due to their lower cost. Some advantages of brushless over brushed DC motors include more torque per weight, more torque per watt (increased efficiency) and increased reliability Sanchez et al. (2011); Solomon & Famouri (2006); Y. (2003).

Quadrotor used in described experiments is equipped with a standard brushed DC motor. Experimental results show that quadratic relationship between rotor speed (applied voltage) and resulting thrust is valid for certain range of voltages. Moving close to saturation point (i.e. 11V-12V), the quadratic relation of thrust and rotor speed deteriorates. Experimental results are shown in figure 7 and in the table 1.

| Voltage [V] | Rotation speed $\Omega$ [rpm] | Induced speed $v_i$ [m/s] | Thrust [N] |
|---|---|---|---|
| 4.04 | 194.465 | 1.5 | 0.16 |
| 5.01 | 241.17 | 2 | 0.29 |
| 5.99 | 284.105 | 2.45 | 0.44 |
| 6.99 | 328.82 | 2.7 | 0.58 |
| 8.00 | 367.357 | 3.2 | 0.72 |
| 8.98 | 403.171 | 3.5 | 0.94 |
| 10.02 | 433.540 | 3.8 | 1.16 |
| 10.99 | 464.223 | 4.05 | 1.34 |
| 12.05 | 490.088 | 4.3 | 1.42 |

Table 1. Data collected from the experiments

In order to use thrust equation 18, certain coefficients need to be known. Some of them like rotor radius $R$ and cord length $c$ can be measured. Others, like the mechanical angle $\Theta_0$ have to be calculated. Solving thrust equation 18 for $\mu = 0$ and $\lambda_c = 0$ (i.e. static conditions) yields:

$$F_V = \frac{1}{2}\rho a c \omega^2 R^3 \left(\Theta_{\frac{3}{4}} - \lambda_i\right) \tag{23}$$

where $\Theta_{\frac{3}{4}}$ is a mechanical angle at the $\frac{3}{4}$ of the blade length $R$ 13. $\Theta_{tw}$ can later be assessed from the blade construction. Rearranging equation 23 yields an equation for solving the mechanical angle problem 24.

$$\Theta_{\frac{3}{4}} = \frac{3}{2}\left(\frac{2F_V}{\rho a c \Omega^2 R^3} + \lambda_i\right) \tag{24}$$

Using experimental data from table 1 it is easy to calculate rotor angle $\Theta_{\frac{3}{4}}$. For given set of data the average $\lambda_i = \frac{v_i}{\Omega R} = 0.0766$. Therefore the mechanical angle $\Theta_{\frac{3}{4}} = 11.6291^o$, which is well between the expected boundaries.

Obtained data is piecewise linearized, in order to clearly demonstrate the differences between various voltage ranges. From Fig. 7 it can be seen how thrust declines near the point of saturation. This is important to notice, when deriving valid algorithms for quadrotor stabilization and control. Linearizaton coefficients are given in table 2.

| Voltage [V] | Linear gain [N/V] |
|---|---|
| [0 − 3] | 0 |
| [3 − 8] | 0.1433 |
| [8 − 11] | 0.2070 |
| [11 − 12] | 0.08 |

Table 2. Piecewise linearization coefficients

### 3.2.2 Applying aerodynamics to rotor dynamic model

To apply aerodynamic coefficient 18 to the static thrust experimental results, one needs to multiply experimental results with dynamic-to-static aerodynamic coefficient ratio 25.

$$T(\mu, \lambda_c, \lambda_i) = \frac{C_T(\mu, \lambda_c, \lambda_i)}{C_T(0,0,0)} T(0,0,0) \tag{25}$$

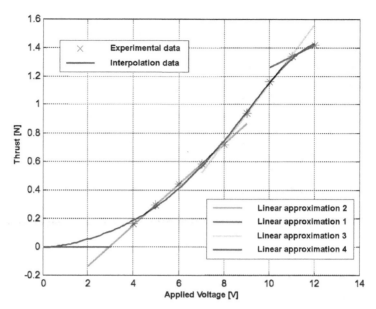

Fig. 7. Static rotor thrust experimental results with interpolation function and piecewise linear approximations

For the calculation of the aerodynamic coefficient $C_T$ it is crucial to know three airspeed coefficients $\mu$, $\lambda_c$ and $\lambda_i$. Two of them, $\mu$, $\lambda_c$, can easily be obtained from the available motion data $V_{xy}$, $V_z$ and $\Omega R$. $\lambda_i$ however, is very hard to know, because it is impossible to measure the induced velocity $v_i$.

One way to solve this problem is to calculate the induced velocity coefficient $\lambda_i$ from the two aerodynamic principals, momentum and blade element theories. The macroscopic momentum equation 10 and the microscopic blade element equation 17 provide the same rotor thrust using different physical approach:

$$T = \frac{1}{4}\rho a R^3 \Omega^2 c \left[\frac{2}{3}\alpha_{meh}\left(1+\frac{3}{2}\mu^2\right) - \lambda_i - \lambda_c\right] = 2\rho R^2 \pi \lambda_i \sqrt{(\lambda_c + \lambda_i)^2 + \mu^2 + \frac{\lambda_c^2}{7,67}} \quad (26)$$

When squared, equation 26 can be easily solved as a quadrinome:

$$\lambda_i^4 + p_3\lambda_i^3 + p_2\lambda_i^2 + p_1\lambda_i + p_0 = 0$$

$$p_0 = -c_1 c_2^2$$

$$p_1 = 2c_1 c_2$$

$$p_2 = \left(1 + \frac{1}{7.67}\right)\lambda_c^2 + \mu^2 - c_1 \quad (27)$$

$$p_3 = 2\lambda_c$$

$$c_1 = \frac{a^2 s^2}{64}$$

$$c_2 = 2\Theta_0 \left(1 + 1.5\mu^2\right)/3 - \lambda_c$$

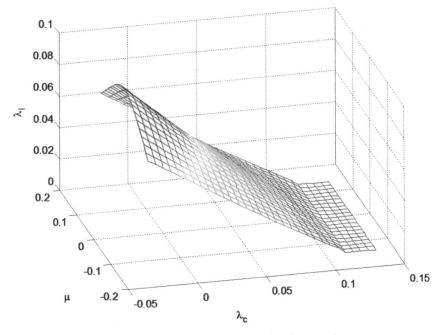

Fig. 8. 3D representation of $\lambda_i$ change during horizontal and vertical movement

The results of solving this quadrinome can be shown in a 3D graph 8. Although equations 27 look straightforward to solve, it still requires a substantial amount of processor capacity. This is why an offline calculation is proposed. This way, the calculated data can be used during simulation without the need for online computation. By using calculated values of the induced velocity, it is easy to calculate the dynamic thrust coefficient from equation 18. The 3D representation of final results is shown in figure 9.

Due to an increase of airflow produced by quadrotor movement, the induced velocity decreases. This can be seen in figure 8. Although both movements tend to increase induced velocity, only the vertical movement decreases the thrust coefficient. As a result, during takeoff the quadrotor looses rotor thrust, but during horizontal movement that same thrust is increased and enables more aggressive maneuvers.

### 3.2.3 Quadrotor model

A complete quadrotor model, incorporating previously mentioned effects is shown in figure 10. A control input block feeds the voltage signals to calculate statics thrust, which is easily interpolated from the available experimental data, using an interpolation function as shown in figure 7.

Static rotor thrust is applied to equation 25 along with aerodynamic coefficient $C_T(\mu, \lambda_c, \lambda_i)$. Induced velocity and aerodynamic coefficient are calculated using inputs from the current flight data (i.e. $\lambda_c, \mu$). This data is supplied from the Quadrotor Dynamics block. The calculation can be done offline, so that a set of data points from figure 9 can be used to

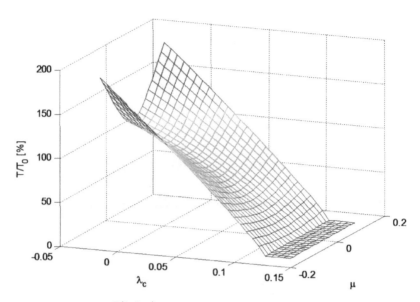

Fig. 9. 3D representation of $\frac{T(\lambda_i, \lambda_c, \mu)}{T(0,0,0)}$ ratio during horizontal and vertical movement

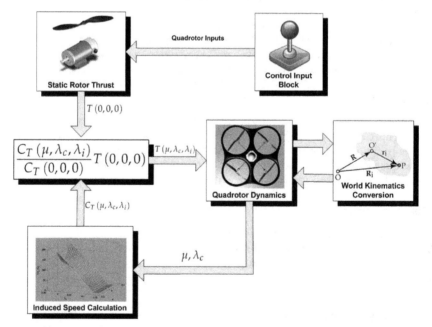

Fig. 10. Quadrotor model

interpolate true aerodynamic coefficient. This speeds up the simulation, as opposed to solving the quadrinome problem online.

A combination of the results provided from these two blocks using equation 25 gives the true aerodynamic rotor thrust. The same procedure is used to calculate the induced speed from the data shown in figure 8. Once the exact induced speed is known it can be applied to horizontal coefficients 19 and torque coefficients 21. In this way, quadrotor dynamics block can calculate quadrotors angular and linear dynamics using equations 6 and 3.

Dynamics data is finally fed into the kinematics block, that calculates quadrotor motion in world coordinate system using transformation matrices 2 and 7.

## 4. Conclusion

As the unmanned aerial research community shifts its efforts towards more and more aggressive flying maneuvers as well as mobile manipulation, the need for a more complete aerodynamic quadrotor model, such as the one presented in this chapter arises.

The chapter introduces a nonlinear mathematical model that incorporates aerodynamic effects of forward and vertical flights. A clear insight on how to incorporate these effects to a basic quadrotor model is given. Experimental results of widely used brushed DC motors are presented. The results show negative saturation effects observed when using this type of DC motors, as well as the phenomenon of thrust variations during quadrotor's flight.

The proposed model incorporates aerodynamic effects using offline precalculated data, that can easily be added to existing basic quadrotor model. Furthermore, the model described in the paper can incorporate additional aerodynamic effects like the In Ground Effect.

## 5. References

Adigbli, P., Grand, C., Mouret, J.-B. & Doncieux, S. (2007). Nonlinear attitude and position control of a micro quadrotor using sliding mode and backstepping techniques, *7th European Micro Air Vehicle Conference (MAV07)*, Toulouse.

Anderson, S. B. (1981). Historical overview of v/stol aircraft technology, *NASA Technical Memorandum* .

Bouabdallah, S., Noth, A. & Siegwart, R. (2004). Pid vs lq control techniques applied to an indoor micro quadrotor, *Proc. of The IEEE International Conference on Intelligent Robots and Systems (IROS)*.

Bouabdallah, S. & Siegwart, R. (2005). Backstepping and sliding-mode techniques applied to an indoor micro quadrotor, *Proc. of The IEEE International Conference on Robotics and Automation (ICRA)*.

Bramwell, A., Done, G. & Balmford, D. (2001). *Bramwell's helicopter dynamics*, American Institute of Aeronautics and Astronautics.

Chaturvedi, N., Sanyal, A. & McClamroch, N. (2011). Rigid-body attitude control, *Control systems magazine* .

Gessow, A. & Myers, G. (1952). *Aerodynamics of the helicopter*, F. Ungar Pub. Co.

Hoffmann, G. M., Huang, H., Wasl, S. L. & Tomlin, E. C. J. (2007). Quadrotor helicopter flight dynamics and control: Theory and experiment, *In Proc. of the AIAA Guidance, Navigation, and Control Conference*.

Jazar, R. (2010). *Theory of Applied Robotics: Kinematics, Dynamics, and Control (2nd Edition)*, Springer.

Korpela, C. M., Danko, T. W. & Oh, P. Y. (2011). Mm-uav: Mobile manipulating unmanned aerial vehicle, *In Proc. of the International Conference on Unmanned Aerial Systems (ICUAS)*.

Mellinger, D., Michael, N. & Kumar, V. (2010). Trajectory generation and control for precise aggressive maneuvers with quadrotors, *Proceedings of the International Symposium on Experimental Robotics*.

Orsag, M. & Bogdan, S. (2009). Hybrid control of quadrotor, *Proc.of the 17. Mediterranean Conference on Control and Automation, (MED)*.

Orsag, M., Poropat, M. & S., B. (2010). Hybrid fly-by-wire quadrotor controller, *AUTOMATIKA: Journal for Control, Measurement, Electronics, Computing and Communications* .

Sanchez, A., García Carrillo, L. R., Rondon, E., Lozano, R. & Garcia, O. (2011). Hovering flight improvement of a quad-rotor mini uav using brushless dc motors, *J. Intell. Robotics Syst.* 61: 85–101.
URL: *http://dx.doi.org/10.1007/s10846-010-9470-3*

Shuster, M. D. (1993). Survey of attitude representations, *Journal of the Astronautical Sciences* 41: 439–517.

Solomon, O. & Famouri, P. (2006). Dynamic performance of a permanent magnet brushless dc motor for uav electric propulsion system - part i, *Proc.of IEEE Industrial Electronics, (IECON)*.

Stingu, E. & Lewis, F. (2009). Design and implementation of a structured flight controller for a 6dof quadrotor using quaternions, *Proc.of the 17. Mediterranean Conference on Control and Automation, (MED)*.

Varga, M. & Bogdan, S. (2009). FuzzyâĂŞlyapunov based quadrotor controller design, *Proc. of the European Control Conference,(ECC)*.

Warwick, G. (2007). Army looking at three configuration concepts for large cargo rotorcraft, *Flight International via www.flightglobal.com* .

Y., P. (2003). Brushless dc (bldc) motor fundamentals, *Microchip Application Notes, AN885* .

# 8

# Advanced Graph Search Algorithms for Path Planning of Flight Vehicles

Luca De Filippis and Giorgio Guglieri
*Politecnico di Torino*
*Italy*

## 1. Introduction

Path planning is one of the most important tasks for mission definition and management of manned flight vehicles and it is crucial for Unmanned Aerial Vehicles (UAVs) that have autonomous flight capabilities. This task involves mission constraints, vehicle's characteristics and mission environment that must be combined in order to comply with the mission requirements. Nevertheless, to implement an effective path planning strategy, a deep analysis of various contributing elements is needed. Mission tasks, required payload and surveillance systems drive the aircraft selection, but its characteristics strongly influence the path. As an example, quad-rotors have hovering capabilities. This feature permits to relax turning constraints on the path (which represents a crucial problem for fixed-wing vehicles). The type of mission defines the environment for planning actions, the path constraints (mountains, hills, valleys, ...) and the required optimization process. The need for off-line or real-time re-planning may also substantially revise the path planning strategy for the selected type of missions. Finally, the computational performances of the Remote Control Station (RCS), where the mission management system is generally running, can influence the algorithm selection and design, as time constraints can be a serious operational issue.

This chapter aims to cover three main topics:

- Describe the most important algorithms developed for path planning of flying vehicles in order to compare them and depict their merits and drawbacks.
- Focus on graph search algorithms in order to define their main characteristics and provide a complete overview of the most important methods developed.
- Present a new graph search algorithm (called Kinematic A*) that has been developed on the base of the well-known A* algorithm and aims to fill the relation gap between the path planned with classical graph search solutions and the aircraft kinematic constraints.

The chapter is structured as follow:

- General description of the most important path planning algorithms:
  - Introduction to first approaches to path planning: manual path planning and Dubins curves. Also some simple applications developed by this research group are presented.

- General description of probabilistic and graph search algorithms.
- General description of potential field and model predictive algorithms.
- Introduction to some generic optimization algorithms.
- Study on graph search algorithms:
  - General description of commonality and differences between methods composing this family. Basic algorithm structure identification and introduction to the general features of these methods.
  - First graph search solutions focusing on the A* algorithm.
  - Introduction to dynamic-graph search and to the principal developed methods.
  - "Any heading" algorithms description focusing on Theta*.
  - Brief comparison between Theta* and A* on paths planned with the tools developed by this research group, focusing on the main improvements introduced with Theta*.
- Kinematic A*:
  - State space definition: in order to implement Kinematic A*, redefinition of the state space is needed.
  - Kinematic model description: the system of differential equation modelling the aircraft kinematic behaviour.
  - Introduction of wind in the kinematic model in order to take into account this disturbance on the path.
  - Formulation of the optimization problem solved with the graph search approach.
  - Constraints definition identifying the set of states evaluated to find the optimal path.
  - Algorithm description.
- Results presentation in order to identify new algorithm merits and drawbacks:
  - Algorithm test on a square map collecting four obstacles placed close to the four corners. A* path comparison with the Kinematic A* one planned with and without wind.
  - Algorithm test on a square map with one obstacle obstructing the path. This test is made to verify the algorithm search performances.
- Conclusion and future work description.

## 2. The path planning task

Generally, path planning aims to generate a real-time trajectory to a target, avoiding obstacles or collisions (assuming reference flight-conditions and providing maps of the environment), but also optimizing a given functional under kinematic and/or dynamic constraints. Several solutions were developed matching different planning requirements: performances optimization, collision avoidance, real-time planning or risk minimization, etc. Several algorithms were designed for robotic systems and ground vehicles. They took hints from research fields like physics for potential field algorithms, mathematics for probabilistic approaches, or computer science for graph search algorithms. Each family of algorithms has been tailored for path planning of UAVs, and future work will enforce the development of new strategies.

## 2.1 Manual path planning and Dubins curves

First studies on path planning of unmanned aircrafts evidenced task complexities, strict safety requirements and reduced technological capabilities that imposed as unique solution manual approaches for path planning of UASs. The waypoint sequences where based on the environment map and on the mission tasks, taking into account some basic kinematic constraints. The flight programs were then loaded on the aircraft flight control system (FCS) and the path tracking were monitored in real time. These approaches were overtaken researching on this problem, but some of them are still used for industrial applications where the plan complexity requires a human supervision at all stages. In these cases computer tools driving the waypoints allocation, the path feasibility verification and the waypoints-sequence conversion in formats compatible with the aircraft FCS assist the human agent.

The above-mentioned path-planning procedures were investigated (De Filippis et al., 2009) and some simple tools were developed in Matlab/Simulink and integrated into a single software package. This software (named PCube) handles geotiff and Digital Elevation Models to generate waypoint sequences compatible with the programming scripts of Micropilot commercial autopilots. The tool has a basic Graphical User Interface (GUI) used to manage the map and the path planning sequence. This tool can be used to:

- generate point and click waypoint sequences,
- choose predefined path shapes (square, rectangular and butterfly shapes),
- generate automatic grid type waypoint sequences (grid patterns for photogrammetric use).

If manual planning is the first and basic approach to path planning, it motivated research of more accurate solutions. In this direction optimization of the path with respect to some performance parameters was the challenge. Many approaches from optimal theory were studied and adapted to path planning and the Dubins curves are one of the most used and attractive solutions for their conceptual and implementation simplicity.

In a bi-dimensional space a couple of points each one associated to a unitary vector is given such that a vehicle is supposed to pass from these points with its trajectory tangent to the vector in that point. Dubins considered a non-holonomic vehicle moving at constant speed with limited turning capabilities and tried to find the shortest path between the two points under such constraints. He demonstrated that assuming constant turning radiuses this path exists and analysing each possible case a set of geodesic curves can be defined (Dubins, 1957). The same work was moved forward through successive studies on holonomic vehicles (Reeds & Shepp, 1990).

Dubins curves are used in PCube to take into account the UAVs turn performances and average flight speed, reallocating waypoints violating the constraints. For grid type patterns, the path generation is optimized for optical type payloads, specifying image overlaps and focal length. The package also allows the manipulation of maps and flight paths (i.e. sizing, scaling and rotation of mapped patterns). 3D surface and contour level plots are available for enhancing the visualization of the flight path. Coordinates and map formats can also be converted in different standards according to user specifications.

An example of manual planning using the point and click technique on a highland area is shown in **Figure 1**. Where the waypoint sequence defined by the user has been modified exploiting the Dubins curves. Manual path planning can generate paths with very simple logics when the optimization constraints do not affect the task and more complex solutions were developed and implemented.

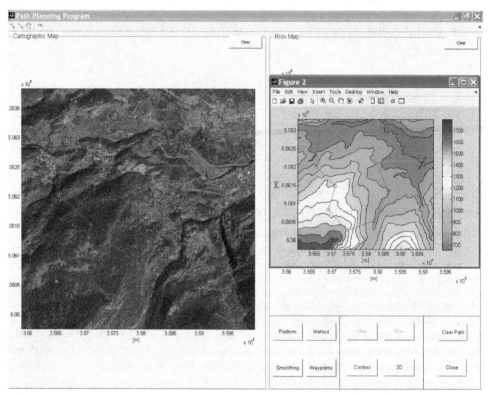

Fig. 1. Manula path planning with PCube Graphical User Interface (GUI).

## 2.2 Probabilistic and graph search algorithms

The problem of path planning is just an optimization problem made complex by the concurring parameters to be optimized on the same path. These parameters sometimes jar each other and they have to be balanced with respect to the mission tasks. All the more advanced algorithms developed for path planning try to identify the object of the optimization and reformulate the problem to cope with the prominent task, finding different approaches to optimize the parameter connected with this task. Many of them were developed for other applications and were modified to match with the problem of path planning. It's the case of the probability algorithms.

These algorithms generate a probability distribution connected with the parameter to be optimized and they implement statistic techniques to find the most probable path that optimizes this parameter (Jun & D'Andrea, 2002). Many implementations are related with the risk distribution of some kind of threat on a map (obstacles, forbidden areas, wind, etc.) and the algorithm creates probabilistic maps (Bertuccelli & How 2005) or look-up tables (that can be updated in real-time) modelling this distribution with various theories and logics (Pfeiffer, B et al., 2008). Markov processes are commonly used to introduce probabilistic uncertainties on the problem of path planning and Markov decision processes (MDPs) are defined all the approaches connecting this uncertainty with the taken action. These techniques are useful for all the cases where the optimization parameters are uncertain and can change in time and space, like conditions of flight, environment, and mission tasks.

Graph search algorithms are then interesting techniques coming from computer science. They were developed to find optimal plans to drive data exchanges on computer networks. These algorithms are commonly defined "greedy" as they generate a local optimal solution that can be quite far from the global optimal one. These algorithms are widely used in different fields thanks to their simplicity and small computational load and in the last five decades they evolved from basic approaches as Djikstra and Bellman-Ford algorithms to more complex solutions as D* Lite and Theta*. All of them differ in some aspects related to arc-weights definition and cost-function, but they are very similar in the implementation philosophy.

The main drawback of probabilistic and graph search algorithms resides in the lack of correlation between the aircraft kinematics and the planned path. Commonly, after the path between nodes of the graph has been generated with the minimum path algorithms, it has to be smoothed in order to be adapted to the vehicle flight performances. Indeed greedy algorithms provide a path constituted by line segments connected with edges that can't be followed by any type of flight vehicle. In order to obtain a more feasible and realistic path, refinement algorithms have to be used. This kind of algorithms can be very different in nature, starting from geometric curve definition algorithms also line flow smoothing logic can be used, but in any case at the end of this process a more realistic path is obtained, which better matches with autopilot control characteristics and flight performances.

Successive research on path planning algorithms brought to development of potential field based solutions. First potential field implementations came out to solve obstacles avoidance and formation flight problems, but in the last few years trajectory optimization under some performance constrains has been investigated.

## 2.3 Potential field and model predictive algorithms

Potential field algorithms come from robotic science and have been adapted to UASs simply modifying the kinematic models and the obstacles models. The environment is modelled to generate attractive forces toward the goal and repulsive ones around the obstacles (Dogan, 2003). The potential field model can be magnetic or electric (Horner & Healey, 2004), but the methods derived from aerodynamics provide the best choice in generation of trajectories for flight (Waydo & Murray, 2003). The vehicle motion is forced to follow the energy minimum respecting some dynamic constraints connected with its characteristics (Ford & Fulkerson, 1962). Two important aerodynamic field methods can be mentioned here: one obtained modelling path through propagation of pressure waves and another based on streamline modelling the motion field. The first method has been implemented supposing the fluid expanding from the target position through the starting one and modelling objects in the environment as obstacles. The second method instead models the environment like an aerodynamic field where obstacles are represented with singularities characterized by outgoing flow direction and target position like attracting singularities. The trajectory is chosen between all the streamlines defined in the field, to minimize the potential field gradient.

As a matter of fact these algorithms give smoothed and flyable paths, avoiding static and dynamic obstacles according with the field complexity. In the last years they have been widely investigated and interesting applications have been published. Even tough they are a promising solution for path planning and collision avoidance their application to some problems seemed hard due to their tendency to local minima on complex potential models.

The last and more advanced family of methods presented here, is based on technique coming from control science and applied to path planning and collision avoidance in the last decades. These algorithms apply model predictive control techniques to path planning problems linking a simplified model of the vehicle to some optimization parameters.

These algorithms solve in open loop an optimization problem constrained with a set of differential equations over a finite time horizon. The fundamental idea is to generate a control input that respects vehicle dynamics, environment characteristics and optimization constrains inside the defined time step and to repeat this process each step up to reach the goal. Sensors data can be integrated to update the model states so that these algorithms are used for collision avoidance in presence of active obstacles and particular harsh environments.

The big merit of model predictive solutions is the inclusion inside the optimization problem of the vehicle kinematics and dynamics in order to generate flyable trajectories. Model Predictive Control (MPC) or Receding Horizon Control (RHC) are the first techniques developed for industrial processes control that have been adapted to path planning (Ma & Castanon, 2006). Relation between control theory and path planning underlines another important characteristic of these methods. Indeed using the same logic for control and path planning opens the possibility to generate an integrated system that provides trajectories and control signals. On the other hand because complex sets of differential equations solved iteratively to generate the path are used in these methods, computation speed has been a real issue for these algorithms to spread. Also, as more as the problem complexity increases, as more the optimization space becomes complex and convergence to the optimal solution becomes an issue. Though, successive evolution of the model predictive technique is the Mixed-Integer Linear Programming (MILP). This algorithm applies the same logics of the model predictive one but allows inclusion of integer variables and discrete logics in a continuous linear optimization problem. Variables are used to model obstacles and to generate collision avoidance rules, while dynamics can be modelled with continuous constrains.

As it was stated previously, path planning is an optimization process then classical optimization techniques must be described to give a complete overview of the main tools developed to cope with this problem.

## 2.4 Generic optimization algorithms

Mathematical methods to solve optimization problems, known as indirect methods, are the most important and referenced techniques in this field. Algorithms based on Pontryagin minimum principle and Lagrange multipliers have been widely used to reduce optimization problems to a boundary condition one (Chitsaz & LaValle, 2007). Sequential Gradient Restoration Algorithm (SGRA) represents an indirect method used for several problems like space trajectories optimization (Miele & Pritchard, 1969, Miele, 1970). These techniques are elegant and reliable thanks to decades of research and application to thousands of different problems. They require a complex problem formulation and simplification in order to reach the required mathematical structure that ensures convergence. In some cases where complex and non-linear problems need to be solved these methods can result impracticable and other optimization techniques are needed (Sussmann & Tang, 1991).

Genetic algorithms are nowadays the most attractive solution in problems where constraints and optimization variables are the issue (Carroll,1996). They are based on the concept of natural selection, modelling the solutions like a population of individuals and evaluating evolution of this population over an environment represented by the problem itself. Using Splines or random threes to model the trajectory, these algorithms can reallocate the waypoint sequence to generate optimum solutions under constraints on complex environments (Nikolos et al., 2003). Being interesting and flexible, the evolutionary algorithms are spreading on different planning problems, but their solving complexity is paid with a heavy computational effort.

Finally, more advanced optimization techniques inspired to biological behaviours must be mentioned. These techniques recall biological behaviours to find the optimal solution to the problem. The key aspect of these solutions is the observation of biological phenomena and the adaptation to path planning problems. These algorithms permit to improve the system flexibility to changes in mission constraints and environmental conditions and with respect to genetic approaches these algorithms optimize the solution through a cooperative search.

# 3. The graph search algorithms for path planning

Graph search algorithms were developed for computer science to find the shortest path between two nodes of connected graphs. They were designed for computer networks to develop routing protocols and were applied to path planning through decomposition of the path in waypoint sequences. The optimization logics behind these algorithms attain the minimization of the distance covered by the vehicle, but none of its performances or kinematic characteristics is involved in the path search.

## 3.1 General overview

Basic elements common to each graph search method are (LaValle, 2006):

- a finite or countably infinite state space that collects all the possible states or nodes of the graph ($X$),
- an actions space that collects for each state the set of action that can be taken to move from a state to the next ($U$),
- a state transition function:

$$f: \quad \forall \quad x \in X \quad and \quad u \in U \quad f(x,u) = x' \quad x' \in X \tag{1}$$

- an initial state $x_I \in X$ ,
- a goal state $x_G \in X$ ,

Classical graph search algorithms applied to path planning tasks then have other common elements:

- the state space is the set of cells obtained meshing the environment in discrete fractions,
- the action space is the set of cells reachable from a given cell,
- the transition function checks the neighbours of a given cell to determine whether motion is possible (i.e. for an eight connected mesh the transition function checks the eight neighbours of a given cell),
- the cost function evaluates the cost to move from a given cell to one of its neighbours.
- the initial state is the starting cell where the aircraft is supposed to be,
- the goal state is the goal cell where the aircraft is supposed to arrive.

Classical graph search algorithms treat each cell as a graph node and they search the shortest path with "greedy" logics. The algorithm applies the transition function to the current cell to move to the next one and it analyses systematically the state space from the starting cell trying to reach the goal one. Each analysed cell can be:

- Unexpanded: a cell that the algorithm has not been reached yet. When the algorithm reaches an unexpanded cell the cost to come to that cell is computed and the cell is stored in a list called *open list*.
- Expanded (a cell already reached):
  - Alive: a cell that the algorithm could reach from another neighbouring cell. A cell alive is yet in the open list. The algorithm computes the new cost to come and substitutes the new cost associated to the cell whether it is lower then the previous one.
  - Dead: a cell that the algorithm already reached and its cost to come cannot be reduced further. These cells are stored in a list called *closed list*.

For each cell together with its coordinates and the cost to come, the algorithm stores in the lists also the parent coordinates. The parent is the cell left to reach a current one (i.e. $x_0$ used in $f(x_0,u_0)$ is the $x$ parent assuming that $x$ is the current cell and $x' = f(x,u)$ is the $x$ neighbour).

Main structure of any classical graph search algorithms is:

- Insert the starting cell in open list
- Searching cycle (this cycle breaks when the goal cell is reached or the open list is empty):
  - Check that the open list is not empty
    - True: go on
    - False: cycle break
  - Sort the open list with respect to the cost to come
  - Take the cell with the lower cost
  - Check that this cell is not the target one

- • True: go on
- • False: cycle break
- Add this cell to the closed list
- Cancel this cell from the open list
- Cell expansion cycle (this cycle breaks when each new cell has been evaluated)
  - • Use the transition function to find a new cell
  - • Check inclusion of the new cell in the closed list
    - • True: jump the state
    - • False: go on
  - • Check inclusion of the new cell in the open list
    - • True:
      - • Evaluate the cost to come
      - • Check if the new cost is lower then the previous one:
        - • True: substitute the new cost and the cell parent
        - • False: jump the state
    - • False:
      - • Evaluate the cost to come
      - • Add the cell to the open list
- End of the new state evaluation cycle
- End of the searching cycle.

The algorithm expands systematically the cells up to reach the goal and the different solutions composing this family of algorithms differ each other because of the logics driving the expansion. However the algorithm breaks when the goal cell is reached without providing any guaranty of global optimality on the solution. More advanced algorithms include more complex cost functions driving the expansion in such a way to provide some guarantees of local optimality of the solution, but the "greedy" optimization logics characterizing these path planning techniques has in this one of its drawbacks.

From late 50s wide research activity was performed on graph-search algorithms within computer science, trying to support the design of computer networks. Soon after, the possibility of their application in robotics resulted evident and new solutions were developed to implement algorithms tailored for autonomous agents. As a consequence, research on graph-search methods brought new solutions and still continues nowadays. Therefore, an accurate analysis is required to understand advantages and drawbacks of each proposed approach, in order to find possible improvements.

### 3.2 From Dijkstra to A*

The Dijkstra algorithm (Dijkstra, 1959) is one of the first and most important algorithms for graph search and permits to find the minimum path between two nodes of a graph with positive arc costs (Chandler et al, 2000). The structure of this algorithm is the one reported in the previous section and it represents the basic code for all the successive developments. An evolution of the Dijkstra algorithm is the Bellman-Ford (Bellman, 1958) algorithm; this method finds the minimum path on oriented graphs with positive, but also negative costs (Papaefthymiou & Rodriguez, 1991). Another method arose by the previous two is the Floyd-Warshall algorithm (Floyd, 1962, Warshall, 1962), that finds the shortest path on a weighted graph with positive and negative weights, but it reduces the number of evaluated nodes compared with Dijkstra.

The A* algorithm is one of the most important solvers developed between 50s and 70s, explicitly oriented to motion-robotics (Hart et al., 1968). A* improved the logic of graph search adding a heuristic component to the cost function. Together with the evaluation of the cost to come (i.e. the distance between the current node and a neighbour), it also considers the cost to go (i.e. an heuristic evaluation of the distance between a neighbour and the goal cell). Indeed the cost function (F) exploited by the A* algorithm is obtained summing up two terms:

- The cost to go H: a heuristic estimation of the distance from the neighbouring cell $x'$ to the goal $x_G$.
- The cost to come G: the distance between the expanded cell $x$ and the neighbouring one $x'$.

The G-value is 0 for the starting cell and it increases while the algorithm expands successive cells. The H-value is used to drive the cells expansion toward the goal, reducing this way the amount of expanded cells and improving the convergence. Because in many cases is hard to determine the exact cost to go for a given cell, the H-function is an heuristic evaluation of this cost that has to be monotone or consistent. In other words, at each step the H-value of a cell has not to overestimate the cost to go and H has to vary along the path in such a way that:

$$H(x',x_G) \leq H(x,x_G) + G(x,x') \qquad (2)$$

### 3.3 Dynamic graph search

The graph-search algorithms developed between 60s and 80s were widely used in many fields, from robotics to video games, assuming fixed and known positions of the obstacles on the map. This is a logic assumption for many planning problems, but represents a limit when robots move in unknown environments. This problem excited research on algorithms able to face with map modifications during the path execution. Particularly, results on sensing robots, able to detect obstacles along the path, induced research on algorithms used to re-plan the trajectory with a more effective strategy than static solvers were able to implement.

Dynamic re-planning with graph search algorithms was introduced. D* (Dynamic A*) was published in 1993 (Stentz, 1993) and it represents the evolution of A* for re-planning . When changes occur on the obstacle distribution some of the cell costs to come changes. Dynamic algorithms update the cost for these cells and replan only the portion of path around them keeping the remaining path unchanged. This way D* expands less cells than A* because it has not to re-plan the whole path through the end. D* focused was the evolution of D*, published by the same authors and developed to improve its characteristics (Stentz, 1995). This algorithm improved the expansion, reducing the amount of analysed nodes and the computational time.

Then, research on dynamic re-planning brought to the development of Lifelong Planning A* (LPA*) and D* Lite (Koenig & Likhachev, 2001, 2002). They are based on the same principles of D* and D* focused, but they recall the heuristic cost component of A* to drive the cell expansion process. They are very similar and can be described together. LPA* and D* Lite exploit an incremental search method to update modified nodes, recalculating only the start distances (i.e. distance from the start cell) that have changed or have not been calculated before. These algorithms exploit the change of *consistency* of the path to replan. When obstacles move, graph cells are updated and their cost to come changes. The algorithm records the cell cost to come before modifications and compares the new cost with the old one to verify consistency. The change in consistency of the path drives the algorithm search.

## 3.4 Any heading graph search

Dynamic algorithms allowed new applications of graph search methods to path planning of robotic systems. More recently, other drawbacks and possible improvements were discovered. Particularly, one of the most important drawbacks of A* and the entire dynamic algorithms resides on the heading constraints connected with the graph structure. The graph obtained from a surface map is a mesh of eight-connected cells with undirected edges. Moving from a given cell to the next means to move along the graph edge. The edges of these graphs are the straight lines connecting the centre of the current cell with the one of the neighbour. As a matter of fact the edges between cells of an eight connected graph can have slope $a$ such that:

$$a = n \cdot \frac{\pi}{4} \quad 0 \le n \le 8 \quad n \in N \tag{3}$$

Then the paths obtained with A* and its successors is made of steps with heading defined in equation [3]. This limit is demonstrated prevents these algorithms to find the real shortest path between goal and start cells in many cases (it is easy to imagine a straight line connecting the start with goal cell having heading different from the ones of equation [3]). A* and dynamic algorithms generate strongly suboptimal solutions because of this limit, that comes out in any application to path planning. Suboptimal solutions are paths with continuous heading changes and useless vehicle steering (increasing control losses) that require some kind of post processing to become feasible. Different approaches were developed to cope with this problem, based on post-processing algorithms or on improvements of the graph-search algorithm itself. Very important examples are Field D* and Theta*. These algorithms refined the graph search obtaining generalized paths with almost "any" heading.

To exploit Field D*, the map must be meshed with cells of given geometry and the algorithm propagates information along the edges of the cells (Ferguson & Stentz, 2006). Field D* evaluates neighbours of the current cell like D*, but it also considers any path from the cell to any point along the perimeter of the neighbouring cell. A functional defines the point on the perimeter characterising the shortest path. With this method a wider range of headings can be achieved and shorter paths are obtained.

Theta* represents the cutting edge algorithm on graph search, solving with a simple and effective method the heading constraint issue (Nash et al., 2007). It evaluates the distance from the parent to one of the neighbours for the current cell so that the shortest path is obtained. When the algorithm expands a cell, it evaluates two types of paths: from the current cell to the neighbour (like in A*) and from the current-cell parent to the neighbour. As a conclusion, paths obtained by the Theta* solver are smoother and shorter than those generated by A*.

Apparently, Theta* is the most promising solution for path planning. As a matter of fact, some other graph search algorithms were not considered here, as this chapter would provide a general overview on the main concepts converging in development of these path-planning methods. By the way all the algorithms described have the common drawback of missing any kind of vehicle kinematic constraints in the path generation. The algorithm presented in the following chapter (Kinematic A*) has been developed to bridge this gap and open investigations in this direction.

### 3.5 Tridimensional path planning with A* and Theta*

The application of A* and Theta* to 3D path planning for mini and micro UAVs was extensively investigated (De Filippis et al., 2010, 2011). The A*-basic algorithm was improved and applied to tri-dimensional path planning on highlands and urban environments. Then this algorithm has been compared with Theta* for the same applications in order to investigate merits and drawbacks of these solutions.

Here is reported the comparison between a path planned with A* with the same one planned with Theta* in order to show the improvements introduced adopting the last algorithm. **Figure 2** is the tri-dimensional view of the two paths implemented for this example.

Map characteristics:

- Cells number: 9990000.
- $\Delta$lat: 1 m.
- $\Delta$long: 1 m.
- $\Delta$Z: 1 m.
- Environment matrix dimensions: 300 x 300 x 111 (lat x long x Z).

| Path 1 | | |
|---|---|---|
|  | A* | Theta* |
| Path length | 386.5 m | 372.4 m |
| Computation time | 3.1 s | 3.6 s |
| Number of heading changes | 327 | 6 |
| Number of altitude changes | 0 | 0 |
| Number of waypoints | 327 | 6 |

Table 1. Example parameters.

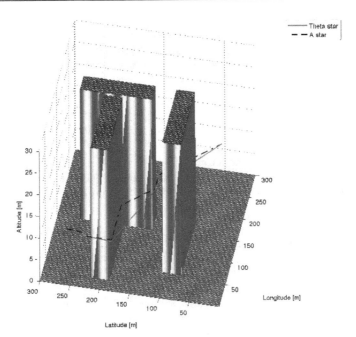

Fig. 2. Comparison between Theta* and A* (3D view).

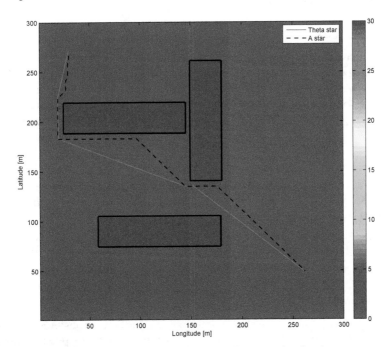

Fig. 3. Comparison between Theta* and A* (Longitude-Latitude plane).

Table 1 collects the map parameters and the algorithm performances while Figure 3 is the longitude-latitude path view. The paths are planned without altitude changes so the last picture is sufficient to depict differences between them. The path obtained with Theta* is slightly shorter then the A* one, but huge difference is in the number of waypoints composing the path and in the amount of heading changes. These parameters testify the previous statements identifying in Theta* an interesting algorithm among the classical graph search solutions here mentioned.

## 4. Kinematic A*

The main drawback of applying classical graph search algorithms to path planning problems resides in the lack of correlation between the path and the vehicle kinematic constraints. In this section, a new path-planning algorithm (Kinematic A*) is presented, implementing the graph search logics to generate feasible paths and introducing basic vehicle characteristics to drive the search.

Kinematic A* (KA*) includes a simple kinematic model of the vehicle to evaluate the moving cost between the waypoints of the path in a tridimensional environment. Movements are constrained with the minimum turning radius and the maximum rate of climb. Furthermore, separation from obstacles is imposed, defining a volume along the path free from obstacles (tube-type boundaries), as inside these limits the navigation of the vehicle is assumed to be safe.

The main structure of the algorithm will be presented in this section, together with the most important subroutines composing the path planner.

### 4.1 From cells to state variables

Classical graph search algorithms solve a discrete optimization problem linking the cost function evaluation to the distance between cells. These cells discretize the motion space representing the discrete state of the system. The states space is finite and discrete containing the positions of the cell centres. The optimization problem requires finding the sequence of states minimizing the total covered distance between the starting and the target cell.

Kinematic A* introduces a vehicle model to generate the states and evaluate the cost function. Each state is made of the model variables and is discrete because the command space is made of discrete variables. So the optimization problem is transformed in finding the discrete sequence of optimal commands generating the minimum path between the starting and the target state.

In the following sections then the concept of cells or nodes of the graph, representing the discrete set of states defining the optimization problem is substituted with the concept of states of the vehicle model and the optimization problem is reformulated.

### 4.2 The kinematic model

In the following description S is the state of the aircraft at the current position. S is the vector of the model state variables. This simple model is used to generate the possible movements from a given state to the next, i.e. the evolution of S from the current condition to the next.

The model is a set of four differential equations describing the aircraft motion in Ground reference frame (G frame). This is not the typical Nort-East-Down (NED) frame used to write navigation equations in aeronautics. The Ground frame is typical of ground robotic applications that inspired this work. The G-frame origin is placed in the aircraft center of mass. The X and Y axes are aligned with the longitude and latitude directions respectively. Then the Z axis points up completing the frame.

In the G frame distances are measured in meters and two control angles ($\chi$ and $\gamma$) act as gains on rate of turn and rate of climb along the path:

- $\chi$ is the angle between the X axis and the projection of the speed vector (V) on the X-Y plane, the variation of this angle is connected with the rate of turn.
- $\gamma$ is the angle between the speed vector and its projection on the X-Z plane (see Figure 4), this angle controls the rate of climb.

The model is obtained considering the aircraft flying at constant speed and the Body frame (B frame) aligned with the Wind frame (W frame). The rate of turn is assumed bounded with the minimum turn radius and the rate of climb with the maximum climb angle.

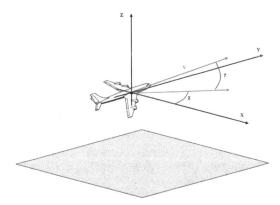

Fig. 4. The Ground Reference frame (G frame).

The speed vector is constant and aligned with the $X_B$ axis. Using the Euler transformation matrix from the body to the ground frame the speed components in G frame are obtained. Combining these differential equations with the turning-rate the aircraft model becomes:

$$
\begin{cases}
\dot{X} = V\cos(\chi)\cos(\gamma_{max} \cdot w) \\
\dot{Y} = V\sin(\chi)\cos(\gamma_{max} \cdot w) \quad |u| \le 1 \\
\dot{Z} = V\sin(\gamma_{max} \cdot w) \quad\quad |w| \le 1 \\
\dot{\chi} = \dfrac{V}{R_{min}} \cdot u
\end{cases}
\tag{4}
$$

where:

X,Y,Z = aircraft positions vector P on the ground frame [m].
V       = aircraft speed [m/s].

$R_{min}$  =maximum turning radius [m].

$\chi$       = turning angle.

$\gamma_{max}$  = maximum climbing angle.

u,w  = command parameters.

To generate the set of possible movements discrete command values ($u_i$ and $w_i$) are chosen and the system of equations [4] is integrated in time with the initial conditions given by the current state S:

$$
\begin{cases}
X(0) = X_s \\
Y(0) = Y_s \quad u_i = [\ -1 \quad -0.5 \quad 0 \quad 0.5 \quad 1] \\
Z(0) = Z_s \quad w_i = [\ -1 \quad -0.5 \quad 0 \quad 0.5 \quad 1] \\
\chi(0) = \chi_s
\end{cases}
\tag{5}
$$

If the command values are constant along the integration time ($\Delta t$), the equations in [4] become:

$$
\begin{cases}
X_i = X_s + \left( \dfrac{R_{min}}{u_i} \right) \cdot \cos(\gamma_{max} \cdot w_i) \cdot \left[ \sin\left( \chi_s + \dfrac{V}{R_{min}} \cdot u_i \cdot \Delta t \right) - \sin(\chi_s) \right] \\[4mm]
Y_i = Y_s - \left( \dfrac{R_{min}}{u_i} \right) \cdot \cos(\gamma_{max} \cdot w_i) \cdot \left[ \cos\left( \chi_s + \dfrac{V}{R_{min}} \cdot u_i \cdot \Delta t \right) - \cos(\chi_s) \right] \\[4mm]
Z_i = Z_s + V \cdot \sin(\gamma_{max} \cdot w_i) \cdot \Delta t \\[4mm]
\chi_i = \chi_s + \dfrac{V}{R_{min}} \cdot u_i \cdot \Delta t
\end{cases}
\tag{6}
$$

providing the evolution of S for each controls space. On Figure 5 25 trajectories are represented. They are obtained combining the two vectors $u$ and $w$ presented in [5] and substituting each command couple (5 $u_i$ values x 5 $w_i$ values) in [6]. For each couple the system of equation is integrated over the time step with initial conditions and parameters equal to:

$P_s$  = [0 0 0 0],

V   = 25 [m/s],

$R_{min}$ = 120 [m],

$\gamma_{max}$  = 4 [deg],

$\Delta t$   = 8 [s].

Once $\Delta t$, aircraft speed, minimum turning radius and maximum climbing angle are chosen according with the aircraft kinematic constraints the equations in [6] can be solved at each cycle for the current state and the algorithm can generate the set of possible movements looking for the optimal path.

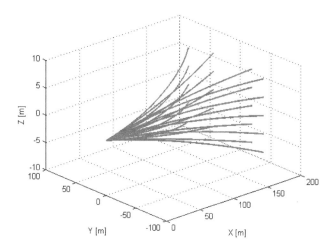

Fig. 5. Sequences of states for a time horizon of 8 seconds.

### 4.3 The kinematic model with wind

The kinematic model can be improved taking into account the wind effect on the states evolution. Summing to the aircraft speed the wind components in G frame and assuming these components constant on Δt the system of equations [6] become:

$$
\begin{cases}
X_i = X_s + \left(\dfrac{R_{min}}{u_i}\right) \cdot \cos(\gamma_{max} \cdot w_i) \cdot \left[\sin\left(\chi_s + \dfrac{V}{R_{min}} \cdot u_i \cdot \Delta t\right) - \sin(\chi_s)\right] + W_x \cdot \Delta t \\[2mm]
Y_i = Y_s - \left(\dfrac{R_{min}}{u_i}\right) \cdot \cos(\gamma_{max} \cdot w_i) \cdot \left[\cos\left(\chi_s + \dfrac{V}{R_{min}} \cdot u_i \cdot \Delta t\right) - \cos(\chi_s)\right] + W_y \cdot \Delta t \\[2mm]
Z_i = Z_s + V \cdot \sin(\gamma_{max} \cdot w_i) \cdot \Delta t + W_z \cdot \Delta t \\[2mm]
\chi_i = \chi_s + \dfrac{V}{R_{min}} \cdot u_i \cdot \Delta t
\end{cases}
\tag{7}
$$

where:

$[W_x \; W_y \; W_z]$ = wind speed components in G frame [m/s].

In Figure 6 and Figure 7 a state evolution with wind is compared with the same without wind. The state and parameters used as an example are:

$P_s = [0\ 0\ 0\ 0]$,
$V = 25\ [m/s]$,
$R_{min} = 120\ [m]$,
$\gamma_{min} = 4\ [deg]$,
$\Delta t = 8\ [s]$.

$W_x = 5 \ [m/s]$
$W_y = 5 \ [m/s]$
$W_z = 0 \ [m/s]$

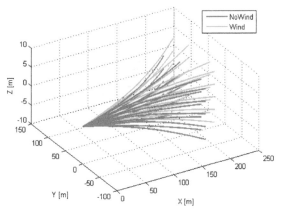

Fig. 6. Comparison of states evolution (3D view) with and without wind ($W_x$=5 m/s, $W_y$=5 m/s, $W_z$=0).

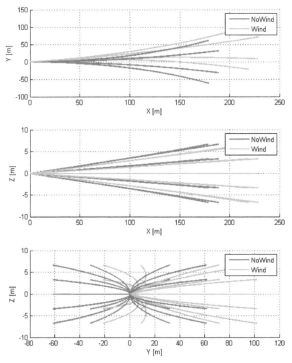

Fig. 7. Comparison of states evolution (2D views) with and without wind ($W_x$=5 m/s, $W_y$=5 m/s, $W_z$=0).

These pictures show how wind affects the evolution of a given state and in turn the effect on the set of possible movements from the current state to the next.

## 4.4 The problem formulation

The functional J minimized through the optimization process is made up of the costs $F_{ij}$ of each state composing the path. The minimum of J is found summing up the smaller cost $F_{ij}$ of each state. $F_{ij}$ is made of two terms related respectively with the states and the commands. At each step the algorithm generates the set of movements from the current state (shown in Figure 5). Then it evaluates $F_{ij}$ for each new state and chooses the one with the smaller value. The global optimization problem is finding:

$$\min(J) = \sum_{S_0}^{S_t} \min(F_{ij}) = \sum_{S_0}^{S_t} \min(\overline{H}_{ji}^T \cdot \overline{\alpha} \cdot \overline{H}_{ij} + \overline{G}_{ij}^T \cdot \overline{\beta} \cdot \overline{G}_{ij}) \tag{8}$$

The H and G vectors take into account respectively the error on the states and the amount of command due to reach a new state. The matrices α and β are diagonal matrices of gains on the states and on the commands:

$$\overline{H}_i = \begin{bmatrix} X_t - X_i \\ Y_t - Y_i \\ Z_t - Z_i \end{bmatrix} = \begin{bmatrix} \Delta X_i \\ \Delta Y_i \\ \Delta Z_i \end{bmatrix}$$

$$\overline{G}_i = \begin{bmatrix} u_i \\ w_i \end{bmatrix} \tag{9}$$

$$\overline{\alpha} = \begin{bmatrix} \alpha_1 & 0 & 0 \\ 0 & \alpha_2 & 0 \\ 0 & 0 & \alpha_3 \end{bmatrix}$$

$$\overline{\beta} = \begin{bmatrix} \beta_1 & 0 \\ 0 & \beta_2 \end{bmatrix} \tag{10}$$

The H vector is the distance between the new state $[X_i\ Y_i\ Z_i]'$ and the target one $[X_t\ Y_t\ Z_t]'$. On the other hand the G vector evaluates the amount of command needed to reach this new state from the current one. Then choosing the smaller value of F the algorithm selects a new state that reduces the distance from the target minimizing the commands. The gain matrices are used to weight the state variables and the commands in order to tune their importance in F.

To complete the problem formulation, the states in J must be included in the state space respecting the differential equations given in [4] and the initial conditions given in [5]. Then the commands must be chosen in the command space given in [5] in order to minimize the functional J. The state space is constrained by the map limits, the obstacles and the separation requirements and will be described in the following section.

### 4.5 The state space

If the command space of the problem solved with KA* is bounded by the kinematic constraints and is discretized according with the optimization requirements, the states are bounded only on the X and Y sets (longitude and latitude coordinates) because of constraints on the Z state and on the X and Y states themselves:

- The map bounds: these bounds affect the X-Y sets because points outside the map limits can not be accepted as new states.
- The ground obstacles: these constraints bound the X-Y sets if the Z component of the new state is lower then the ground altitude at the same X-Y coordinates, so the relative new state must be rejected.
- The separation constraints: the new state not only has to have a Z component higher then the ground one, but has also to respect the horizontal and vertical separations from the obstacles. The X-Y sets are bounded because states too close to the obstacles must be rejected.

Figure 8 shows the horizontal ($HZ_1$ and $HZ_2$) and vertical ($VZ_1$) separation constraints imposed on the path from the current state S to the next state I. These constraints guarantee the flight safety along the path because possible tracking errors of the guidance system are acceptable and safe inside the boundaries imposed by the separation constraints.

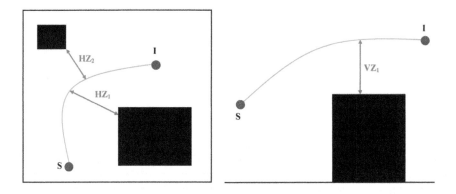

Fig. 8. Horizontal and Vertical separation from an obstacle.

### 4.6 Algorithm description

- Initialize the control variables and parameters.
- Evaluate the F cost for the initial state
- Add this state to the open list.
- Searching cycle (this cycle breaks when the target state is reached or the open list is empty):
  - Check that the open list is not empty
    - True: go on
    - False: cycle break

- Open list search for the state with the lower F value
- Check that this state is not the target one
  - True: go on
  - False: cycle break
- Add this state to the closed list
- Cancel this state from the open list
- New states evaluation cycle (this cycle breaks when each new state has been evaluated)
  - New states generation through the model equations
  - Check inclusion of the new state inside the state space
    - True: go on
    - False: jump the state
  - Check inclusion of this state in the closed list
    - True: jump the state
    - False: go on
  - Check inclusion of this state in the open list
    - True:
      - Evaluate the F cost for the state
      - Check if the new cost is lower then the previous one:
        - True: substitute this new cost and the new current state to the open list
        - False: jump the state
    - False:
      - Evaluate the F cost for the state
      - Add the state to the open list
  - End of the new state evaluation cycle
- End of the searching cycle.

## 5. Results

The following chapter collects two tests with the obstacles placed on the map in such a way to force KA* toward its limits. The new algorithm is compared with A* in order to show the improvements introduced. The paths are generated with and without wind to show its effects on the path and to compare the results. These tests then give the opportunity to show limitations of this new technique in order to stimulate future developments.

The first path is on a map with four obstacles symmetrically placed. They are close to the four corners of a square area and the aircraft is forced to slalom between them. The starting and target points are placed respectively at the bottom-left and top-right corner with different altitudes in order to force the algorithm to plan a descent meeting the four obstacles along the flight. Finding the path for this test is easier then finding it for the next one. The algorithm is able to follow the minimum of the cost function without analyzing too many states and it converges rapidly.

The map of the second path has just one wide obstacle placed in the middle. The obstacle is placed slightly closer to the right border of the map in order to obstruct the path to the aircraft that is supposed to move from the bottom-left to the top-right corner. The path search for this test is harder then the previous one because the algorithm has to analyze many states to find the optimum path. Following a monotonic decrease of the cost function along the path search is impossible for this case. The obstacle in the middle forces the aircraft far from the target point. The aircraft has to go around the obstacle to reach the target; this induces a cost increase to move from a state to the next that makes the optimization harder.

The component of wind introduced to implement the following tests is considered constant in time and space on the whole map. This approach clearly does not mean to solve the problems due to wind disturbances in the path optimization. This is a complex and hard problem due to wind model complexity, effects of the wind on the aircraft performances and dynamics, turbulent components effects, etc. Face properly this problem requires specific studies and techniques, but it is useful to introduce this simple study at this level in order to show potential developments of this path planning technique for future applications.

Then the aircraft parameters chosen to implement the tests must be motivated. The small area of the map, induced to chose accordingly the aircraft parameters needed for the model. The reference vehicle is a mini UAV with reduced cruise speed, turning radius and climbing performances but agile enough to perform the required paths. Particularly speed is chosen so that the trajectories needed to avoid the obstacles would be feasible and the turning radius is calculated considering coordinated turns:

$$R_{min} = \sqrt{\frac{V^4}{g^2 \cdot \left(\left(\frac{1}{\cos(\varphi_{max})}\right)^2 - 1\right)}} \tag{11}$$

where:

$R_{min}$ = minimum turning radius.
$V$ = aircraft cruise speed.
$g$ = gravitational acceleration.
$\varphi_{max}$ = maximum bank angle.

Finally for each test a table collecting all the data is reported. All the reported paths are obtained with the MATLAB version 7.11.0 (R2010b), running on MacBook Pro with Intel Core 2 Duo (2 X 2.53 GHz), 4 Gb RAM and MAC OS X 10.5.8. The table contains:

- map dimensions,
- obstacles dimensions,
- obstacles center position,
- starting point,
- target point,
- aircraft parameters,
- optimization parameters,
- obstacles separation parameters,
- wind speed,
- computation time,

- path length,
- number of waypoints.

## 5.1 Four obstacles

**Figure 9** shows the obstacles position on the map and the three paths (KA* with wind, KA* without wind, A*) in tridimensional view.

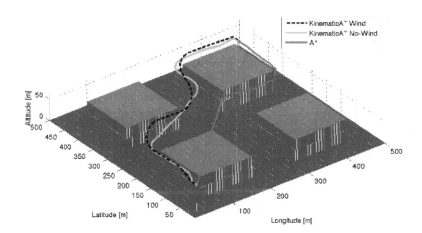

Fig. 9. Four obstacles test (3D view).

**Table 2** collects the numerical data used to implement this test. The time step between two states is set to 2 seconds in order to have a sufficiently discretized path without increasing too much the computation time. Horizontal and vertical obstacles separations are set to 15 m and 10 m respectively. This should guarantee sufficient safety without limiting the aircraft agility between the obstacles.

The constant wind along the $Y$ ground direction has 5 m/s intensity. This value is sufficiently high and it affects deeply the path as the computation time, the number of waypoints and the path shape testify. The wind pushes the aircraft toward the target, reducing the computation time with respect to the case without wind.

In order to compare the KA* performances with the A* one, it can be noticed that the computation time for the two algorithms is almost the same for the path with wind, but it is increased without wind. This is due to the reduced speed of the aircraft without wind and to the higher number of possible movements from one state to the next. With wind the feasible movements between states are strongly reduced because of the wind disturbance. Flying at lower speed the algorithm is forced to analyze much more states and for each position much more possible movements are feasible. Then the optimization process takes more time. Finally shall be noticed how KA* generates a path with a really small number of waypoints with respect to A*. This permits to obtain more handy waypoints lists without need of post processing, ready to be loaded on the flight control system.

| Map dimension | | | |
|---|---|---|---|
| | X | 500 | [m] |
| | Y | 500 | [m] |
| | Z | 80 | [m] |
| | ΔXY | 1 | [m] |
| | ΔXZ | 1 | [m] |
| | | | |
| Obstacles dimension | | | |
| | X | 250 | [m] |
| | Y | 125 | [m] |
| | Z | 50 | [m] |
| | | | |
| Obstacles-center position (1) | | | |
| | X | 125 | [m] |
| | Y | 125 | [m] |
| | | | |
| Obstacles-center position (2) | | | |
| | X | 125 | [m] |
| | Y | 375 | [m] |
| | | | |
| Obstacles-center position (3) | | | |
| | X | 375 | [m] |
| | Y | 375 | [m] |
| | | | |
| Obstacles-center position (4) | | | |
| | X | 375 | [m] |
| | Y | 125 | [m] |
| | | | |
| Starting point | | | |
| | X | 20 | [m] |
| | Y | 20 | [m] |
| | Z | 60 | [m] |
| | | | |
| Target point | | | |
| | X | 480 | [m] |
| | Y | 480 | [m] |
| | Z | 30 | [m] |
| | | | |
| Aircraft parameters | | | |
| | Speed | 10 | [m/s] |
| | Min turning radius | 25 | [m] |
| | Max climbing angle | 4 | [deg] |

| Optimization parameters | | | |
|---|---|---|---|
| | Time step | 2 | [s] |
| | α | 10 | |
| | β | 1 | |
| | | | |
| Obstacles separation | | | |
| | Horizontal | 15 | [m] |
| | Vertical | 10 | [m] |
| | | | |
| Wind Speed | | | |
| | X | 0 | [m/s] |
| | Y | 5 | [m/s] |
| | Z | 0 | [m/s] |
| | | | |
| Computation time | | | |
| | KA* with Wind | 2.1827 | [s] |
| | KA* without Wind | 9.9657 | [s] |
| | A* | 2.3674 | [s] |
| | | | |
| Path length | | | |
| | KA* with Wind | 772 | [m] |
| | KA* without Wind | 769 | [m] |
| | A* | 740 | [m] |
| | | | |
| WayPoints | | | |
| | KA* with Wind | 34 | |
| | KA* without Wind | 40 | |
| | A* | 591 | |

Table 2. Four-obstacles test parameters.

In **Figure 10** the paths on the Longitude-Latitude plane are presented. The path obtained with A* pass over the bottom-left obstacle and very close to the top-right one. Planning sharp heading changes to reach the target. This is typical of classical graph search algorithms that do not take into account the vehicle kinematic constraints. The path obtained with KA* on the other hand is smooth and obstacles separation constraints is evident. Comparing the path with wind with the one without wind between the obstacles on the left is evident the disturbance induced by the wind that pushes the path closer to the top-left obstacle.

In **Figure 11** on the X-axis is plotted the distance covered from start to target point and on the Y-axis the aircraft altitude. Again A* plans sharp altitude changes and particularly sharp descends to reach the target. These changes are unfeasible with real aircrafts. As a matter of fact in general the A* path requires deep post processing and waypoints reallocation to make the path flyable. Analyzing in detail though the algorithm plans a path passing over the bottom-left obstacle and then descending close to the top-right one. Being this descent unfeasible for the aircraft a complete re-planning is needed to reallocate the waypoints sequence. This is one of many cases evidencing that classical graph search algorithms used for tridimensional path planning can generate unfeasible paths because of the strong longitudinal constraints of aircrafts and they need high intrusive post processing algorithms to modify the waypoint sequence.

In **Figure 12** the time history of the turning rate (connected with the $u$ command) is plotted. The comparison between the command sequence with and without wind puts in evidence that the path needs more aggressive commands to compensate disturbances introduced from wind, but the average value remains limited thanks to the $G$ value in the cost function that takes care of the amount of command needed to perform the path. On the other hand in **Figure 13** the climbing angle (due to the $w$ command) is plotted. Also in this case the average amount of command is limited. Limiting turning rates and climbing angles required to follow the path is important. The main path-planning task is to generate a trajectory driving the aircraft from start to target in safe conditions. If tracking the path planned requires aggressive maneuvers, the aircraft performances will be completely absorbed by this task. However in many cases tracking the path is just a low-level task prerogative to accomplish with the high-level mission task (i.e in a save and rescue mission tracking the path could be one of the tasks together with many others. As an example it could be required also to avoid collision with dynamic obstacles along the flight, to deploy the payload and collect data, to interact with other aircrafts involved in the mission). If the aircraft must exploit its best performances to track the path it will not be able to accomplish also with the other mission tasks and this is not acceptable.

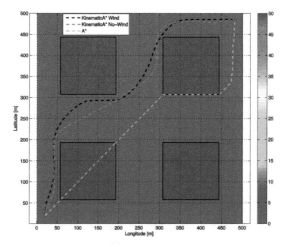

Fig. 10. Longitude-Latitude view (four obstacles test).

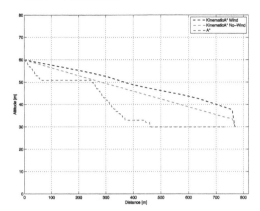

Fig. 11. Distance-Altitude view (four obstacles test).

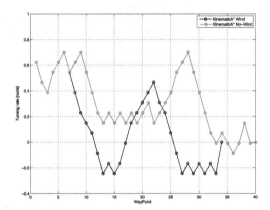

Fig. 12. Turning rate (four obstacles test).

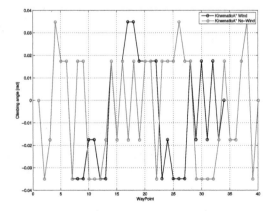

Fig. 13. Climbing angle (four obstacles test).

## 5.2 One obstacle

**Figure 14** is the tridimensional view of the three paths (KA* with wind, KA* without wind, A*) generated with this test. The picture shows that the obstacle obstructs almost completely the path to the aircraft on the right, leaving just a small aisle to reach the target.

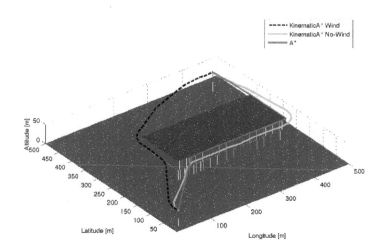

Fig. 14. One obstacle test (3D view).

All the data collected in **Table 3** are almost the same of the previous test. The environment, the aircraft, starting and target point, obstacles separation and optimization parameters do not change. Just the number and distribution of the obstacles is changed, together with the wind speed. The last parameter is changed to investigate the effects of diagonal wind on the path. The wind intensity is reduced to avoid reaching conditions too harsh for the flight. Two [m/s] of wind along the $X$ and $Y$ ground axes are introduced and the effects on the path are evidenced by the search performances.

The computation time between the path with and without wind is strongly different. As in the previous case wind forces the aircraft to move faster with respect to the ground, but the big difference between the paths now is due to the different path followed to reach the target. As shown in **Figure 14** the path with wind goes to the left of the obstacle and reaches the target directly. This is due to the negative wind speed along the $X$-axis that opposes the tendency of the aircraft to go straight from start to target (as the first part of the path without wind shows). The aircraft is pushed to the left forcing it to find a different path to reach the goal. In this way the computation time is strongly reduced because crossed the obstacle the aircraft can go straight to the target.

On the other hand KA* with wind and A* look for a way to reach the goal crossing the obstacle to the right. This is due to the $H$ component in the cost function that drives the

| Map dimension | | | |
|---|---|---|---|
| | X | 500 | [m] |
| | Y | 500 | [m] |
| | Z | 80 | [m] |
| | ΔXY | 1 | [m] |
| | ΔXZ | 1 | [m] |
| | | | |
| Obstacles dimension | | | |
| | X | 300 | [m] |
| | Y | 125 | [m] |
| | Z | 50 | [m] |
| | | | |
| Obstacles-center position | | | |
| | X | 300 | [m] |
| | Y | 250 | [m] |
| | | | |
| Starting point | | | |
| | X | 20 | [m] |
| | Y | 20 | [m] |
| | Z | 40 | [m] |
| | | | |
| Target point | | | |
| | X | 450 | [m] |
| | Y | 450 | [m] |
| | Z | 50 | [m] |
| | | | |
| Aircraft parameters | | | |
| | Speed | 10 | [m/s] |
| | Min turning radius | 25 | [m] |
| | Max climbing angle | 4 | [deg] |
| | | | |
| Optimization parameters | | | |
| | Time step | 2 | [s] |
| | α | 10 | |
| | β | 1 | |
| | | | |
| Obstacles separation | | | |
| | Horizontal | 15 | [m] |
| | Vertical | 10 | [m] |
| | | | |
| Wind | | | |
| | X | -2 | [m/s] |
| | Y | 2 | [m/s] |

| | | Z | 0 | [m/s] |
|---|---|---|---|---|
| Computation time | | | | |
| | KA* with Wind | | 0.4854 | [s] |
| | KA* without Wind | | 32.476 | [s] |
| | A* | | 6.5682 | [s] |
| Path length | | | | |
| | KA* with Wind | | 689 | [m] |
| | KA* without Wind | | 796 | [m] |
| | A* | | 776 | [m] |
| WayPoints | | | | |
| | KA* with Wind | | 37 | |
| | KA* without Wind | | 41 | |
| | A* | | 698 | |

Table 3. One-obstacle test parameters.

search along the diagonal between the start and the target point. In this way the algorithms look for the optimal path following the diagonal up to meeting the obstacle. Then the search continues choosing to turn on the right because in that direction the $F$-value is decreasing. Because of this process the computation time is higher and also the covered distance is more then the one with wind.

Analyzing this behavior an important limit of this optimization technique comes out: greedy algorithms become slow when the optimum search does not provide a continuing monotonic decrease of the cost function. Because of this tendency this first version of KA* must be improved in order to accelerate the convergence to the optimal solution in cases where the continuing descent to the minim is not guaranteed.

**Figure 15** shows on the Longitude-Latitude plane the different paths planned in this test. In this case, the post processing phase for the A* path would be less intrusive because of the slight altitude variation and of the few heading changes, but the 90 degrees heading change on the right of the obstacle is clearly unfeasible. This is evident comparing the turning radius planned by KA* with the sharp angle planned with A*. Some of these sharp heading changes can be easily corrected with a smoothing algorithm in post processing, but some of them can require a complete waypoints reallocation (as shown in the previous example). The important advantage of KA* is to generate a feasible path respecting the basic aircraft kinematic constraints with low computation workload.

**Figure 16** again has on the $X$-axis the distance covered from start to target and on the $Y$-axis the aircraft altitude. For this test the altitude changes are smother then the one in the previous test, but here also it is possible to see the stepwise approach to climb of the A* path compared with the smooth climbing maneuver planned with KA*. About altitude variations the relation with the integration time step must be mentioned. Because of the slower behavior of an aircraft to altitude variations with respect to heading changes, when KA* is used on environments requiring strong altitude variations to avoid the obstacles, the integration time must be increased. The integration time provides to the algorithm the capability to forecast the possible movements from the current state to the next. Then, if the aircraft has to climb to avoid an obstacle flying above it, a longer integration horizon is needed in order to plan in time the climbing maneuver reducing the computation time.

In **Figure 17** the turn rate (related to the $u$ command) is plotted. In this case the two command sequences cannot be compared because of the different paths followed. Anyway it is possible to see the strong turning rate imposed to the aircraft reaching the bottom-right corner of the obstacle. Reaching that corner the aircraft has to turn in order to go toward the target respecting the obstacle separation constrains, than strong heading changes are needed. In order to limit turning radiuses and climbing angles along the path and generate smooth and flyable trajectories for the aircraft the $u$ and $w$ command vectors provided to the model to generate the possible movements are limited to half of the maximum turning rate and climbing angle. Finally **Figure 18** shows the climb angle (related to the $w$ command) time history as for the previous test. Here the climb angle is always small for both the paths and the aircraft climbs slowly to the target altitude.

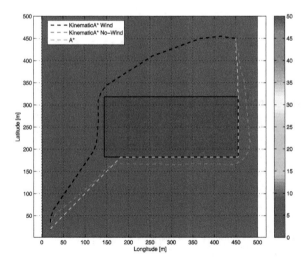

Fig. 15. Longitude-Latitude view (one obstacle test).

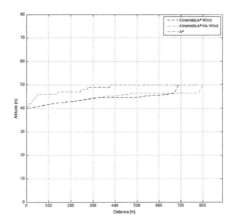

Fig. 16. Distance-Altitude view (one obstacle test).

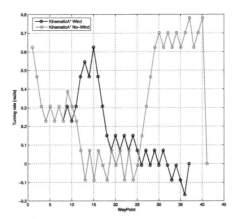

Fig. 17. Turn rate (one obstacle test).

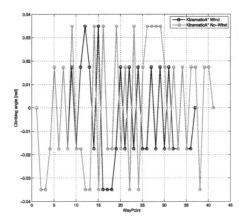

Fig. 18. Climb angle (one obstacle test).

# 6. Conclusions and future work

Kinematic A* has been developed to fill the relation gap between the aircraft kinematic constraints and the logics used in graph search approaches to find the optimal path. The simple aircraft model generates the state transitions in the state space and drives the search toward feasible directions. This approach has been tested on several cases. The algorithm generates feasible paths respecting limits on vehicle turning rates and climbing angles but the tests evidence also some issues that have to be investigated to improve the algorithm.

As a matter of fact the following conclusion can be taken:

- The goal to obtain paths respecting the basic aircraft kinematic constraints has been reached. KA* generates smooth and safe paths respecting the imposed constraints. No sharp heading changes or strong altitude variations typical of classical graph search paths are shown. Tests put in evidence that some paths obtained with classical graph search algorithms cannot be adapted in post processing with waypoints reallocation, to reach the aircraft kinematic constraints. Some heading changes or altitude variations affect deeply the whole path and a complete re-planning is needed when these unfeasible trajectories are included in the full path. This point addresses the development of an algorithm that like KA* may be able to match the graph search logics with the aircraft kinematic constraints.

- Obstacles separation represents an important improvement with respect to other graph search solutions. In classic graph search formulations obstacles separation was implemented just imposing to skip a given amount of cells around the obstacles whether the algorithm should try to expand them. In KA* the obstacle separation is more elegant; the algorithm skips the states with positions too close to the obstacles modifying accordingly the full planned path.

- Introducing the model to expand the states and perform the graph search is the fundamental novelty introduced with KA*, but is paid with an algorithm increased complexity. A longer computation time was expected but several tests demonstrated just slight time increases to obtain the solution. This is important to save the merit of low computation effort characteristic of graph search algorithms.

- Tests show another important KA* merit: the lower waypoints number on the path generated by KA* with respect to A*. KA* algorithm naturally generates just the amount of waypoints needed to reach the goal and because of this the waypoints filtering and reallocation needed for A* can be skipped.

- In spite of the simplified wind model, the effects of this disturbance are hardly relevant also for the preliminary tests reported in this paper. Wind modifies the state space and forces the algorithm to obtain solutions very different from the one without wind. Because of this further and deeper investigations are required to better understand this problem and improve accordingly the implementation.

- Analyzing heading changes and altitude variations needed to follow the KA* paths it is evident the strong effect that wind has on the path following performances. Paths with wind require harder heading and altitude variations pushing the aircraft to reach its limits in order to follow the path. This aspect shall be taken into account in the following work in order to study deeper the problem and find modifications on the cost function that can improve the algorithm performances.

- The second test puts in evidence an issue still present in KA*: the optimality of the paths planned. The wind effect on the model drives the states expansion along directions that are not taken into account otherwise. This way a shorter and computationally lighter solution is found. This means that the algorithm search must be improved modifying the cost function in order to investigate possible optimality proofs for the generated path. However this is a hard task because the graph search logics in it self makes optimality proof possible just for limited and simple problem formulations.
- Another issue outlined by the tests that must be analyzed in the following work is the exponential increase of computational time when the algorithm cannot follow a monotonic cost function decrease along the states expansion. Part of the problem is due to the graph structure and needs a deep state space analysis to be improved. On the other side, the cost function then needs to be modified, investigating solutions able to drive differently the graph search.
- Finally tests show that the different aircraft behavior on longitudinal with respect to lateral-directional plane affects largely the model time step selection. Particularly, longer time steps are needed when strong altitude variations are required to cross the obstacles. As a consequence, the time step must be tuned in order to improve the algorithm performances according with the test cases.

## 7. References

Bellman, R. (1958). On a Routing Problem. *Quarterly of Applied Mathematics*, Vol. 16, No 1, pp. 87–90.

Bertuccelli, L.F. & How, J.P. (2005). Robust UAV Search for Environmentas with Imprecise Probability Maps. *Proceedings of IEEE Conference of Decision and Control*, Seville, Dec 2005.

Boissonnat, J.D., Cèrèzo, A. & Leblond, J. (1994). Shortest Paths of Bounded Curvature in the Plane. *Journal of Intelligent and Robotic Systems*, Vol. 11, No. 1-2.

Carroll, D.L. (1996). Chemical Laser Modeling With Genetic Algoritms. *AIAA Journal*, Vol. 34, No.2, pp.338-346.

Chandler, P.R., Rasmussen, S. & Patcher, M. (2000). UAV Cooperative Path Planning. *Proceedings of AIAA Guidance, Navigation and Control Conference*, Denver, USA.

Chitsaz, H., LaValle, S.M. (2007). Time-optimal Paths for a Dubins airplane. *Proceedings of IEEE Conference on Decision and Control*, New Orleans, USA.

De Filippis, L., Guglieri, G., Quagliotti, F. (2009). Flight Analysis and Design for Mini-UAVs. *Proceedings of XX AIDAA Congress*, Milano, Italy.

De Filippis, L., Guglieri, G., Quagliotti, F. (2010). A minimum risk approach for path planning of UAVs. *Journal of Intelligent and Robotic Systems*, Springer, pp. 203-222.

De Filippis, L., Guglieri, G. & Quagliotti, F. (2011). Path Planning strategies for UAVs in 3D environments. *Journal of Intelligent and Robotic Systems*, Springer, pp. 1-18.

Dijkstra, E. W. (1959). A note to two problems in connexion with graphs. *Numerische Mathematik*, Vol:1, pp. 269–271.

Dogan, A. (2003). Probabilistic path planning for UAVs. *proceeding of 2nd AIAA Unmanned Unlimited Systems, Technologies, and Operations - Aerospace, Land, and Sea Conference and Workshop & Exhibition*, San Diego, California, September 15-18.

Dubins, L.E. (1957). On Curves of Minimal Length With a Constraint on Average Curvature and with Prescribed Initial and Terminal Positions and Tangents. *American Journal of Mathematics*, No. 79.

Ferguson, D. & Stentz, A. (2006). Using interpolation to improve path planning: The Field D* algorithm. *Journal of Field Robotics*, 23(2), 79-101.

Floyd, R. W. (1962). Algorithm 97: Shortest Path. *Communications of the ACM*, 5(6), pp. 345.

Ford, L. R. Jr. & Fulkerson, D. R. (1962). *Flows in Networks*. Princeton University Press.

Hart, P., Nilsson, N. & Raphael, B. (1968). A formal basis for the heuristic determination of minimum cost paths. *IEEE Transactions on Systems Science and Cybernetics*, SCC-4(2), pp. 100-107.

Horner, D., P. & Healey, A., J. (2004). Use of artificial potential fields for UAV guidance and optimization of WLAN communications. *Autonomous Underwater Vehicles*, 2004 IEEE/OES, vol., no., pp. 88- 95, 17-18.

Jun, M. & D'Andrea, R. (2002). Path Planning for Unmanned Aerial Vehicles in Uncertain and Adversarial Environments. *Models, Applications and Algorithms*, Kluwer Academic Press.

Koenig, S. & Likhachev, M. (2001). Incremental A*. *Proceedings of the Natural Information Processing Systems*.

Koenig, S. & Likhachev, M. (2002). D* Lite. *Proceedings of the AAAI Conference on Artificial Intelligence*, pp. 476-483.

LaValle, S. M. (2006). *Planning Algorithms*. Cambridge University Press.

Ma, X. & Castanon, D.A. (2006). Receding Horizon Planning for Dubins Traveling Salesman Problems. *Proceedings of IEEE Conference on Decision and Control*, San Diego, USA.

Miele, A. & Pritchard, R.E. (1969). Gradient Methods in Control Theory. Part 2 – Sequential Gradient-Restoration Algorithm. *Aero-Astronautics Report* ,n° 62, Rice University.

Miele, A. (1970). Gradient Methods in Control Theory. Part 6 – Combined Gradient-Restoration Algorithm. *Aero-Astronautics Report*, n° 74, Rice University.

Nash, A., Daniel, K., Koenig, S. & Felner, A. (2007). Theta*: Any-angle path planning on grids. *Proceedings of the AAAI Conference on Artificial Intelligence*, pp. 1177-1183.

Nikolos, I.K., Tsourveloudis, N.C. & Valavanis, K.P. (2003). Evolutionary Algorithm Based Offline/Online Path Planner for UAV Navigation. *IEEE Transactions on Systems, Man and Cybernetics - part B: Cybernetics*, Vol. 33, No. 6.

Papaefthymiou, M. & Rodriguez., J. (1991). Implementing Parallel Shortest-Paths Algorithms. *DIMACS Series in Discrete Mathematics and Theoretical Computer Science*.

Pfeiffer, B., Batta, R., Klamroth, K. & Nagi, R. (2008). Path Planning for UAVs in the Presence of Threat Zones Using Probabilistic Modelling. In: *Handbook of Military Industrial Engineering*, Taylor and Francis, USA.

Reeds, J.A & Shepp, L.A. (1990). Optimal Path for a Car That Goes Both Forwards and Backwards. *Pacific Journal of Mathematics*, Vol. 145, No. 2.

Stentz, A. (1993). Optimal and efficient path planning for unknown and dynamic environments. *Carnegie Mellon Robotics Institute Technical Report*, CMU-RI-TR-93-20.

Stentz, A. (1995). The focussed D* algorithm for real-time replanning. *Proceedings of the International Joint Conference on Artificial Intelligence*, pp.1652-1659.

Sussmann, H.J. & Tang, W. (1991). Shortest Paths for the Reeds-Shepp Car: a Worked Out Example of the Use of Geometric Technique in Nonlinear Optimal Control. *Report SYCON-91-10*, Rutgers University.

Warshall, S. (1962). A theorem on Boolean matrices. *Journal of the ACM*, 9(1), pp. 11–12.

Waydo, S. & Murray, R.M. (2003). Vehicle Motion Planning Using Stream Functions. *Proceedings of 2003 IEEE International Conference on Robotics and Automation*, September.

# 9

# GNSS Carrier Phase-Based Attitude Determination

Gabriele Giorgi[1] and Peter J. G. Teunissen[2,3]

[1] *Technische Universität München*
[2] *Curtin University of Technology*
[3] *Delft University of Technology*
[1]*Germany*
[2]*Australia*
[3]*The Netherlands*

## 1. Introduction

The GNSS (Global Navigation Satellite Systems) are a valid aid in support of the aeronautic science. GNSS technology has been successfully implemented in aircraft design, in order to provide accurate position, velocity and heading estimations. Although it does not yet comply with aviation integrity requirements, GNSS-based aircraft navigation is one of the alternative means to traditional dead-reckoning systems. It can provide fast, accurate, and driftless positioning solutions. Additionally, ground-based GNSS receivers may be employed to aid navigation in critical applications, such as precision approaches and landings.

One of the main issues in airborne navigation is the determination of the aircraft attitude, i.e., the orientation of the aircraft with respect to a defined reference system. Many sensors and technologies are available to estimate the attitude of a aircraft, but there is a growing interest in GNSS-based attitude determination (AD), often integrated at various levels of tightness to other types of sensors, typically Inertial Measurements Units (IMU). Although the accuracy of a stand-alone GNSS attitude system might not be comparable with the one obtainable with other modern attitude sensors, a GNSS-based system presents several advantages. It is inherently driftless, a GNSS receiver has low power consumption, it requires minor maintenance, and it is not as expensive as other high-precision systems, such as laser gyroscopes.

GNSS-based AD employs a number of antennas rigidly mounted on the aircraft's structure, as depicted in Figure 1. The orientation of each of the baselines formed between the antennas is determined by computing their relative positions. The use of GNSS carrier phase signals enables very precise range measurements, which can then be related to angular estimations. However, carrier phase measurements are affected by unknown integer ambiguities, since only their fractional part is measured by the receiver. The process of reconstructing the number of whole cycles from a set of measurements affected by errors goes under the name of ambiguity resolution (AR). Only after these ambiguities are correctly resolved to their correct integer values, will reliable baseline measurements and attitude estimations become available. This chapter focuses on novel AR and AD methods. Recent advances in GNSS-based attitude determination have demonstrated that the two problems can be formulated in an integrated manner, i.e., aircraft attitude and the phase ambiguities can be considered as the unknown parameters of a common ambiguity-attitude estimation method. In this integrated approach, the AR and AD problems are solved together by means of the theory of Constrained Integer Least-Squares (C-ILS). This theory extends the well-known least-squares theory (LS), by having geometrical constraints as well as integer constraints imposed on parameter subsets. The novel AR-AD estimation problem is discussed and its various properties are analyzed. The method's complexity is addressed by presenting new numerical algorithms that largely reduce the required processing load. The main objective of this chapter is to provide evidence that:

- GNSS carrier-phase based attitude determination is a viable alternative to existing attitude sensors

- Employing the new ambiguity-attitude estimation method enhances ambiguity resolution performance

- The new method can be implemented such that it is suitable for real-time applications

The structure of this contribution is as follows. Section 2 gives the observation and stochastic model which cast the set of GNSS observations, with special focus on the derivation of the GNSS-based attitude model. Section 3 reviews the most common attitude parameterization and estimation methods, mainly focusing on those widely used in aviation applications. Section 4 introduces a new ambiguity-attitude estimation method, which enhances the existing approach for attitude determination using GNSS signals. Section 5 presents flight-test results, which provide practical evidence of the novel method's performance. Finally, section 6 draws several conclusions.

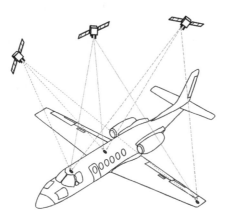

Fig. 1. GNSS data collected on multiple antennas installed on the fuselage and wings allow the estimation of an aircraft's orientation (attitude).

## 2. The GNSS-based attitude model

A GNSS receiver works by tracking satellites in view and storing the data received. Each GNSS satellite broadcasts a coded message with information about its orbit, the time of transmission, and few other parameters necessary for the correct processing at receiver side (Misra & Enge, 2001). By collecting signals from three or more satellites a GNSS receiver determines its own position with a triangulation procedure, exploiting the knowledge about both the satellites positions and the slant distance (range) by each satellite in view. The range measurements are obtained by detecting the time of arrival of the signal, from which the range can be inferred. This measurement is affected by several error sources: the satellite and receiver clocks are not perfectly synchronized; the signal travels through the atmosphere, which causes delays; the direct signal may be affected by unwanted reflections (multipath) that cannot be perfectly eliminated by careful antenna design. If not properly modeled, each of these effects will limit the achievable GNSS accuracy. The observed pseudorange or code observable is therefore modeled as

$$\begin{aligned}
P^s_{r,f}(t) = {} & \rho^s_r(t, t - \tau^s_r) + I^s_{r,f} + T^s_r + dm^s_{r,f} + c\left[dt_r(t) - dt^s(t - \tau^s_r)\right] \\
& + c\left[d_{r,f}(t) + d^s_f(t - \tau^s_r)\right] + \varepsilon^s_{P,r,f}
\end{aligned} \tag{1}$$

where the superscript $s$ indicates the satellite and the subscripts $r$ and $f$ indicate the receiver and the frequency, respectively. The different terms are:

| | |
|---|---|
| $P$ | code observation [m] |
| $\tau$ | signal travel time [s] |
| $\rho$ | geometrical distance between receiver and satellite [m] |
| $I$, $T$ | ionospheric and tropospheric delays [m] |
| $dm$ | multipath error [m] |
| $c$ | speed of light : 299 792 458 $[\frac{m}{s}]$ |
| $dt$ | clock errors [s] |
| $d$ | instrumental delays [s] |
| $\varepsilon_P$ | remaining unmodeled code errors [m] |

The magnitude of errors involved in these observations - decimeter or meter level - would not allow high-precision applications, such AD, which require cm- or mm-level accuracy in the final positioning product. Therefore, another set of observations is considered: the phase of the tracked signal, modeled as

$$\begin{aligned}
\Phi^s_{r,f}(t) = {} & \rho^s_r(t, t - \tau^s_r) - I^s_{r,f} + T^s_r + \delta m^s_{r,f} + c\left[dt_r(t) - dt^s(t - \tau^s_r)\right] \\
& + c\left[\delta_{r,f}(t) + \delta^s_f(t - \tau^s_r)\right] + \lambda_f[\varphi^s_{r,f}(t_0) - \varphi^s_f(t_0)] + \lambda_f z^s_{r,f} + \varepsilon^s_{\Phi,r,f}
\end{aligned} \tag{2}$$

with $\varphi$ the phase of the generated carrier signal (original or replica) in cycles, $t_0$ the time of reference for phase synchronization, and $\lambda_f$ the wavelength of frequency $f$. The phase reading is characterized by different atmospheric delays (the ionosphere causes an anticipation of phase instead of a delay), different instrumental biases (indicated with $\delta$), different multipath and an additional bias which is represented by the unknown number of whole cycles that cannot be detected by the tracking loop, since only the fractional part is measured. These are the integer ambiguities $z$. In case of GNSS, the precision of the phase measurements far exceeds the one of code observations: typically the phase observable is two orders of magnitude more accurate than the code measurement.

The many sources of error in (1) and (2) can be mitigated in relative positioning models. First, we form the so-called single difference (SD) code and carrier phase observations by taking the differences between observations simultaneously collected at two antennas tracking the same satellite:

$$P^s_{r_2,f}(t) - P^s_{r_1,f}(t) = P^s_{r_{12},f} = \rho^s_{r_{12},f} + I^s_{r_{12},f} + T^s_{r_{12}} + dm^s_{r_{12},f} + cdt_{r_{12}} + cd_{r_{12},f} + \varepsilon^s_{P,r_{12},f}$$

$$\Phi^s_{r_2,f}(t) - \Phi^s_{r_1,f}(t) = \Phi^s_{r_{12},f} = \rho^s_{r_{12},f} - I^s_{r_{12},f} + T^s_{r_{12}} + \delta m^s_{r_{12},f} + cdt_{r_{12}} + c\delta_{r_{12},f} + \lambda_f \varphi^s_{r_{12},f}(t_0)$$
$$+ \lambda_f z^s_{r_{12},f} + \varepsilon^s_{\Phi,r_{12},f}$$

$$\tag{3}$$

where subscript $r_{12}$ indicates the difference between two antennas: $\Box_{r_{12}} = \Box_{r_2} - \Box_{r_1}$. The phase value $\varphi^s_f(t_0)$, relative to the common satellite, is eliminated. The instrumental delays and clock errors of the satellite are usually considered constant over short time spans, since the travel time difference with respect to any two points on the Earth surface is small (Teunissen & Kleusberg, 1998).

The terms $cdt_{r_{12}}$, $cd_{r_{12},f}$ and $\delta_{r_{12},f}$ refer to the relative clock errors and relative instrumental delays between the two receivers. A perfect synchronization between receivers implies the cancellation of the clock biases, and a correct calibration would reduce the impact of instrumental delays. In the case of a single receiver connected to two antennas, these two sources of relative error could cancel out with a proper calibration.

The receiver clock errors and hardware delays in the single difference equations (3) are common for all the satellites tracked at the same frequency. Therefore these terms can be eliminated by forming a double difference (DD) combination, obtained by subtracting two SD measurements from two different satellites:

$$P^{s_{12}}_{r_{12},f} = \rho^{s_{12}}_{r_{12},f} + I^{s_{12}}_{r_{12},f} + T^{s_{12}}_{r_{12}} + dm^{s_{12}}_{r_{12},f} + \varepsilon^{s_{12}}_{P,r_{12},f}$$

$$\tag{4}$$

$$\Phi^{s_{12}}_{r_{12},f} = \rho^{s_{12}}_{r_{12},f} - I^{s_{12}}_{r_{12},f} + T^{s_{12}}_{r_{12}} + \delta m^{s_{12}}_{r_{12},f} + \lambda_f z^{s_{12}}_{r_{12},f} + \varepsilon^{s_{12}}_{\Phi,r_{12},f}$$

It has been assumed that the real-valued initial phase of the receiver replica does not vary for different tracked GNSS satellites.

The differential atmospheric delays depend on the distance between antennas. For sufficiently short baselines - typically shorter than a kilometer - the signals received by the antennas have traveled approximately the same path, thus the atmospheric delays becomes highly correlated. The differencing operation makes these errors negligible with respect to the measurement white noise for the baselines typically employed in AD applications, which rarely exceeds a few hundred meters.

Note that the relation between observations and baseline coordinates is nonlinear, since these are contained in the range term

$$\rho^s_r = \|r^s(t_r - \tau^s_r) - r_r(t_r)\| \tag{5}$$

with $r^s$ and $r_r$ the satellite and receiver antenna position vectors, respectively. By assuming the atmospheric delays negligible and applying the Taylor expansion to expression (4) one obtains the linearized relations

$$\triangle P^{S_{12}}_{r_{12},f} = -(u^{S_{12}}_r)^T \triangle r_{12} + \varepsilon^{S_{12}}_{P,r_{12},f}$$
$$\triangle \Phi^{S_{12}}_{r_{12},f} = -(u^{S_{12}}_r)^T \triangle r_{12} + \lambda_f z^{S_{12}}_{r_{12},f} + \varepsilon^{S_{12}}_{\Phi,r_{12},f} \tag{6}$$

where the observables are now 'observed minus computed' terms, and the unknowns are expressed as increments with respect to a computed approximate value. $\triangle r_{12}$ is the baseline vector - the difference between the absolute antennas positions - whereas $u^{S_{12}}_r = u^{S_2}_r - u^{S_1}_r$ is the difference between unit line-of-sight vectors of different satellites. Also note that the multipath terms have been lumped into the remaining unmodeled errors $\varepsilon^{S_{12}}_{P,r_{12},f}$ and $\varepsilon^{S_{12}}_{\Phi,r_{12},f}$.

Consider now two antennas simultaneously tracking the same $m + 1$ satellites at $N$ frequencies. The vector of DD observations of type (6) are cast in the linear(ized) functional model (Teunissen & Kleusberg, 1998)

$$y = Az + Gb + \varepsilon \quad ; \quad z \in \mathbb{Z}^{mN}, \quad b \in \mathbb{R}^3 \tag{7}$$

with $y$ the $2mN$-vector of code and carrier phase observations, $z$ the unknown integer-valued ambiguities and $b$ the vector of real-valued baseline coordinates. $A$ and $G$ are the design matrices

$$A = \begin{bmatrix} 0 \\ \Lambda \end{bmatrix} \otimes I_m \qquad G = e_N \otimes \begin{bmatrix} U \\ U \end{bmatrix} \tag{8}$$

with $\Lambda$ the diagonal matrix of $N$ carrier wavelengths and $U$ the $m \times 3$ matrix of DD unit line-of-sight vectors. Symbol $\otimes$ denotes the Kronecker product (Van Loan, 2000).

Model (7) describes the linear relationship between GNSS observables and the parameters of the two antennas. However, a single baseline is generally not sufficient to estimate the full orientation of an aircraft with respect to a given reference frame. At least three non-aligned antennas are necessary to guarantee that each rotation of the aircraft can be tracked unambiguously. It is straightforward to generalize the model formulation (7) to cast $n$ DD baseline observations, obtained with $n + 1$ GNSS antennas (Teunissen, 2007a):

$$Y = AZ + GB + \Xi \quad ; \quad Z \in \mathbb{Z}^{mN \times n}, \quad B \in \mathbb{R}^{3 \times n} \tag{9}$$

This formulation is obtained by casting the observations at each baseline in the columns of the $2mN \times n$ matrix $Y$. Consequently, $Z = [z_1, \ldots, z_n]$ is the matrix whose $n$ columns are the integer ambiguity $mN$-vectors, and $B = [b_1, \ldots, b_n]$ is the $3 \times n$ matrix that contains the $n$ real-valued baseline vectors. We exploited here once again the short baseline hypothesis: the same matrix of line-of-sight vectors $U$ is used for all baselines.

Besides describing the functional relationship between observables and unknowns, a proper modeling should also capture the observation noise, i.e., the measurement error. The error is relative to the receiver, to the satellite, to the frequency and to the type of observations (code or phase). The variance-covariance (v-c) matrix of a vector of DD observations $y$ collected at baseline $b$ will be denoted as $D(y) = Q_{yy}$, with $D(\cdot)$ the dispersion operator. For the multibaseline model (9), the description of measurement errors requires a further step: the observations are cast into a $2mNn$ vector by applying the *vec* operator, which stacks the columns of a matrix. The v-c matrix $Q_{YY}$ that characterizes the error statistic of $vec(Y)$ is

$$D(vec(Y)) = Q_{YY} \tag{10}$$

A simple expression for $Q_{YY}$ is obtained by assuming that each of the baselines is characterized by the same v-c matrix $Q_{yy}$:

$$D(vec(Y)) = Q_{YY} = P_n \otimes Q_{yy} \tag{11}$$

with $P_n$ the $n \times n$ matrix that takes care of the correlation which is introduced by having a common antenna:

$$P_n = \frac{1}{2} [I_n + e_n e_n^T] = \begin{bmatrix} 1 & 0.5 & \cdots & 0.5 \\ 0.5 & 1 & & \vdots \\ \vdots & & \ddots & 0.5 \\ 0.5 & \cdots & 0.5 & 1 \end{bmatrix} \tag{12}$$

Expressions (9) and (11) define the *GNSS multibaseline model* that we use in this contribution as the foundation of our GNSS-based attitude estimation theory.

With the available code and phase observations it is possible to estimate the set of baseline coordinates. These can then be used to provide the aircraft attitude, but *only* when a further condition is realized: the positions of the antennas installed aboard the given aircraft are known, rigid and do not change over time (or, if change occurs, it is perfectly known and predictable). This is so because it is necessary to have a one-to-one relationship between aircraft attitude and baselines attitude. As an example, consider two antennas mounted on the two extremities of a flexible mast: it is not possible to separate the rotations of the mast from its deformations by only observing the variations of the mutual position between the two antennas.

The rigidity assumption is formalized in the following way. Consider two orthonormal frames, defined by the basis $\{u_1, u_2, u_3\}$ and $\{u_1', u_2', u_3'\}$. Let us assume that the second frame is integrally fixed with the aircraft. An arbitrary vector $x$ can be equivalently described by using either reference system:

$$\begin{aligned} x &= \left(x^T u_1\right) u_1 + \left(x^T u_2\right) u_2 + \left(x^T u_3\right) u_3 \\ x' &= \left(x^T u_1'\right) u_1' + \left(x^T u_2'\right) u_2' + \left(x^T u_3'\right) u_3' \end{aligned} \tag{13}$$

The relation between the components of vectors $x$ and $x'$ is completely defined by the mutual orientation of the two reference systems. The linear transformation $x = Rx'$ allows for a one-to-one relationship. Matrix $R$, hereafter referred to as *rotation matrix* or *attitude matrix*, belongs to the class of orthonormal matrices $O$: its column vectors $r_i$ are normal and their product null: $r_i^T r_j = \delta_{i,j}$, with $\delta_{i,j}$ the Kronecker's delta ($\delta_{i,j} = 1$ if $i = j$, 0 otherwise). These constraints are necessary for the admissibility of transformation $x = Rx'$. In absence of deformations, the scalar product between any two vectors should be invariant with respect to the transformation:

$$x'^T y' = x^T R^T R y = x^T y \tag{14}$$

whereas the vectorial product is invariant under rotations about the axis defined by $x' \times y'$:

$$x' \times y' = (Rx) \times (Ry) = |R| R (x \times y) \tag{15}$$

These conditions are fulfilled for orthonormal rotation matrices with determinant equal to one.

Model (9) can then be reformulated by means of the linear transformation $B = RF$, where $F$ is used to cast the set of known local baseline coordinates and $R$ is the orthonormal ($R^T R = I_q$) matrix that rotates $B$ into $F$. The complete GNSS attitude model reads then (Teunissen, 2007a; 2011)

$$Y = AZ + GRF + \Xi;$$
$$Z \in \mathbb{Z}^{mN \times n},$$
$$R \in \mathbb{O}^{3 \times q}$$
$$D(vec(Y)) = Q_{YY}$$

(16)

Parameter $q$ is introduced in order to make model (16) of general applicability. The $n$ baselines may be aligned or coplanar, impeding the estimation of a full $3 \times 3$ matrix $R$. Therefore, $q$ defines the span of matrix $F$. For baseline sets formed by aligning $n + 1$ antennas we set $q = 1$, whereas configurations of coplanar antennas are defined by $q = 2$. With four or more non-coplanar antennas, $q = 3$.

The GNSS attitude model (16) is a nonlinear model. Although the relation between observables and unknowns remain linear, the orthonormal constraint is of a nonlinear nature, and profoundly affects the estimation process. This is investigated in section 4. First, the following section gives an overview of common attitude parameterization and estimation methods.

## 3. Attitude parameterization and estimation

The orthonormality of $R$ ($R^T R = I_q$) imposes $\frac{q(q+1)}{2}$ constraints on its components $r_{ij}$. The full matrix $R$ can then be parameterized with a properly chosen set of variables, whose number can be as little as two (if $q = 1$) or three (if $q \leq 2$). To this purpose, several representations may be used, and few are briefly reviewed in the following.

From a set of code and phase observations cast as in (16), the problem of extracting the components of the attitude representation involves, as shown in section 4, a nonlinear least squares problem. Its formulation and solution are the second topic discussed in this section.

### 3.1 Attitude parameterization

Several attitude parameterizations are available in the literature, see e.g., Shuster (1993) and references therein. The most common parameterizations are briefly reviewed in the following.

### 3.1.1 Direction cosine matrix

The transformation between two basis of orthonormal frames reads

$$\{u'_1, u'_2, u'_3\} = R\{u_1, u_2, u_3\} \implies u'_i = \sum_{j=1}^{3} r_{ij} u_j$$

(17)

with $r_{ij}$ the entries of $R$. The scalar product between any two unit vectors of the two frames is

$$u_i'^T u_j = \sum_{k=1}^{3} r_{ik}\left(u_k^T u_j\right) = r_{ij} = \cos\left(\widehat{u_i' u_j}\right) \tag{18}$$

Hence, the attitude matrix can be expressed by nine direction cosines, i.e., the nine cosines of the angles formed by the three unit vectors of the first frame and the three unit vectors of the second frame:

$$R = \begin{bmatrix} u_1'^T u_1 & u_1'^T u_2 & u_1'^T u_3 \\ u_2'^T u_1 & u_2'^T u_2 & u_2'^T u_3 \\ u_3'^T u_1 & u_3'^T u_2 & u_3'^T u_3 \end{bmatrix} \tag{19}$$

This representation fully defines the mutual orientation of the two frames, by using a set of nine parameters (see Figure 2). Each configuration can be described without incurring any singularity, at the cost of having a larger number of parameters than other representations.

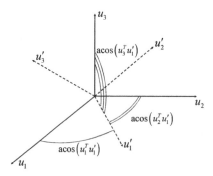

Fig. 2. The main axis $u_1'$ is completely defined by the knowledge of the three direction cosines $u_1^T u_1'$, $u_2^T u_1'$ and $u_3^T u_1'$.

### 3.1.2 Euler angles

Consider counterclockwise rotations about one of the main axis of a frame $\{u_1, u_2, u_3\}$. Then the rotation matrix $R$ is obtained through one of the following expressions:

$$R(u_1, \phi) = \begin{bmatrix} 1 & 0 & 0 \\ 0 & C_\phi & S_\phi \\ 0 & -S_\phi & C_\phi \end{bmatrix}$$

$$R(u_2, \phi) = \begin{bmatrix} C_\phi & 0 & -S_\phi \\ 0 & 1 & 0 \\ S_\phi & 0 & C_\phi \end{bmatrix} \tag{20}$$

$$R(u_3, \phi) = \begin{bmatrix} C_\phi & S_\phi & 0 \\ -S_\phi & C_\phi & 0 \\ 0 & 0 & 1 \end{bmatrix}$$

Any arbitrary rotation can always be decomposed as a combination of three consecutive rotations about the main axis $u_1$, $u_2$ or $u_3$, represented by one of the relations in (20). Figure 3 shows the example of a 321 rotation: the first rotation is about the third main axis $u_3$ with magnitude $\psi$, the second is about the (new) second main axis $u_2'$ with magnitude $\theta$, the last about the (new) first main axis $u_1''$ with magnitude $\phi$. The rotation matrix that defines the transformation between the frames $\{u_1, u_2, u_3\}$ and $\{u_1''', u_2''', u_3'''\}$ is built as

$$\{u_1, u_2, u_3\} \underset{R(u_3,\psi)}{\overset{\psi}{\Longrightarrow}} \{u_1', u_2', u_3'\} \underset{R(u_2,\theta)}{\overset{\theta}{\Longrightarrow}} \{u_1'', u_2'', u_3''\} \underset{R(u_1,\phi)}{\overset{\phi}{\Longrightarrow}} \{u_1''', u_2''', u_3'''\} \tag{21}$$

Therefore, $R_{321}(\psi, \theta, \phi) = R(u_1, \phi) R(u_2, \theta) R(u_3, \psi)$. Twelve combinations of rotations are possible, whose choice depends on the application. As an example, the sequence 321 is commonly used to describe the orientation of an aircraft, where the angles $\psi, \theta, \phi$ are named heading, elevation and bank, respectively.

It is easy to see that the Euler angles representation is not unique: e.g, the combination 321 is equivalently expressed as $R_{321}(\psi, \theta, \phi)$ or $R_{321}(\psi + \pi, \pi - \theta, \phi + \pi)$. This ambiguity is usually avoided by imposing $-90° < \theta \leq 90°$. The main advantage of the Euler angles representation is its straightforward physical interpretation, of importance for human-machine interfaces. The disadvantage lies in fact that the construction of the attitude matrix requires the evaluation of trigonometric functions, of higher computational load than other parameterizations. Also, the derivatives of the components of the rotation matrix are nonlinear (trigonometric), and affected by singularities.

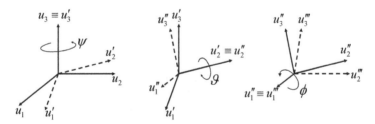

Fig. 3. The three consecutive rotations that rotate the frame $\{u_1, u_2, u_3\}$ into the frame $\{u_1''', u_2''', u_3'''\}$. The first one is about the main axis $u_3$ and magnitude $\psi$, the second is about the main axis $u_2'$ and magnitude $\theta$ and the third is about the main axis $u_1''$ and magnitude $\phi$.

### 3.1.3 Quaternions

A quaternion is an order-4 vector whose components can be used to define the mutual rotation between reference systems:

$$\bar{q} = (q_1, q_2, q_3, q_4)^T \tag{22}$$

$q_4$ is named the scalar (real) component of the quaternion, whereas $(q_1, q_2, q_3)^T$ forms the imaginary (or vectorial) part. The components of a quaternion must respect the constraint $\bar{q}^T \bar{q} = 1$. Physically, the four components of $\bar{q}$ define the magnitude and axis of the rotation necessary to rotate one reference system into the other, see Figure 4. The attitude matrix $R$ is parameterized in terms of quaternions as

$$R(\bar{q}) = R(q, q_4) = \left(q_4^2 - \|q\|^2\right) I_3 + 2qq^T + 2q_4 \left[q^+\right]$$

$$= \begin{bmatrix} q_1^2 - q_2^2 - q_3^2 + q_4^2 & 2(q_1q_2 + q_3q_4) & 2(q_1q_3 - q_2q_4) \\ 2(q_1q_2 - q_3q_4) & -q_1^2 + q_2^2 - q_3^2 + q_4^2 & 2(q_2q_3 + q_1q_4) \\ 2(q_1q_3 + q_2q_4) & 2(q_2q_3 - q_1q_4) & -q_1^2 - q_2^2 + q_3^2 + q_4^2 \end{bmatrix} \quad (23)$$

with

$$[q^+] = \begin{bmatrix} 0 & q_3 & -q_2 \\ -q_3 & 0 & q_1 \\ q_2 & -q_1 & 0 \end{bmatrix} \quad (24)$$

This parameterization is non ambiguous and it does not involve any trigonometric function, so that the computational burden is lower than with other representations. The quaternion representation is of common use in attitude estimation and control applications, since it guarantees high numerical robustness.

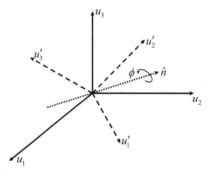

Fig. 4. The frame $\{u_1', u_2', u_3'\}$ can be rotated to equal the orientation of frame $\{u_1, u_2, u_3\}$ by means of a single rotation of magnitude $\phi$ about axis $\hat{n}$. The four components of a quaternion are proportional to the entries of the normal vector $\hat{n}$ and to the magnitude $\phi$.

### 3.2 Attitude estimation

As it will be shown in the next sections, the least-squares solution of model (16), requires the solution of a constrained least-squares problem of the type:

$$\check{R} = \arg \min_{R \in \mathcal{O}^{3 \times q}} \left\| vec\left(\hat{R} - R\right) \right\|_Q^2 \quad (25)$$

with $\|\cdot\|_Q^2 = (\cdot)^T Q^{-1} (\cdot)$. The shape of $Q$ drives the choice of the solution technique to be adopted for solving (25).

If $Q$ is a diagonal matrix, problem (25) becomes an Orthogonal Procrustes Problem (OPP), see Schonemann (1966). This class of constrained least-squares problem have been thoroughly analyzed, and fast algorithms have been devised to quickly extract the minimizer $\check{R}$, see (Davenport, 1968; Shuster & Oh, 1981). Various fast methods for the solution of an OPP have been introduced - and widely used in practice - based on the Singular Value Decomposition

(SVD) or the EIGenvalues decomposition (EIG), such as the QUaternion ESTimator (QUEST) (Shuster, 1978; Shuster & Oh, 1981), the Fast Optimal Attitude Matrix (FOAM) (Markley & Landis, 1993), the EStimator of the Optimal Quaternion (ESOQ) (Mortari, 1997) or the Second ESOQ (ESOQ2) (Mortari, 2000) algorithms, which have been extensively compared in Markley & Mortari (1999; 2000) and Cheng & Shuster (2007).

For nondiagonal matrices $Q$, the extraction of the orthonormal attitude matrix $\check{R}$ has to be performed through nonlinear estimation techniques. A first numerical scheme for the solution of (25) is derived by applying the Lagrangian multipliers method. The Lagrangian function is

$$L(R) = vec\left(\hat{R} - R\right)^T Q^{-1} vec\left(\hat{R} - R\right) - tr\left[[\lambda]_q \left[R^T R - I_q\right]\right] \qquad (26)$$

with $[\lambda]_q$ the $q$ by $q$ matrix of Lagrangian multipliers:

$$[\lambda]_1 = \lambda \quad ; \quad [\lambda]_2 = \begin{bmatrix} \lambda_1 & \frac{1}{2}\lambda_3 \\ \frac{1}{2}\lambda_3 & \lambda_2 \end{bmatrix} \quad ; \quad [\lambda]_3 = \begin{bmatrix} \lambda_1 & \frac{1}{2}\lambda_4 & \frac{1}{2}\lambda_5 \\ \frac{1}{2}\lambda_4 & \lambda_2 & \frac{1}{2}\lambda_6 \\ \frac{1}{2}\lambda_5 & \frac{1}{2}\lambda_6 & \lambda_3 \end{bmatrix} \qquad (27)$$

The last term of (26) gives the $\frac{q(q+1)}{2}$ constraining functions that follows from the orthonormality of $R$: $q$ constraints are given by the normality (unit length) of the columns of $R$, whereas $\frac{q(q-1)}{2}$ constraints are given by the orthogonality of the columns of $R$.

The gradient of the Lagrangian function (26), together with the $\frac{q(q+1)}{2}$ constraining functions, defines the nonlinear system to be solved:

$$\begin{cases} \frac{1}{2}\nabla L(R) = \left[Q^{-1} - [\lambda]_q \otimes I_3\right] vec\,(R) - Q^{-1} vec\left(\hat{R}\right) = 0 \\ vec\left(R^T R - I_q\right) = 0 \end{cases} \qquad (28)$$

Due to the symmetry of matrix $\left[R^T R - I_q\right]$, only its upper (or lower) triangular part has to be considered in (28). The Newton-Raphson method can then be applied to iteratively converge to the sought orthonormal matrix of rotations.

This method is computationally heavier than other iterative schemes, since it requires the explicit computation of larger-sized matrices than other methods given in the following.

A second viable solution scheme is obtained by re-parameterizing the attitude matrix with the vector of Euler angles $\mu = (\psi, \theta, \phi)^T$. Following the reparameterization, matrix $R(\mu)$ implicitly fulfills the constraint $R^T R = I_q$, and problem (25) is rewritten as

$$\check{\mu} = \arg\min_{\mu \in \mathbb{R}^3} \|h(\mu)\|_I^2 \qquad (29)$$

with $h(\mu) = Q^{-\frac{1}{2}} vec\left(\hat{R} - R(\mu)\right)$. The nonlinear least-squares problem (29) is solved by applying iterative methods, e.g., the Newton method. This approach (Euler angles parameterization) works with a minimal set of unknowns - the Euler angles - and it can quickly converge to the sought minimizer if an accurate initial guess is used. The disadvantage is that trigonometric functions have to be evaluated, increasing the computational load.

A third viable approach is devised by employing the quaternions parameterization of $R$ and to solve for (25):

$$\breve{q} = \arg \min_{q \in \mathbb{R}^4, \|q\|=1} \left\| vec \left( \hat{R} - R(\bar{q}) \right) \right\|_Q^2 \tag{30}$$

The orthonormality of $R$ is guaranteed by the normality of the quaternion: this introduces a single constraint in the minimization problem (30). A Lagrangian function is formed as

$$L'(\bar{q}) = vec \left( \hat{R} - R(\bar{q}) \right)^T Q^{-1} vec \left( \hat{R} - R(\bar{q}) \right) - \lambda \left( \bar{q}^T \bar{q} - 1 \right) \tag{31}$$

and the (nonlinear) system to be solved is

$$\begin{cases} \frac{1}{2} \nabla L'(R(\bar{q})) = J_{R(\bar{q})}^T Q^{-1} vec \left( \hat{R} - R(\bar{q}) \right) - \lambda \bar{q} = 0 \\ \bar{q}^T \bar{q} - 1 = 0 \end{cases} \tag{32}$$

with $J_{R(\bar{q})}$ the Jacobian of $vec(R(\bar{q}))$.

The three iterative solutions given above rigorously solve for problem (25), but are generally slower than the methods available for diagonal $Q$ matrices (SVD, EIG, QUEST, FOAM, ESOQ, and ESOQ2). Figure 5a illustrates the mean number of floating-point operations for different attitude estimation methods, per number of baselines employed. $10^4$ samples $\hat{R}$ have been generated via Monte Carlo simulations for a given fully-populated $Q$ matrix. The gray bars span between the maximum and minimum numbers obtained for each algorithm. The off-diagonal elements of $Q$ are disregarded when applying the SVD, EIG, QUEST, FOAM, ESOQ, and ESOQ2 methods. These techniques outperform each iterative method: the number of required floating-point operations is generally two to three orders of magnitude lower. Among the iterative methods, the Lagrangian multiplier technique generally requires the highest number of operations, making it the least efficient method, while the Euler angle method and the Quaternion parameterization provide better overall results. Figure 5b shows the corresponding mean, maximum and minimum computational times marked during the simulations. The Lagrangian parameterization method generally takes the longest time to converge, whereas the quaternion and Euler angle methods show better results. Note that higher number of floating operations does not directly translate into longer computational times, because modern processor architectures efficiently operate by means of multi-threading and parallel processing.

## 4. Reliable attitude-ambiguity estimation methods

This section reviews the solution of the GNSS attitude model (16). This can be presented by addressing two consecutive steps: float estimation and ambiguity resolution.

### 4.1 Float ambiguity-attitude solution

We indicate with *float* the solution of (16) obtained by disregarding the whole set of constraints, i.e., the integerness of $Z$ and the orthonormality of $R$:

$$\{\hat{Z}, \hat{R}\} = \arg \min_{Z \in \mathbb{R}^{mN \times n}, R \in \mathbb{R}^{3 \times q}} \left\| vec \left( Y - AZ - GRF \right) \right\|_{Q_{YY}}^2 \tag{33}$$

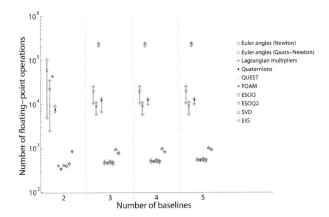

(a) Floating point operations.

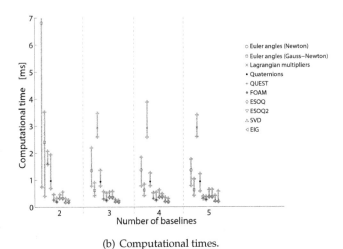

(b) Computational times.

Fig. 5. Mean, maximum and minimum numbers of floating-point operations (left) and computational times (right) per number of baseline, for each of the attitude estimation method analyzed.

This float solution follows from solving the system of normal equations

$$
M \begin{pmatrix} vec(\hat{R}) \\ vec(\hat{Z}) \end{pmatrix} = \begin{bmatrix} FP_n^{-1} \otimes G^T Q_{yy}^{-1} \\ P_n^{-1} \otimes A^T Q_{yy}^{-1} \end{bmatrix} vec(Y)
$$

$$
M = \begin{bmatrix} FP_n^{-1}F^T \otimes G^T Q_{yy}^{-1}G & FP_n^{-1} \otimes G^T Q_{yy}^{-1}A \\ P_n^{-1}F^T \otimes A^T Q_{yy}^{-1}G & P_n^{-1} \otimes A^T Q_{yy}^{-1}A \end{bmatrix}
$$

(34)

where the v-c matrix $Q_{YY}$ is written as in (11). Inversion of the normal matrix $M$ gives the v-c matrix of the float estimators $\hat{R}$ and $\hat{Z}$:

$$\begin{bmatrix} Q_{\hat{R}\hat{R}} & Q_{\hat{R}\hat{Z}} \\ Q_{\hat{Z}\hat{R}} & Q_{\hat{Z}\hat{Z}} \end{bmatrix} = M^{-1} \tag{35}$$

The float estimators are explicitly derived as

$$\hat{R} = \left[\overline{G}^T Q_{yy}^{-1} \overline{G}\right]^{-1} \overline{G}^T Q_{yy}^{-1} Y P_n^{-1} F^T \left[F P_n^{-1} F^T\right]^{-1} \tag{36}$$
$$\hat{Z} = \left[A^T Q_{yy}^{-1} A\right]^{-1} A^T Q_{yy}^{-1} \left[Y - \overline{G}\hat{R}F\right]$$

with $\overline{G} = \left[I - A\left[A^T Q_{yy}^{-1} A\right]^{-1} A^T Q_{yy}^{-1}\right] G$. Next to the above float solution, we can also define the following *conditional* float solution for the attitude matrix:

$$\hat{R}(Z) = \arg \min_{R \in \mathbb{R}^{3 \times q}} \|vec\,(Y - AZ - GRF)\|_{Q_{YY}}^2 \tag{37}$$

In this case the ambiguity matrix is assumed completely known. The solution $\hat{R}(Z)$ can be computed form the float solutions $\hat{R}$ and $\hat{Z}$ as:

$$vec(\hat{R}(Z)) = vec(\hat{R}) - Q_{\hat{R}\hat{Z}} Q_{\hat{Z}\hat{Z}}^{-1} vec(\hat{Z} - Z) \tag{38}$$

Application of the variance propagation law gives

$$Q_{\hat{R}(Z)\hat{R}(Z)} = Q_{\hat{R}\hat{R}} - Q_{\hat{R}\hat{Z}} Q_{\hat{Z}\hat{Z}}^{-1} Q_{\hat{Z}\hat{R}} = \left[F P_n^{-1} F^T\right]^{-1} \otimes \left[G^T Q_{yy}^{-1} G\right]^{-1} \tag{39}$$

There is a very large difference in the precision of the float solution $\hat{R}$ and the precision of the conditional float solution $\hat{R}(Z)$. This can be demonstrated by comparing expression (39) with $Q_{\hat{R}\hat{R}}$ in (35), whose relation with the design matrices can be made explicit as

$$Q_{\hat{R}\hat{R}} = \left[F P_n^{-1} F^T\right]^{-1} \otimes \left[\overline{G}^T Q_{yy}^{-1} \overline{G}\right]^{-1} \tag{40}$$

Matrix $\left[G^T Q_{yy}^{-1} G\right]^{-1}$ is characterized by much smaller entries than $\left[\overline{G}^T Q_{yy}^{-1} \overline{G}\right]^{-1}$. This is demonstrated as follows. Matrices $A$, $G$ and $Q_{yy}$ may be partitioned as

$$A = \begin{bmatrix} 0 \\ \Lambda \end{bmatrix} \otimes I_m \qquad G = e_N \otimes \begin{bmatrix} u \\ u \end{bmatrix} \qquad Q_{yy} = I_N \otimes \begin{bmatrix} \sigma_P^2 Q & 0 \\ 0 & \sigma_\Phi^2 Q \end{bmatrix} \tag{41}$$

where we assumed, for simplicity, the same code and phase standard deviations for each observation, independent from the combination of satellites, receivers and frequency. $\Lambda$ is the diagonal matrix of carrier wavelengths, whereas $Q$ is the matrix that introduces correlation due to the DD operation.

It follows that

$$\left[\overline{G}^T Q_{yy}^{-1} \overline{G}\right]^{-1} = \frac{\sigma_P^2}{N} \left[U^T Q^{-1} U\right]^{-1}$$
$$\left[G^T Q_{yy}^{-1} G\right]^{-1} = \frac{1}{N} \frac{\sigma_\Phi^2}{\frac{\sigma_\Phi^2}{\sigma_P^2}+1} \left[U^T Q^{-1} U\right]^{-1} \approx \frac{\sigma_\Phi^2}{N} \left[U^T Q^{-1} U\right]^{-1} \tag{42}$$

The ratio between the entries of matrix $Q_{\hat{R}\hat{R}}$ and $Q_{\hat{R}(Z)\hat{R}(Z)}$ is then proportional to the ratio $\frac{\sigma_\Phi^2}{\sigma_P^2}$. In GNSS applications, this phase-code variance ratio is in the order of $10^{-4}$. This clearly demonstrates the importance of ambiguity resolution: if we can integer-estimate $Z$ with sufficiently high probability, then the attitude matrix $R$ can be estimated with a precision that is comparable with the high precision of $\hat{R}(Z)$.

## 4.2 Ambiguity resolution

The second step consists of the resolution of the carrier phase integer ambiguities. The solution of model (16) is obtained through the following C-ILS minimization problem:

$$\{\check{Z}, \check{R}\} = \arg \min_{Z \in \mathbb{Z}^{mN \times n}, R \in \mathbb{O}^{3 \times q}} \|vec\,(Y - AZ - GRF)\|_{Q_{YY}}^2 \tag{43}$$

Both sets of constraints are now imposed: the matrix of ambiguities $\check{Z}$ is integer valued and the matrix $\check{R}$ belongs to the class of $3 \times q$ orthonormal matrices $\mathbb{O}^{3 \times q}$. The C-ILS solution $\check{Z}$ can be computed from the float solutions as (Teunissen & Kleusberg, 1998):

$$\check{Z} = \arg \min_{Z \in \mathbb{Z}^{mN \times n}} \underbrace{\left( \|vec(\hat{Z} - Z)\|_{Q_{\hat{Z}\hat{Z}}}^2 + \|vec(\hat{R}(Z) - \check{R}(Z))\|_{Q_{\hat{R}(Z)\hat{R}(Z)}}^2 \right)}_{C(Z)} \tag{44}$$

with

$$\check{R}(Z) = \arg \min_{R \in \mathbb{O}^{3 \times q}} \|vec(\hat{R}(Z) - R)\|_{Q_{\hat{R}(Z)\hat{R}(Z)}}^2 \tag{45}$$

The cost function $C(Z)$ is the sum of two terms. The first weighs the distance between a candidate integer matrix $Z$ and the float solution $\hat{Z}$, weighted by the v-c matrix $Q_{\hat{Z}\hat{Z}}$. The second weighs the distance between the conditional (on the candidate $Z$) attitude matrix $\hat{R}(Z)$ and the orthonormal matrix $\check{R}(Z)$ that follows from the solution of (45). Therefore, the computation of cost function $C(Z)$ also involves a term that weighs the distance of the conditional attitude matrix from its orthogonal projection. This second term greatly aids the search for the correct ambiguities: integer candidates $Z$ that produce matrices $\hat{R}(Z)$ too far from their orthonormal projection contribute to a much higher value of the cost function.

Since the minimization problem (44) is not solvable analytically due to the integer nature of the parameter involved, an extensive search in a subset of the space of integer matrices $\mathbb{Z}^{mN \times n}$ has to be performed. The definition of an efficient and fast solution scheme for problem (44) is not a trivial task. In order to highlight the intricacies of such formulation, we first give an approximate solution, obtained by neglecting the orthonormal constraint.

### 4.2.1 The LAMBDA method

Consider first the integer minimization problem (44) without the orthonormality constraint on $R$. Then the second term of $C(Z)$ reduces to zero and the integer minimization problem becomes

$$\check{Z}^u = \arg \min_{Z \in \mathbb{Z}^{mN \times n}} \|vec(\hat{Z} - Z)\|_{Q_{\hat{Z}\hat{Z}}}^2 \tag{46}$$

This is the usual approach of doing GNSS integer ambiguity resolution. Due to the absence of the orthonormality constraint on $R$ one may expect lower success rates, i.e., lower probability of identifying the correct ambiguity matrix $Z$. However, the ILS problem (46) is of lower complexity than (44) and a very fast implementation of it is available: the LAMBDA (Least-squares AMBiguity Decorrelation Adjustment) (Teunissen, 1995) method, see, e.g., Boon & Ambrosius (1997); Cox & Brading (2000); Huang et al. (2009); Ji et al. (2007); Kroes et al. (2005). It consists of two steps, namely decorrelation and search.

The integer minimizer has to be extensively searched within a subset of the whole space of integers:

$$\Omega^u\left(\chi^2\right) = \{Z \in \mathbb{Z}^{mN \times n} \mid \|vec(\hat{Z} - Z)\|_{Q_{\hat{Z}\hat{Z}}}^2 \le \chi^2\} \tag{47}$$

$\Omega^u$ is the so-called search space, a region of the space of integer matrices that contains only those candidates $Z$ for which the squared norm (46) is bounded by the value $\chi^2$. This can be set by choosing an integer matrix $Z_c$ and taking $\chi^2 = \|vec(\hat{Z} - Z_c)\|_{Q_{\hat{Z}\hat{Z}}}^2$. Rounding the float solution, $Z_c = [\hat{Z}]$, is an option, as well as bootstrapping an integer matrix, as in Teunissen (2000; 2007b).

Searching for the integer minimizer in $\Omega^u$ proves inefficient due to the weight matrix $Q_{\hat{Z}\hat{Z}}$. Geometrically, the search space defines a hyperellipsoid centered in $\hat{Z}$ and whose shape and orientation are driven by the entries of matrix $Q_{\hat{Z}\hat{Z}}$. The difficulty of the search lies in the fact that the search space is highly elongated, as detailed in Teunissen & Kleusberg (1998). The reason is that the ambiguities are highly correlated. While the set wherein the independent ambiguities (e.g., three ambiguities for a single baseline scenario) can be chosen is rather large, the set of admissible values for the remaining ambiguities is very small. This causes major halting problems during the search, since many times the selected subset of independent ambiguities does not yield admissible integer matrix candidates. This issue is tackled and solved in the LAMBDA method with a decorrelation step. The decorrelation of matrix $Q_{\hat{Z}\hat{Z}}$ is achieved by an admissible transformation matrix $T$. In order to preserve the integerness, such matrix has to fulfill the following two conditions: $T$ as well as its inverse $T^{-1}$ need to have integer entries. The matrix of transformed ambiguities $Z'$ and corresponding v-c matrix are then obtained as

$$Z' = TZ \quad ; \quad Q_{\hat{Z}'\hat{Z}'} = TQ_{\hat{Z}\hat{Z}}T^T \tag{48}$$

The decorrelation procedure is described in Teunissen & Kleusberg (1998). The v-c matrix is iteratively decorrelated by a sequence of admissible transformations $T_i$, until matrix

$$Q_{\hat{Z}'\hat{Z}'} = \left(\prod_i T_i\right) Q_{\hat{Z}\hat{Z}} \left(\prod_i T_i\right)^T = TQ_{\hat{Z}\hat{Z}}T^T \tag{49}$$

cannot be further decorrelated. Note that due to the integer conditions on $T$, a full decorrelation cannot generally be achieved. Figure 6 shows three steps of the decorrelation process for a two-dimensional example. Figure 6a shows the original (elongated) ellipse associated to $Q_{\hat{Z}\hat{Z}}$, Figure 6b shows an intermediate decorrelation step, and Figure 6c shows the final decorrelated search space.

After the decorrelation step, the actual search is performed by operating the $LDL^T$ factorization of matrix $Q_{\hat{Z}'\hat{Z}'}$, so that the quadratic form in (46) can be written as a

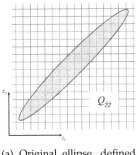

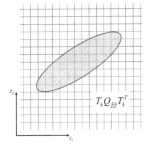

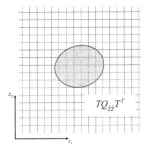

(a) Original ellipse, defined by $Q_{\hat{z}\hat{z}}$.

(b) Intermediate decorrelated ellipse, defined by $T_k Q_{\hat{z}\hat{z}} T_k^T$.

(c) Final decorrelated ellipse, defined by $T Q_{\hat{z}\hat{z}} T^T$.

Fig. 6. Initial, intermediate and decorrelated search space defined by the (transformed) v-c matrix of the ambiguities.

summation:

$$\left\| vec(\hat{Z}' - Z') \right\|^2_{Q_{\hat{z}'\hat{z}'}} = \left\| vec(\hat{Z}' - Z') \right\|^2_{LDL^T} = \sum_{i=1}^{mNn} \frac{\left( \hat{z}'_{i|I} - z'_i \right)^2}{\sigma^2_{i|I}} \leq \chi^2 \tag{50}$$

where the scalars $\hat{z}'_{i|I}$ and $\sigma^2_{i|I}$ are the conditional float ambiguity estimator and its corresponding conditional variance, respectively. These are conditioned to the previous $I = 1, \ldots, i-1$ values, and directly follow from the entries of matrices $L$ and $D$. More details on the way the search is actually performed can be found in de Jonge & Tiberius (1996).

Due to the decorrelation step, the extensive search for the integer minimizer $\check{Z}^u$ is performed quickly and efficiently, making the LAMBDA method perfectly suitable for real-time applications.

### 4.2.2 The MC-LAMBDA method

The MC-LAMBDA method is an extension of the LAMBDA method that applies to the geometrically-constrained problem (44). The MC-LAMBDA method shares the same working principle of the LAMBDA method: first the search space is decorrelated, then the search for the integer minimizer is performed. However, an extensive search within a (decorrelated) search space is generally not efficient as it is with the LAMBDA method, as explained in the following.

The search space is now defined as

$$\Omega^c \left( \chi^2 \right) = \{ Z \in \mathbb{Z}^{mN \times n} \mid C(Z) \leq \chi^2 \} \tag{51}$$

The cost function $C(Z)$ takes, for the same candidate $Z$, much larger values than the first quadratic term in (44), due to the matrix $Q_{\hat{R}(Z)\hat{R}(Z)}$, whose inverse has entries two orders of magnitude larger than the entries of $Q_{\hat{z}\hat{z}}^{-1}$ (Giorgi et al., 2011; Teunissen, 2007a). For this reason it is not trivial to set a proper value of $\chi^2$, since the cost function $C(Z)$ is highly sensitive to the choice of $Z$ (Giorgi, 2011; Giorgi et al., 2011). This problem becomes more marked for weaker models (single frequency, low number of satellites tracked, high noise levels). Obviously, larger values of $\chi^2$ imply longer computational times due to the larger number of candidates to be evaluated. Also, the constrained least-squares problem (45) has to be solved for each of the integer candidates in $\Omega^c\left(\chi^2\right)$, thereby further increasing the computational load.

The aforementioned issues are solved with a novel numerical efficient search scheme for the solution of (44). This is achieved by employing easier-to-evaluate bounding functions and introducing new search algorithms.

First, consider two functions, $C_1(Z)$ and $C_2(Z)$, that satisfy the following inequalities:

$$C_1(Z) \le C(Z) \le C_2(Z) \tag{52}$$

These functions provide a lower and an upper bound for the cost function $C(Z)$. The choice for these bounding functions is driven by two requirements: their evaluation should be less time consuming than the evaluation of $C(Z)$, and each bound should be sufficiently tight. Several alternatives have been studied in (Giorgi, 2011; Giorgi et al., 2012; Nadarajah et al., 2011; Teunissen, 2007a;c), based on

-   the eigenvalues of matrix $Q_{\hat{R}(Z)\hat{R}(Z)}^{-1}$

-   the analytical solution of Wahba's problem (Wahba, 1965)

-   a tighter geometrical bound based on Procustes problem (Schonemann, 1966)

-   a QR factorization (Gram-Schmidt process)

For example, the first method listed exploits the inequalities $\xi_m \|\cdot\|_I^2 \le \|\cdot\|_Q^2 \le \xi_M \|\cdot\|_I^2$, with $\xi_m$ and $\xi_M$ the smallest and largest eigenvalues of $Q_{\hat{R}(Z)\hat{R}(Z)}^{-1}$, respectively. After some manipulation, the two bounding functions read

$$C_1(Z) = \left\|vec(\hat{Z} - Z)\right\|_{Q_{\hat{Z}\hat{Z}}}^2 + \xi_m \sum_{i=1}^{q} \left(\|\hat{r}_i(Z)\| - 1\right)^2$$

$$\tag{53}$$

$$C_2(Z) = \left\|vec(\hat{Z} - Z)\right\|_{Q_{\hat{Z}\hat{Z}}}^2 + \xi_M \sum_{i=1}^{q} \left(\|\hat{r}_i(Z)\| + 1\right)^2$$

where $\hat{r}_i(Z)$ are the column vectors of $\hat{R}(Z)$.

Two efficient search methods have been developed to reduce the computational burden associated to an extensive search. Independently from the bounding functions used, these novel search schemes allow for a quick minimization of $C(Z)$.

Consider first the lower bound $C_1(Z)$. The search space associated to $C_1(Z)$ is

$$\Omega_1\left(\chi^2\right) = \{Z \in \mathbb{Z}^{mN \times n} \mid C_1(Z) \le \chi^2\} \supset \Omega^c\left(\chi^2\right) \tag{54}$$

Obviously, the search space $\Omega^c\left(\chi^2\right)$ is contained within $\Omega_1\left(\chi^2\right)$. One may proceed, for example, by choosing $\chi^2 = \left\|vec(\hat{Z} - Z')\right\|_{Q_{\hat{Z}\hat{Z}}}^2$ with $Z'$ a given integer matrix (both rounding the float solution and bootstrapping an integer matrix are viable choices). Then, we can enumerate all the integers matrices contained in $\Omega_1\left(\chi^2\right)$ and compute $C(Z)$ for each candidate (if any, since set $\Omega_1\left(\chi^2\right)$ may also turn out empty), in order to also evaluate $\Omega^c\left(\chi^2\right)$. If this set turns out non-empty, then one has simply to extract the minimizer $\check{Z}$ by sorting the integer matrices according to the values of $C(Z)$. However, there is no guarantee that $\Omega^c\left(\chi^2\right)$ is non-empty. If the search space $\Omega^c\left(\chi^2\right)$ is empty, the size of $\Omega_1\left(\chi^2\right)$ is increased and the process repeated iteratively until the minimizer $\check{Z}$ is found. This search scheme, illustrated with the flow chart in Figure 7, is named *Expansion* approach, since the size of the search space is iteratively 'expanded'.

An alternative approach is devised by considering the upper bound $C_2(Z)$. Its search space is

$$\Omega_2\left(\chi^2\right) = \{Z \in \mathbb{Z}^{mN \times n} \mid C_2(Z) \le \chi^2\} \subset \Omega^c\left(\chi^2\right) \tag{55}$$

which is contained in the set $\Omega^c\left(\chi^2\right)$. Consider the following iterative procedure. First, the scalar $\chi^2$ is set such that it guarantees the non-emptiness of $\Omega_2\left(\chi^2\right)$, and therefore $\Omega^c\left(\chi^2\right)$ is non-empty either. This can be done by choosing $\chi^2 = C_2(Z')$ for an integer matrix $Z'$, which can be the rounded float solution, a bootstrapped solution, or an integer matrix obtained by other means (see for further options Giorgi et al. (2008)). Then, the search proceeds by looking for an integer candidate in the set $\Omega_2\left(\chi^2\right)$, aiming to find a matrix $Z_1$ that provides a smaller value for the upper bound $C_2(Z_1) = \chi_1^2 < \chi^2$. When it is found, the set is shrunk to $\Omega_2\left(\chi_1^2\right)$ and the search continues by looking for another integer candidate $Z_2$ capable of reducing the value $C_2(Z_2) = \chi_2^2 < \chi_1^2$. This process is repeated until the minimizer of $C_2(Z)$, say $\check{Z}_2$, is found. Since this may differ from the minimizer of $C(Z)$, the search space $\Omega^c\left(\bar{\chi}^2\right)$, with $\bar{\chi}^2 = C_2(\check{Z}_2)$, is evaluated and the sought-for integer minimizer $\check{Z}$ extracted. This iterative search scheme is named *Search and Shrink* approach, and it is detailed in the flow chart of Figure 8.

Both the *Expansion* and the *Search and Shrink* approaches implement the search for integer minimizer (44) in a fast and efficient way, such that the algorithm can be used for real-time applications.

The MC-LAMBDA method achieves very high success rates. The success rate is defined as the probability of providing the correct set of integer ambiguities. The inclusion of geometrical constraints, which follow from the a priori knowledge of the antennas relative positions aboard the aircraft, largely aids the ambiguity resolution process, allowing for higher success rates in weaker models, such as with the single-frequency and/or high measurement noise scenarios. These performance improvements associated to the MC-LAMBDA method with respect to classical methods (such as the LAMBDA) are analyzed in the following section with actual data collected during two different flights tests.

## 5. Flight test results

The performance of the MC-LAMBDA method is analyzed with data collected on two flight-tests performed with a Cessna Citation jet aircraft. The aircraft attitude is extracted from unaided, single-epoch, single-frequency ($N = 1$) GNSS observations, in order to demonstrate the method capabilities in the most challenging scenario, i.e., stand-alone, high observation noise and low measurements redundancy. Also, single-epoch performance is extremely important for dynamic platforms, where a quick recovery from changes of tracked satellites, cycle slips and losses of lock is necessary to avoid undesired loss of guidance. The

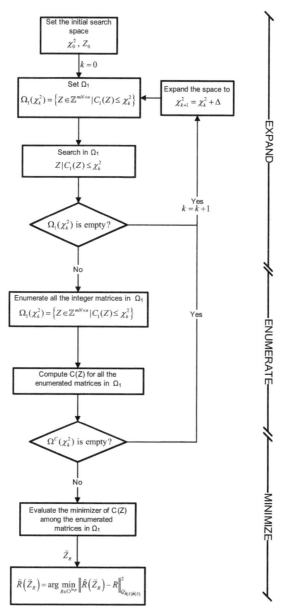

Fig. 7. The *Expansion* approach: flow chart.

single-frequency case is of interest for many aerospace applications, where limits on weight and power consumption must often be respected.

In both tests the same receiver (Septentrio PolaRx2@) was connected to three antennas, placed on the middle of the fuselage, on the wing and on the nose (see Figure 9). In the first

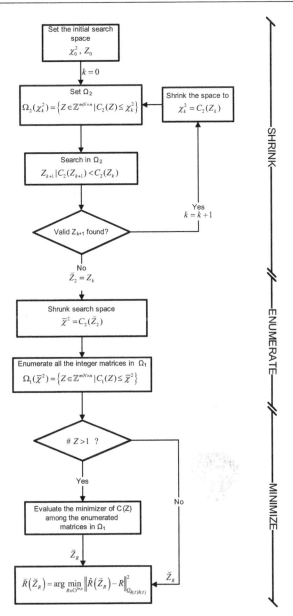

Fig. 8. The *Search and Shrink* approach: flow chart.

test analyzed $(T - I)$ the nose antenna was placed on the extremity of a boom, whereas in second test $(T - II)$ it was directly placed on the aircraft body. The two tests largely differ by the flight dynamic. Test $T - I$ was conducted with aggressive maneuvering and few zero-gravity parabolas, whereas $T - II$ was performed as part of a gravimetry campaign, with very few smooth maneuvers, as shown in Figure 10. During test $T - II$, the aircraft

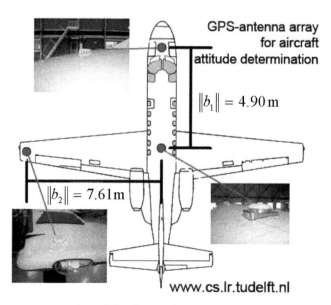

Fig. 9. The antennas set-up onboard the Cessna Citation II.

was also equipped with an Inertial Navigation System (INS), whose output is used to test the GNSS-based attitude estimation accuracy. Figure 11 reports the number of tracked satellites for the duration of the two tests. The PDOP (Precision Dilution of Precision) is also shown.

The matrix of local body-frame baseline coordinates for the two tests are

$$F_{T-I} = \begin{bmatrix} 5.45 & -0.34 \\ 0 & 7.60 \end{bmatrix} \text{ [m]} \qquad F_{T-II} = \begin{bmatrix} 4.90 & -0.39 \\ 0 & 7.60 \end{bmatrix} \text{ [m]} \tag{56}$$

The receiver collected GPS-L1 data for about 6000 epochs (zero cut-off angle 1Hz sampling), between 11:42 and 13:20 UTC, 2nd June 2005 on the first test, and 15000 epochs (zero cut-off angle, 1 Hz sampling), between 11:00 and 14:23 UTC, 1st November 2007 on the second test.

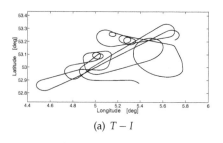
(a) $T - I$

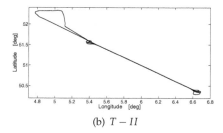
(b) $T - II$

Fig. 10. Ground traces of the two test flights.

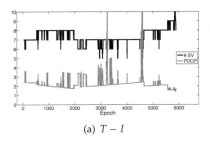

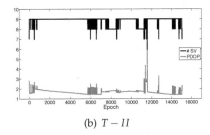

(a) $T - I$                                              (b) $T - II$

Fig. 11. Number of satellites tracked and corresponding PDOP values.

## 5.1 Instantaneous ambiguity resolution

The success rate marked by the LAMBDA and MC-LAMBDA methods applied to both flight tests is reported in Table 1 (Giorgi et al., 2011). The single-epoch performance of the

| $T - I$ | | $T - II$ | |
|---|---|---|---|
| LAMBDA | MC-LAMBDA | LAMBDA | MC-LAMBDA |
| 5.8 | 81.5 | 24.7 | 88.1 |

Table 1. $T - I$ and $T - II$ tests: unaided single-epoch, single-frequency success rate (%) for the LAMBDA and the MC-LAMBDA methods, two-baseline processing.

unconstrained method are rather unsatisfactory. The correct set of integer ambiguities is resolved only for 5.8% of time in test $T - I$ and in 24.7% of time in test $T - II$. The difference is due to the higher number of satellites tracked in the second test.

Instead, application of the MC-LAMBDA method yields a strong performance improvement. The constrained method is capable of providing the correct integer solution for more than 80% of the epochs in test $T - I$ and more than 88% in test $T - II$.

Both airborne tests confirm the very large improvement that is obtained by strengthening the underlying model with the inclusion of geometrical constraints. It is stressed that all the ambiguity resolution performance reported are obtained by processing the GNSS signals without any a priori information or assumption about the attitude or the aircraft motion. Also, mask angles, elevation-dependent models, dynamic models or any kind of filtering are not applied.

## 5.2 Attitude determination

| | | |
|---|---|---|
| Heading | $\sigma_\psi$ [deg] | 0.07 |
| Elevation | $\sigma_\theta$ [deg] | 0.20 |
| Bank | $\sigma_\phi$ [deg] | 0.12 |

Table 2. $T - II$ test: standard deviations of the differences between GPS and INS attitude angles output.

High single-epoch success rates yield precise epoch-by-epoch attitude solutions for the larger part of the flights duration. The attitude angles based on the correctly fixed integer ambiguities in test $T - I$ are shown in Figure 12. The high dynamics of the flight is evident from the steep variations of the attitude. In particular, Figure 13 shows a zero-gravity maneuver: the aircraft promptly pitched up, gained some altitude, and performed an ample arc to create a virtual absence of gravity on board.

Figure 14 shows the GNSS-based attitude angles for the test $T - II$. The INS solutions are also reported in the figures, in order to provide a comparison between the two systems. Table 2 reports the standard deviations of the differences between the INS and GNSS-based attitude estimations. Taking the precise INS output as benchmark solution, it can be inferred that the accuracy obtained is within the expected range, given the baseline lengths employed. The heading angle is estimated with the highest precision, whereas the elevation estimation is characterized by the highest noise levels. This is due to the relative geometry of the antennas and to the fact that the vertical components of the GNSS-based baseline estimations are inherently less accurate that the horizontal components. The bank angle is estimated with higher precision than the elevation angle, being driven by the longer baseline $Body - Wing$.

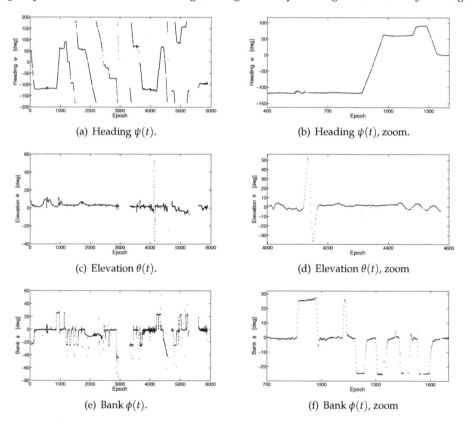

(a) Heading $\psi(t)$.

(b) Heading $\psi(t)$, zoom.

(c) Elevation $\theta(t)$.

(d) Elevation $\theta(t)$, zoom

(e) Bank $\phi(t)$.

(f) Bank $\phi(t)$, zoom

Fig. 12. $T - I$ test: time series of the three attitude angles as estimated via GNSS. On the right, a closer look at the estimates.

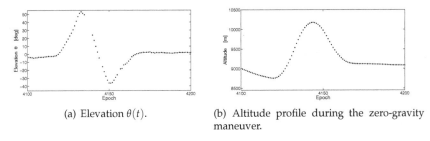

(a) Elevation $\theta(t)$.

(b) Altitude profile during the zero-gravity maneuver.

Fig. 13. $T - I$ test: zero-gravity maneuver.

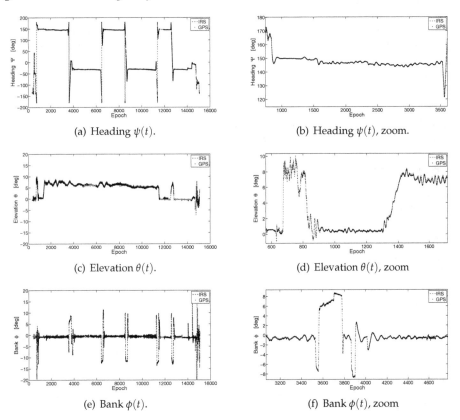

(a) Heading $\psi(t)$.

(b) Heading $\psi(t)$, zoom.

(c) Elevation $\theta(t)$.

(d) Elevation $\theta(t)$, zoom

(e) Bank $\phi(t)$.

(f) Bank $\phi(t)$, zoom

Fig. 14. $T - II$ test: time series of the three attitude angles as estimated via GNSS and provided by the INS. On the right, a closer look at the estimates.

## 6. Summary and conclusions

Ambiguity resolution can be effectively enhanced by means of a rigorous formulation of the ambiguity-attitude estimation problem. In order to infer the aircraft's orientation from the GNSS antenna positions, each antenna location on the aircraft body has to be precisely known. This geometrical information can be embedded in the ambiguity resolution step, thus strengthening the underlying functional model - i.e., additional information is added to the functional model - and enhancing the whole estimation process. The higher ambiguity resolution performance comes at the cost of an increased computational complexity. In order to overcome the issue, a number of solutions are presented, which allow for fast and reliable solutions without requiring extensive computational loads. A fast implementation of the geometrically constrained problem is obtained by modifying a well-known method for ambiguity resolution: the LAMBDA (Least-squares AMBiguity Decorrelation Adjustment) method. This method is nowadays the standard for carrier-phase based applications, and it is being implemented in a number of receivers employed for high-precision navigation applications. The complexity of the constrained estimation method requires the development of novel strategies to extract the solution in a timely manner. This is achieved by properly modifying the LAMBDA method to address the specific ambiguity-attitude estimation problem: the Multivariate Constrained (MC)-LAMBDA method. Through the use of two novel search schemes the sought-for set of carrier phase ambiguities can be efficiently estimated.

The method is tested on actual data collected on two different flight tests. Each test indicates the feasibility of employing GNSS as attitude sensor, an application that might be increasingly adopted in the aviation industry, either stand-alone for non-critical applications, or in combinations with other sensors for safety-critical applications.

## 7. Acknowledgment

The second author is the recipient of an Australian Research Council Federation Fellowship (project number FF0883188). This support is gratefully acknowledged.

## 8. References

Boon, F. & Ambrosius, B. A. C. (1997). Results of Real-Time Applications of the LAMBDA Method in GPS Based Aircraft Landings, *Proceedings KIS97* pp. 339–345.

Cheng, Y. & Shuster, M. D. (2007). Robustness and Accuracy of the QUEST Algorithm, *Advances in the Astronautical Sciences* 127: 41–61.

Cox, D. B. & Brading, J. D. (2000). Integration of LAMBDA Ambiguity Resolution with Kalman Filter for Relative Navigation of Spacecraft, *NAVIGATION* 47(3): 205–210.

Davenport, P. B. (1968). A Vector Approach to the algebra of Rotations with Applications, *NASA Technical Note D-4696, Goddard Space Flight Center* .

de Jonge, P. & Tiberius, C. (1996). The LAMBDA Method for Integer Ambiguity Estimation: Implementation Aspects, *LGR Series 12, Publications of the Delft Geodetic Computing Centre, Delft, The Netherlands* .

Giorgi, G. (2011). GNSS Carrier Phase-based Attitude Determination. Estimation and applications., *PhD dissertation, Delft University of Technology, Delft, The Netherlands* .

Giorgi, G., Teunissen, P. J. G. & Buist, P. J. (2008). A Search and Shrink Approach for the Baseline Constrained LAMBDA: Experimental Results, *Proceedings of the International Symposium on GPS/GNSS 2008. A. Yasuda (Ed.), Tokyo University of Marine Science and Technology* pp. 797–806.

Giorgi, G., Teunissen, P. J. G., Verhagen, S. & Buist, P. J. (2011). Instantaneous Ambiguity Resolution in GNSS-based Attitude Determination Applications: the MC-LAMBDA method, *Journal of Guidance, Control, and Dynamics, to be published* .

Giorgi, G., Teunissen, P. J. G., Verhagen, S. & Buist, P. J. (2012). Integer Ambiguity Resolution with Nonlinear Geometrical Constraints., *N. Sneeuw et al. (eds.), VII Hotine-Marussi Symposium on Mathematical Geodesy, International Association of Geodesy Symposia 137, Springer-Verlag* .

Huang, S. Q., Wang, J. X., Wang, X. Y. & Chen, J. P. (2009). The Application of the LAMBDA Method in the Estimation of the GPS Slant Wet Vapour, *Acta Aeronautica et Astronautica Sinica* 50(1): 60–68.

Ji, S., Chen, W., Zhao, C., Ding, X. & Chen, Y. (2007). Single Epoch Ambiguity Resolution for Galileo with the CAR and LAMBDA Methods, *GPS Solutions* 11(4): 259–268.

Kroes, R., Montenbruck, O., Bertiger, W. & Visser, P. (2005). Precise GRACE Baseline Determination Using GPS, *GPS Solutions* 9(1): 21–31.

Markley, F. L. & Landis, F. (1993). Attitude Determination Using Vector Observations: a Fast Optimal Matrix Algorithm, *The Journal of the Astronautical Sciences* 41(2): 261–280.

Markley, F. L. & Mortari, D. (1999). How to Estimate Attitude from Vector Observations, *Presented at AAS/AIAA Astrodynamics Specialist Conference, Paper 99-427* .

Markley, F. L. & Mortari, D. (2000). Quaternion Attitude Estimation Using Vector Observations, *The Journal of the Astronautical Sciences* 48(2-3): 359–380.

Misra, P. & Enge, P. (2001). *Global Positioning System: Signals, Measurements, and Performance,* 2nd edn, Ganga-Jamuna Press, Lincoln MA.

Mortari, D. (1997). ESOQ: A Closed-form Solution to the Wahba Problem, *The Journal of the Astronautical Sciences* 45(2): 195–204.

Mortari, D. (2000). Second Estimator of the Optimal Quaternion, *Journal of Guidance, Control, and Dynamics* 23(5): 885–888.

Nadarajah, N., Teunissen, P. J. G. & Giorgi, G. (2011). Instantaneous GNSS Attitude Determination for Remote Sensing Platforms, *Presented at the XXV International Union of Geodesy and Geophysics General Assembly (IUGG), Melbourne, Australia* .

Schonemann, P. H. (1966). A Generalized Solution of the Orthogonal Procrustes Problem, *Psychometrika* 31(1): 1–10.

Shuster, M. D. (1978). Approximate Algorithms for Fast Optimal attitude Computation, *Proceedings of the AIAA Guidance and Control conference, Palo Alto, CA, US* pp. 88–95.

Shuster, M. D. (1993). A Survey of Attitude Representations , *The Journal of the Astronautical Sciences* 41(4): 439–517.

Shuster, M. D. & Oh, S. D. (1981). Three-Axis Attitude Determination from Vector Observations, *Journal of Guidance and Control* 4(1): 70–77.

Teunissen, P. J. G. (1995). The Least-Squares Ambiguity Decorrelation Adjustment: a Method for Fast GPS Integer Ambiguity Estimation, *Journal of Geodesy* 70(1-2): 65–82.

Teunissen, P. J. G. (2000). The Success Rate and Precision of GPS Ambiguities, *Journal of Geodesy* 74(3): 321–326.

Teunissen, P. J. G. (2007a). A General Multivariate Formulation of the Multi-Antenna GNSS Attitude Determination Problem, *Artificial Satellites* 42(2): 97–111.

Teunissen, P. J. G. (2007b). Influence of Ambiguity Precision on the Success Rate of GNSS Integer Ambiguity Bootstrapping, *Journal of Geodesy, Springer* 81(5): 351–358.

Teunissen, P. J. G. (2007c). The LAMBDA Method for the GNSS Compass, *Artificial Satellites* 41(3): 89–103.

Teunissen, P. J. G. (2011). A-PPP: Array-aided Precise Point Positioning with Global Navigation Satellite Systems, *IEEE Transactions on Signal Processing (submitted for publication)* pp. 1–12.

Teunissen, P. J. G. & Kleusberg, A. (1998). GPS for Geodesy, *Springer, Berlin Heidelberg New York* .

Van Loan, C. F. (2000). The Ubiquitous Kronecker Product, *Journal of Computational and Applied Mathematics* 123: 85–100.

Wahba, G. (1965). Problem 65-1: A Least Squares Estimate of Spacecraft Attitude, *SIAM Review* 7(3): 384–386.

# 10

# A Variational Approach to the Fuel Optimal Control Problem for UAV Formations

Andrea L'Afflitto and Wassim M. Haddad

*Georgia Institute of Technology*
*USA*

## 1. Introduction

The pivotal role of unmanned aerial vehicles (UAVs) in modern aircraft technology is evidenced by the large number of civil and military applications they are employed in. For example, UAVs successfully serve as platforms carrying payloads aimed at land monitoring (Ramage et al., 2009), wildfire detection and management (Ambrosia & Hinkley, 2008), law enforcement (Haddal & Gertler, 2010), pollution monitoring (Oyekan & Huosheng, 2009), and communication broadcast relay (Majewski, 1999), to name just a few.

A formation of UAVs, defined by a set of vehicles whose states are coupled through a common control law (Scharf et al., 2003b), is often more valuable than a single aircraft because it can accomplish several tasks concurrently. In particular, UAV formations can guarantee higher flexibility and redundancy, as well as increased capability of distributed payloads (Scharf et al., 2003a). For example, an aircraft formation can successfully intercept a vehicle which is faster than its chasers (Jang & Tomlin, 2005). Alternatively, a UAV formation equipped with interferometic synthetic aperture radar (In-SAR) antennas can pursue both along-track and cross-track interferometry, which allow harvesting information that a single radar cannot detect otherwise (Lillesand et al., 2007).

Path planning is one of the main problems when designing missions involving multiple vehicles; a UAV formation typically needs to accomplish diverse tasks while meeting some assigned constraints. For example, a UAV formation may need to intercept given targets while its members maintain an assigned relative attitude. Trajectories should also be optimized with respect to some performance measure capturing minimum time or minimum fuel expenditure. In particular, trajectory optimization is critical for mini and micro UAVs ($\mu$UAVs) because they often operate independently from remote human controllers for extended periods of time (Shanmugavel et al., 2010) and also because of limited amount of available energy sources (Plnes & Bohorquez, 2006).

The scope of the present paper is to provide a rigorous and sufficiently broad formulation of the optimal path planning problem for UAV formations, modeled as a system of n 6-degrees of freedom (DoF) rigid bodies subject to a constant gravitational acceleration and aerodynamic forces and moments. Specifically, system trajectories are optimized in terms of control effort, that is, we design a control law that minimizes the forces and moments needed to operate a UAV formation, while meeting all the mission objectives. Minimizing the control effort is equivalent to minimizing the formation's fuel consumption in the case of vehicles equipped with conventional fuel-based propulsion systems (Schouwenaars et al., 2006) and is a suitable indicator of the energy consumption for vehicles powered by batteries or other power sources.

In this paper, we derive an optimal control law which is independent of the size of the formation, the system constraints, and the environmental model adopted, and hence, our framework applies to aircraft, spacecraft, autonomous marine vehicles, and robot formations. The direction and magnitude of the optimal control forces and moments is a function of the dynamics of two vectors, namely the translational and rotational primer vectors. In general, finding the dynamics of these two vectors over a given time interval is a demanding task that does not allow for an analytical closed-form solution, and hence, a numerical approach is required. Our main result involves necessary conditions for optimality of the formations' trajectories.

The contents of this paper are as follows. In Section 2, we present notation and definitions of the physical variables needed to formulate the fuel optimization problem. Section 3 gives a problem statement of the UAV path planning optimization problem, whereas Section 4 provides the necessary mathematical background for this problem. Next, in Section 5, we survey the relevant literature and highlight the advantages related to the proposed approach. Section 6 discusses results achieved by applying the theoretical framework developed in Section 4. In Section 7, we present an illustrative numerical example that highlights the efficacy of the proposed approach. Finally, in Section 8, we draw conclusions and highlight future research directions.

## 2. Notation and definitions

The notation used in this paper is fairly standard. When a word is defined in the text, the concept defined is *italicized* and it should be understood as an "if and only if" statement. Mathematical definitions are introduced by the symbol "$\triangleq$." The symbol $\mathbb{N}$ denotes the set of positive integers, $\mathbb{R}$ denotes the set of real numbers, $\overline{\mathbb{R}}_+$ denotes the set of nonnegative real numbers, $\mathbb{R}^n$ denotes the set of $n \times 1$ column vectors on the field of real numbers, and $\mathbb{R}^{n \times m}$ denotes the set of real $n \times m$ matrices. Both natural and real numbers are denoted by lower case letters, e.g., $j \in \mathbb{N}$ and $a \in \mathbb{R}$, vectors are denoted by bold lower case letters, e.g., $x \in \mathbb{R}^n$, and matrices are denoted by bold upper case letters, e.g., $A \in \mathbb{R}^{n \times m}$. Subsets of $\mathbb{R}^n$ and $\mathbb{R}^{n \times m}$ are denoted by italicized upper case letters, e.g., $A \subseteq \mathbb{R}^n$ and $B \subseteq \mathbb{R}^{n \times m}$. The interior of the set $A$ is denoted by $\text{int}(A)$. The zero vector in $\mathbb{R}^n$ is denoted by $\mathbf{0}_n$, the zero matrix in $\mathbb{R}^{n \times m}$ is denoted by $\mathbf{0}_{n \times m}$, and the identity matrix in $\mathbb{R}^{n \times n}$ is denoted by $\mathbf{I}_n$.

For $x \in \mathbb{R}^n$ we write $x \geq\geq \mathbf{0}_n$ (respectively, $x >> \mathbf{0}_n$) to indicate that every component of $x$ is nonnegative (respectively, positive). We write $|| \cdot ||_p$ for the p-norm of a vector and its corresponding equi-induced matrix norm, e.g., $||x||_p$ and $||A||_p$. The transpose of a vector or of a matrix is denoted by the superscript $(\cdot)^T$, e.g., $x^T$ and $A^T$. The cross product between two vectors $\mathbf{a}$ and $\mathbf{b}$ is denoted by $\mathbf{a} \wedge \mathbf{b}$. Given $x \in \mathbb{R}^3$ such that $x \triangleq [x_1, x_2, x_3]^T$, we define

$$x^{\times} \triangleq \begin{bmatrix} 0 & -x_3 & x_2 \\ x_3 & 0 & -x_1 \\ -x_2 & x_1 & 0 \end{bmatrix}.$$

The inverse of a square matrix $\mathbf{A}$ is denoted by $\mathbf{A}^{-1}$, the transpose of $\mathbf{A}^{-1}$ is denoted by $\mathbf{A}^{-T}$, the determinant of $\mathbf{A}$ is denoted by $\det(\mathbf{A})$, the diagonal of $\mathbf{A}$ is denoted by $\text{diag}(\mathbf{A})$, and the nullspace of a matrix $\mathbf{A}$ is denoted by $\mathcal{N}(\mathbf{A})$.

Functions are always introduced by specifying their domain and codomain, e.g., $\mathbf{h} : A_1 \times A_2 \to B$. The arguments of a function will not be indicated in the text unless necessary, e.g., $\mathbf{h}(\mathbf{x}, \mathbf{y})$ is simply denoted by $\mathbf{h}$. If a function is dependent on some unspecified variables, then its arguments will be replaced by dots, e.g., $\mathbf{h}(\cdot, \cdot)$. The same convention is used for functionals; however, their arguments are embraced by square brackets, i.e., $J[\mathbf{x}, \mathbf{y}]$.

The first derivative with respect to time of a differentiable function $\mathbf{q} : [t_1, t_2] \to \mathbb{R}^n$ is denoted by the a dot on top of the function, e.g., $\dot{\mathbf{q}}(t)$. Given $\mathbf{g} : A \to \mathbb{R}^m$, where $A \subset \mathbb{R}^n$ is an open set, we say that $\mathbf{g}(\cdot)$ *is of class* $C^k$, that is, $\mathbf{g}(\cdot) \in C^k(A)$, if $\mathbf{g}(\cdot)$ is continuous on $A$ with k-continuous derivatives. If $\mathbf{g}(\cdot) \in C^1(A)$, then $\mathbf{g}(\cdot)$ is *continuously differentiable*.

Throughout the paper we use two types of mathematical statements, namely, existential and universal statements. An existential statement has the form: "there exist $\mathbf{x} \in A$ such that condition $\Phi$ is satisfied." A universal statement has the form: "condition $\Phi$ is satisfied for all $\mathbf{x} \in A$." For universal statements we often omit the words "for all" and write: "condition $\Phi$ holds, $\mathbf{x} \in A$."

Time is the only independent variable used in this paper and is denoted by $t$. In this paper, $t \in [t_1, t_2]$, where $[t_1, t_2] \subset \mathbb{R}$ is a fixed time interval and is a priori assigned. A generic member of a formation of $n \in \mathbb{N}$ UAVs is identified by the subscript i and, hence, $i = 1, ..., n$. We define $\mathbf{r}_i : [t_1, t_2] \to \mathbb{R}^3$ as the *position vector* of the center of mass of the i-th vehicle in a given inertial reference frame, $\boldsymbol{\sigma}_i : [t_1, t_2] \to \mathbb{R}^3$ as the *attitude vector* of the i-th vehicle in modified rodrigues parameters (MRPs) (Shuster, 1993), and $\mathbf{x}_i \triangleq [\mathbf{r}_i^T, \boldsymbol{\sigma}_i^T]^T$ as the *state vector* of the i-th vehicle. The *system's configuration* at time $t$ is defined by $[\mathbf{x}_1^T(t), ..., \mathbf{x}_n^T(t)]^T$.

The vector $\mathbf{v}_i : [t_1, t_2] \to \mathbb{R}^3$ denotes the *velocity* of the center of mass of the i-th vehicle, $\boldsymbol{\omega}_i : [t_1, t_2] \to \mathbb{R}^3$ denotes the *angular velocity* of the i-th vehicle in a principal body reference frame, and $\tilde{\mathbf{x}}_i \triangleq [\mathbf{r}_i^T, \mathbf{v}_i^T, \boldsymbol{\sigma}_i^T, \boldsymbol{\omega}_i^T]^T$ is the *augmented state vector* of the i-th vehicle. For all $t \in [t_1, t_2]$, $\mathbf{r}_i(t) = \int_{t_1}^t \mathbf{v}_i(\tau)\, d\tau$ and $\dot{\boldsymbol{\sigma}}_i(t) = \mathbf{R}_{\text{rod}}(\boldsymbol{\sigma}_i(t))\boldsymbol{\omega}_i(t)$, where $\mathbf{R}_{\text{rod}}(\boldsymbol{\sigma}_i(t)) \triangleq \frac{1}{4}(1 - \boldsymbol{\sigma}_i^T(t)\boldsymbol{\sigma}_i(t))\mathbf{I}_3 + \frac{1}{2}\boldsymbol{\sigma}_i^\times(t) + \frac{1}{2}\boldsymbol{\sigma}_i(t)\boldsymbol{\sigma}_i^T(t)$ (Neimark & Fufaev, 1972; Shuster, 1993). We assume $[\mathbf{x}_1^T(t), ..., \mathbf{x}_n^T(t)]^T \in D_{\text{rel}} \subseteq \mathbb{R}^{6n}$ and $[\tilde{\mathbf{x}}_1^T(t), ..., \tilde{\mathbf{x}}_n^T(t)]^T \in D_{\text{abs}} \subseteq \mathbb{R}^{12n}$, $t \in [t_1, t_2]$.

We define $\mathbf{u}_{i,\text{tran}} : [t_1, t_2] \to \Gamma_{i,\text{tran}}$ (respectively, $\mathbf{u}_{i,\text{rot}} : [t_1, t_2] \to \Gamma_{i,\text{rot}}$) as the *translational acceleration* (respectively, the *rotational acceleration*) provided by the control system of the i-th vehicle in the formation, e.g., $\mathbf{u}_{i,\text{tran}}$ is the acceleration provided by the propulsion system and $\mathbf{u}_{i,\text{rot}}$ is the acceleration provided by the ailerons. The vector $\mathbf{u}_{i,\text{tran}}$ (respectively, $\mathbf{u}_{i,\text{rot}}$) is also referred to as the *i-th translational control vector* (respectively, the *i-th rotational control vector*). For a given set of real constants $\rho_{i,1}$, $\rho_{i,2}$, $\rho_{i,3}$, and $\rho_{i,4}$ such that $0 \le \rho_{i,1} < \rho_{i,2}$ and $0 \le \rho_{i,3} < \rho_{i,4}$, $\Gamma_{i,\text{tran}}$ and $\Gamma_{i,\text{rot}}$ are defined as

$$\Gamma_{i,\text{tran}} \triangleq \left\{ \mathbf{a} \in \mathbb{R}^3 : \rho_{i,1} \le ||\mathbf{a}||_2 \le \rho_{i,2} \right\} \cup \{\mathbf{0}_3\},$$

$$\Gamma_{i,\text{rot}} \triangleq \left\{ \mathbf{a} \in \mathbb{R}^3 : \rho_{i,3} \le ||\mathbf{a}||_2 \le \rho_{i,4} \right\} \cup \{\mathbf{0}_3\}.$$

Finally, for a given set $\Gamma \subset \mathbb{R}^p$, $\mathbf{u} : [t_1, t_2] \rightarrow \Gamma$ is an *admissible control in* $\Gamma$ if *i*) $\mathbf{u}(\cdot)$ is continuous at the endpoints of $[t_1, t_2]$, *ii*) $\mathbf{u}(\cdot)$ is continuous for all $t \in (t_1, t_2)$ with the exception of a finite number of times $t$ at which $\mathbf{u}(\cdot)$ may have discontinuities of the first kind, and *iii*) $\mathbf{u}(\tau) = \lim_{t \rightarrow \tau^-} \mathbf{u}(t)$, where $\tau \in [t_1, t_2]$ is a point of discontinuity of first kind for $\mathbf{u}(t)$ (Pontryagin et al., 1962). We assume that $\mathbf{u}_{i,\mathrm{tran}}$ (respectively, $\mathbf{u}_{i,\mathrm{rot}}$) is an admissible control in $\Gamma_{i,\mathrm{tran}}$ (respectively, $\Gamma_{i,\mathrm{rot}}$) for each $i \in \{1, \ldots, n\}$.

## 3. Problem statement

### 3.1 Fuel consumption performance functional

A measure of the effort needed to control the i-th formation vehicle is given by the *performance functional*

$$J[\mathbf{u}_i(\cdot)] \triangleq \int_{t_1}^{t_2} ||\mathbf{u}_i(t)||_2 \, dt, \tag{1}$$

where $\mathbf{u}_i(t) \triangleq [\mathbf{u}_{i,\mathrm{tran}}^T(t), c\mathbf{u}_{i,\mathrm{rot}}^T(t)]^T$ and c is a real constant with units of distance. Without loss of generality we assume that $|c| = 1$. The performance functional $\int_{t_1}^{t_2} ||\mathbf{u}_{i,\mathrm{tran}}(t)||_2 \, dt$ represents a measure of the fuel consumed over the time interval $[t_1, t_2]$ (Schouwenaars et al., 2006). Path planning for UAV formations is sometimes addressed by minimizing the more conservative performance functional $\int_{t_1}^{t_2} ||\mathbf{u}_{i,\mathrm{tran}}(t)||_1 \, dt$ (Blackmore, 2008). It is important to note that $||\mathbf{u}_{i,\mathrm{rot}}(t)||_2$ is much smaller than $||\mathbf{u}_{i,\mathrm{tran}}(t)||_2$ for conventional aircraft and, hence, its contribution to the performance functional (1) is negligible. However, this assumption does not hold for the case of $\mu$UAVs (Bataillé et al., 2009).

The control effort for the entire formation can be captured by the performance measure

$$J_{\mathrm{formation}}[\tilde{\mathbf{u}}(\cdot)] \triangleq \sum_{i=1}^{n} \mu_i J[\mathbf{u}_i(\cdot)], \tag{2}$$

where $\tilde{\mathbf{u}}(t) \triangleq [\mathbf{u}_1^T(t), \ldots, \mathbf{u}_n^T(t)]^T$ and $\mu_i \in [0, 1]$, with $\sum_{i=1}^{n} \mu_i = 1$, which represents the relative importance of minimizing the control effort of the i-th vehicle with respect to the others.

### 3.2 Aircraft dynamic equations

Aircraft are subject to external forces and moments from the environment. Specifically, an aerial vehicle is subject to gravitational forces, aerodynamic forces, and aerodynamic moments. Accelerations induced by external forces and external moments acting on a formation vehicle are denoted by $\mathbf{a} : \mathbb{R}^{12} \rightarrow \mathbb{R}^3$ and $\mathbf{m} : \mathbb{R}^{12} \rightarrow \mathbb{R}^3$, respectively, where $\mathbf{a}(\tilde{\mathbf{x}}_i)$, $\mathbf{m}(\tilde{\mathbf{x}}_i) \in C^1(\mathbb{R}^{12})$.

The unconstrained dynamic equations for the i-th vehicle are given by (Greenwood, 2003)

$$\frac{d}{dt}\tilde{\mathbf{x}}_i(t) = \begin{bmatrix} \mathbf{v}_i(t) \\ \mathbf{a}(\tilde{\mathbf{x}}_i(t)) \\ \mathbf{R}_{\mathrm{rod}}(\sigma_i(t))\boldsymbol{\omega}_i(t) \\ -\mathbf{I}_{\mathrm{in},i}^{-1}\boldsymbol{\omega}_i^{\times}(\boldsymbol{\omega}_i(t))\mathbf{I}_{\mathrm{in},i}\boldsymbol{\omega}_i(t) + \tilde{\boldsymbol{\omega}}_i(\tilde{\mathbf{x}}_i(t)) \end{bmatrix} + \begin{bmatrix} \mathbf{0}_3 \\ \mathbf{u}_{i,\mathrm{tran}}(t) \\ \mathbf{0}_3 \\ \mathbf{u}_{i,\mathrm{rot}}(t) \end{bmatrix}, \tag{3}$$

where $\mathbf{I}_{in,i}$ is the inertia matrix of the i-th vehicle in a principal body reference frame and $\tilde{\omega}_i(\tilde{\mathbf{x}}_i(t)) \triangleq \mathbf{I}_{in,i}^{-1}\mathbf{m}(\tilde{\mathbf{x}}_i(t))$, $t \in [t_1, t_2]$. The boundary conditions for (3) are given by the endpoint constraints discussed in Section 3.3.

### 3.3 Formation constraints

Given $D_1 \subset \mathbb{R}^p$ and $D_2 \subset \mathbb{R}^m$, the function $\mathbf{S} : D_1 \rightarrow D_2$ is a *continuously differentiable manifold* if $\mathbf{S}(\mathbf{y}) = 0$, $m < p$, $\mathbf{S}(\mathbf{y}) \in C^1(D_1)$, and rank $\frac{\partial \mathbf{S}(\mathbf{y})}{\partial \mathbf{y}} = m$ (Pontryagin et al., 1962). Let $\mathbf{S}_1 : D_{abs} \rightarrow \mathbb{R}^{n_1}$ and $\mathbf{S}_2 : D_{abs} \rightarrow \mathbb{R}^{n_2}$ be two continuously differentiable manifolds, and define the *endpoint constraints*

$$\mathbf{S}_1\left( \left[\tilde{\mathbf{x}}_1^T(t_1), ..., \tilde{\mathbf{x}}_n^T(t_1)\right]^T \right) = \mathbf{0}_{r_1},$$
$$\mathbf{S}_2\left( \left[\tilde{\mathbf{x}}_1^T(t_2), ..., \tilde{\mathbf{x}}_n^T(t_2)\right]^T \right) = \mathbf{0}_{r_2}. \tag{4}$$

Endpoint constraints partly impose the formation's configuration at times $t_1$ and $t_2$, and hence, can model point-to-point or rendezvous maneuvers.

*State inequality constraints* are given by

$$\mathbf{f}_{ineq}(\mathbf{x}_1(t), ..., \mathbf{x}_n(t)) \leq \mathbf{0}_{r_3}, \tag{5}$$

where $\mathbf{f}_{ineq} : D_{rel} \rightarrow \mathbb{R}^{n_3}$ and $\mathbf{f}_{ineq}(\mathbf{x}_1, ..., \mathbf{x}_n) \in C^3(\text{int}(D_{rel}))$. *State equality constraints* are given by

$$\mathbf{f}_{eq}(t, \mathbf{x}_1(t), ..., \mathbf{x}_n(t)) = \mathbf{0}_{r_4}, \tag{6}$$

where $\mathbf{f}_{eq} : [t_1, t_2] \times D_{rel} \rightarrow \mathbb{R}^{n_4}$ and $\mathbf{f}_{eq}(t, \mathbf{x}_1, ..., \mathbf{x}_n) \in C^2((t_1, t_2) \times \text{int}(D_{rel}))$. Here we assume that the constraints are *compatible*, that is, for all $t \in [t_1, t_2]$ there exists at least one set of 2n admissible controls $\{\mathbf{u}_{1,tran}(t), ..., \mathbf{u}_{n,tran}(t); \mathbf{u}_{1,rot}(t), ..., \mathbf{u}_{n,rot}(t)\}$ that satisfies (3) – (6).

State constraints given in terms of $\tilde{\mathbf{x}}_1(t), ..., \tilde{\mathbf{x}}_n(t)$ that can be reduced to the form given by (5) and (6) are called *holonomic* constraints. In particular, for n = 2 and $t \in [t_1, t_2]$, the constraint $\mathbf{v}_1(t) = \mathbf{v}_2(t)$ is holonomic since it can be rewritten as $\mathbf{r}_1(t) + \mathbf{r}_1(t_1) = \mathbf{r}_2(t) + \mathbf{r}_2(t_1)$, $t \in [t_1, t_2]$. It is important to note that the constraint $\omega_1(t) \leq \omega_2(t)$, $t \in [t_1, t_2]$, is nonholonomic since $\sigma_i(t) \neq \int_{t_1}^t \omega_i(\tau)\,d\tau + \sigma_i(t_1)$, $t \in [t_1, t_2]$ and i = 1, 2 (Greenwood, 2003).

State constraints can model collision avoidance, keeping the formation far from no-fly zones, or the requirement of pointing payloads toward the same target. It is obvious that (6) is a special case of (5); however, as noted in Section 4.2, this distinction is useful in reducing computational complexity.

### 3.4 Path planning optimization problem

For all i = 1, ..., n and $t \in [t_1, t_2]$ find the control vectors $\mathbf{u}_{i,tran}(t)$ and $\mathbf{u}_{i,rot}(t)$ among all admissible controls in $\Gamma_{i,tran}$ and $\Gamma_{i,tran}$ such that the performance measure (2) is minimized and $\tilde{\mathbf{x}}_i(t)$ satisfies (3) – (6).

## 4. Mathematical background

### 4.1 Slack variables

Inequality constraints (5) can be reduced to equality constraints by introducing $s : [t_1, t_2] \rightarrow \mathbb{R}^{n_3}$ such that $\mathbf{s}(t) \in C^2(t_1, t_2)$ and $\mathbf{f}_{ineq}(t, \mathbf{x}_1(t), ..., \mathbf{x}_n(t)) + \frac{1}{2}\text{diag}(\mathbf{ss}^T) = \mathbf{0}_{r_3}$. The components of $\mathbf{s}$ are called *slack variables*. Thus, (5) can be rewritten as (Valentine, 1937)

$$\widetilde{\mathbf{f}}_{ineq}(\mathbf{s}(t), \mathbf{x}_1(t), ..., \mathbf{x}_n(t)) = \mathbf{0}_{r_3}, \tag{7}$$

where $\widetilde{\mathbf{f}}_{ineq}(\mathbf{s}(t), \mathbf{x}_1(t), ..., \mathbf{x}_n(t)) \triangleq \mathbf{f}_{ineq}(\mathbf{x}_1(t), ..., \mathbf{x}_n(t)) + \frac{1}{2}\text{diag}(\mathbf{ss}^T)$.

### 4.2 Lagrange coordinates

The following theorem is needed for the main results of this paper.

**Theorem 4.1.** *(Pars, 1965) Let $D_q \subseteq \mathbb{R}^{6n-n_4}$ be an open connected set and let $\mathbf{q} : [t_1, t_2] \times \mathbb{R}^{n_3} \times D_{rel} \rightarrow D_q$ be such that $\mathbf{q}(t, \mathbf{s}(t), \mathbf{x}_1(t), ..., \mathbf{x}_n(t)) \in C^2((t_1, t_2) \times \mathbb{R}^{n_3} \times \text{int}(D_{rel}))$. Assume that*

$$\det\left(\frac{\partial\left[\widetilde{\mathbf{f}}^T_{ineq}(\mathbf{s}, \mathbf{x}_1, ..., \mathbf{x}_n)\,\mathbf{f}^T_{eq}(t, \mathbf{s}, \mathbf{x}_1, ..., \mathbf{x}_n)\,\mathbf{q}^T(t, \mathbf{s}, \mathbf{x}_1, ..., \mathbf{x}_n)\right]^T}{\partial\left[\mathbf{s}^T, \mathbf{x}_1^T, ..., \mathbf{x}_n^T\right]^T}\right) \neq 0 \tag{8}$$

*for all $(t, \mathbf{s}, \mathbf{x}_1..., \mathbf{x}_n) \in \mathcal{I} \times \Delta$, where $\mathcal{I} \subset (t_1, t_2)$ and $\Delta \subset \mathbb{R}^{n_3} \times D_{rel}$ are open connected sets. Then q can be rewritten as a function of t, that is, $\mathbf{q} : \mathcal{I} \rightarrow D_q$, and $\mathbf{s}, \mathbf{x}_1, ..., \mathbf{x}_n, \widetilde{\mathbf{x}}_1, ..., \widetilde{\mathbf{x}}_n$ can be rewritten as unique functions of t and q, that is, $\mathbf{s} : \mathcal{I} \times D_q \rightarrow \mathbb{R}^{n_3}$, $\mathbf{x}_i : \mathcal{I} \times D_q \rightarrow \mathbb{R}^6$, and $\widetilde{\mathbf{x}}_i : \mathcal{I} \times D_q \rightarrow \mathbb{R}^{12}$ for all $i = 1, ..., n$ and $(t, \mathbf{s}, \mathbf{x}_1..., \mathbf{x}_n) \in \mathcal{I} \times \Delta$. Furthermore, the components of q are independent and uniquely characterize the system's configuration.*

Under the hypothesis of Theorem 4.1, the components of $\mathbf{q}(t)$ are called *Lagrange coordinates*. As will be shown in Section 4.3, the key advantage of using Lagrange coordinates is that the constraints (5) – (7) are automatically accounted for when rewriting the formation's dynamic equations in terms of $t$ and $\mathbf{q}(t)$ (Pars, 1965). In this paper, we assume that $\mathbf{s}, \mathbf{x}_1, ..., \mathbf{x}_n, \widetilde{\mathbf{x}}_1, ..., \widetilde{\mathbf{x}}_n$ are explicit functions of q only and not $t$, which occurs in most practical applications (Pars, 1965). In practice, given constraints in the form of (6) and (7), q is chosen such that Theorem 4.1 holds. As will be further discussed in Section 4.3, we select $\mathbf{q}(t, \mathbf{s}(t), \mathbf{x}_1(t), ..., \mathbf{x}_n(t))$ as an explicit function of $(\mathbf{s}(t), \mathbf{x}_1(t), ..., \mathbf{x}_n(t))$.

Given $\mathbf{q}(t, \mathbf{s}(t), \mathbf{x}_1(t), ..., \mathbf{x}_n(t))$, $\dot{\mathbf{q}}$ is a function of $\mathbf{s}(t)$, $\mathbf{r}_i(t)$, $\sigma_i(t)$, $i = 1, ..., n$, and their first time derivatives. In practice, however, we measure $\omega_i(t)$ rather than $\sigma_i(t)$, and hence, if the assumptions of Theorem 4.1 hold, we define the *kinematic equation*

$$\mathbf{q}_{dot}(t) \triangleq \mathbf{\Psi}(\mathbf{q}(t))\,\dot{\mathbf{q}}(t) + \psi(\mathbf{q}(t)), \tag{9}$$

where $\omega_i(t)$, $i = 1, 2, ..., n$, explicitly appears in $\mathbf{q}_{dot}(t)$, $\mathbf{\Psi} : D_q \rightarrow \mathbb{R}^{(6n-n_4)\times(6n-n_4)}$ is an invertible continuously differentiable matrix function, and $\psi : D_q \rightarrow \mathbb{R}^{6n-n_4}$ is continuously differentiable. Consequently, $\mathbf{s}, \mathbf{x}_1, ..., \mathbf{x}_n, \widetilde{\mathbf{x}}_1, ..., \widetilde{\mathbf{x}}_n$ can be rewritten as unique functions of q and $\mathbf{q}_{dot}$, that is, $\mathbf{s} : D_q \times \mathbb{R}^{6n-n_4} \rightarrow \mathbb{R}^{n_3}$, $\mathbf{x}_i : D_q \times \mathbb{R}^{6n-n_4} \rightarrow \mathbb{R}^6$, and $\widetilde{\mathbf{x}}_i : D_q \times \mathbb{R}^{6n-n_4} \rightarrow \mathbb{R}^{12}$, $(t, \mathbf{s}, \mathbf{x}_1..., \mathbf{x}_n) \in \mathcal{I} \times \Delta$ (Greenwood, 2003). Here, we assume that $\mathbf{q}_{dot}$ satisfies (23) below.

In the following we assume that the path planning optimization problem can be solved over the time interval $[t_1^*, t_2^*] \supset [t_1, t_2]$ and that the given set of Lagrange coordinates can be defined on the open connected set $\widetilde{\mathcal{I}}$, where $[t_1, t_2] \subset \widetilde{\mathcal{I}} \subset (t_1^*, t_2^*)$. Thus, (4) can be rewritten as

$$\mathbf{S}_1 \left( \left[ \widetilde{\mathbf{x}}_1^T (\mathbf{q}(t_1), \mathbf{q}_{\text{dot}}(\mathbf{q}(t_1))), ..., \widetilde{\mathbf{x}}_n^T (\mathbf{q}(t_1), \mathbf{q}_{\text{dot}}(\mathbf{q}(t_1))) \right]^T \right) = \mathbf{0}_{r_1}, \tag{10}$$

$$\mathbf{S}_2 \left( \left[ \widetilde{\mathbf{x}}_1^T (\mathbf{q}(t_2), \mathbf{q}_{\text{dot}}(\mathbf{q}(t_2))), ..., \widetilde{\mathbf{x}}_n^T (\mathbf{q}(t_2), \mathbf{q}_{\text{dot}}(\mathbf{q}(t_2))) \right]^T \right) = \mathbf{0}_{r_2}. \tag{11}$$

**Example 4.1.** Consider a UAV formation with two vehicles so that n = 2. Assume that

$$\mathbf{f}_{\text{ineq}} (\mathbf{x}_1(t), \mathbf{x}_2(t)) = \begin{bmatrix} ||\mathbf{r}_1(t) - \mathbf{r}_2(t)||_2^2 - r_{\max} \\ r_{\min} - ||\mathbf{r}_1(t) - \mathbf{r}_2(t)||_2^2 \end{bmatrix} << \mathbf{0}_2, \tag{12}$$

$$\mathbf{f}_{\text{eq}} (t, \mathbf{x}_1(t), \mathbf{x}_2(t)) = \sigma_1(t) - \sigma_2(t) = \mathbf{0}_3, \tag{13}$$

$$\mathbf{S}_1 \left( \left[ \widetilde{\mathbf{x}}_1^T(t_1) \, \widetilde{\mathbf{x}}_2^T(t_1) \right]^T \right) = \begin{bmatrix} ||\mathbf{r}_1(t_1) - \mathbf{r}_2(t_1)||_2^2 - \left( \frac{r_{\max}+r_{\min}}{2} \right) \\ \sigma_1(t_1) - \sigma_2(t_1) \end{bmatrix} = \mathbf{0}_4, \tag{14}$$

$$\mathbf{S}_2 \left( \left[ \widetilde{\mathbf{x}}_1^T(t_2) \, \widetilde{\mathbf{x}}_2^T(t_2) \right]^T \right) = \begin{bmatrix} ||\mathbf{r}_1(t_2) - \mathbf{r}_2(t_2)||_2^2 - \frac{2(r_{\max}-r_{\min})}{3} \\ \sigma_1(t_2) - \sigma_2(t_2) \end{bmatrix} = \mathbf{0}_4, \tag{15}$$

where $r_{\min}$ and $r_{\max}$ are real constants such that $0 < r_{\min} < r_{\max}$. Equation (12) ensures that $r_{\min} \leq ||\mathbf{r}_1(t) - \mathbf{r}_2(t)||_2^2 \leq r_{\max}$ and (13) ensures that both vehicles always have the same attitude: $D_{\text{rel}} = \left\{ \left[ \mathbf{x}_1^T(t) \, \mathbf{x}_2^T(t) \right]^T : r_{\min} \leq ||\mathbf{r}_1(t) - \mathbf{r}_2(t)||_2^2 \leq r_{\max}, \sigma_1(t) = \sigma_2(t), t \in [t_1, t_2] \right\}$.

Introducing the slack variables $s_1 : [t_1, t_2] \to \mathbb{R}$ and $s_2 : [t_1, t_2] \to \mathbb{R}$, (12) becomes

$$\widetilde{\mathbf{f}}_{\text{ineq}}(\mathbf{s}(t), \mathbf{x}_1(t), \mathbf{x}_2(t)) = \begin{bmatrix} ||\mathbf{r}_1(t) - \mathbf{r}_2(t)||_2^2 - r_{\max} + \frac{1}{2}s_1^2(t) \\ r_{\min} - ||\mathbf{r}_1(t) - \mathbf{r}_2(t)||_2^2 + \frac{1}{2}s_2^2(t) \end{bmatrix} = \mathbf{0}_2. \tag{16}$$

As noted in Section 3.3, the equality constraint (13) can be embedded into (12) to give

$$\widetilde{\mathbf{f}}_{\text{ineq}}(\mathbf{s}(t), \mathbf{x}_1(t), \mathbf{x}_2(t)) = \begin{bmatrix} ||\mathbf{r}_1(t) - \mathbf{r}_2(t)||_2^2 - r_{\max} + \frac{1}{2}s_1^2(t) \\ r_{\min} - ||\mathbf{r}_1(t) - \mathbf{r}_2(t)||_2^2 + \frac{1}{2}s_2^2(t) \\ \sigma_1(t) - \sigma_2(t) + \frac{1}{2}\text{diag}(\mathbf{s}_3\mathbf{s}_3^T) \\ \sigma_2(t) - \sigma_1(t) + \frac{1}{2}\text{diag}(\mathbf{s}_4\mathbf{s}_4^T) \end{bmatrix} = \mathbf{0}_8,$$

where $\mathbf{s}_j : [t_1, t_2] \to \mathbb{R}^3, j = 3, 4$. Note that in this case, the dimension of $\widetilde{\mathbf{f}}_{\text{ineq}}$ is increased since six additional slack variables have been introduced, which increases computational complexity.

Next, define $r_{i,j} : [t_1, t_2] \to \mathbb{R}$ (respectively, $\sigma_{i,j} : [t_1, t_2] \to \mathbb{R}$) as the j-th component of $\mathbf{r}_i(t)$ (respectively, $\sigma_i(t)$). If $\mathbf{q}(t) = \left[ s_1(t), s_2(t), \mathbf{r}_1^T(t), \sigma_1^T(t), r_{2,1}(t) \right]^T$, then (8) gives

$$\det \left( \frac{\partial \left[ \widetilde{\mathbf{f}}_{\text{ineq}}^T (\mathbf{s}, \mathbf{x}_1, \mathbf{x}_2), \mathbf{f}_{\text{eq}}^T (t, \mathbf{s}, \mathbf{x}_1, \mathbf{x}_2), \mathbf{q}^T (t, \mathbf{s}, \mathbf{x}_1, \mathbf{x}_2) \right]^T}{\partial \left[ \mathbf{s}^T, \mathbf{x}_1^T, \mathbf{x}_2^T \right]^T} \right) = 0.$$

Thus, by Theorem 4.1, the components of $\mathbf{q}$ are not Lagrange coordinates.

Alternatively, if $\mathbf{q}(t) = \left[s_1(t),\ r_{1,1}(t),\ r_{1,2}(t),\ \boldsymbol{\sigma}_1^T(t),\ \mathbf{r}_2^T(t)\right]^T$, then

$$\det\left(\frac{\partial\left[\tilde{\mathbf{f}}_{\text{ineq}}^T(\mathbf{s},\mathbf{x}_1,\mathbf{x}_2),\ \mathbf{f}_{\text{eq}}^T(t,\mathbf{s},\mathbf{x}_1,\mathbf{x}_2),\ \mathbf{q}^T(t,\mathbf{s},\mathbf{x}_1,\mathbf{x}_2)\right]^T}{\partial\left[\mathbf{s}^T,\ \mathbf{x}_1^T,\ \mathbf{x}_2^T\right]^T}\right) = -2s_2(t)\,(r_{1,3}(t) - r_{2,3}(t)),$$

for all $(t,\mathbf{s},\mathbf{x}_1,\mathbf{x}_2) \in (t_1,t_2) \times \mathbb{R}^2 \times \text{int}\,(D_{\text{rel}})$ such that $r_{1,3}(t) \neq r_{2,3}(t)$, and hence, the components of $\mathbf{q}$ are suitable Lagrange coordinates if $r_{\min} < \|\mathbf{r}_1(t) - \mathbf{r}_2(t)\|_2^2$ and $r_{1,3}(t) \neq r_{2,3}(t)$. In this case, (9) gives

$$\mathbf{q}_{\text{dot}}(t) = \begin{bmatrix} \dot{s}_1(t) \\ v_{1,1}(t) \\ v_{1,2}(t) \\ \boldsymbol{\omega}_1(t) \\ \mathbf{v}_2(t) \end{bmatrix} = \begin{bmatrix} \mathbf{I}_3 & \mathbf{0}_{3\times3} & \mathbf{0}_{3\times3} \\ \mathbf{0}_{3\times3} & \mathbf{R}_{\text{rod}}^{-1}(\boldsymbol{\sigma}_1(t)) & \mathbf{0}_{3\times3} \\ \mathbf{0}_{3\times3} & \mathbf{0}_{3\times3} & \mathbf{I}_3 \end{bmatrix} \begin{bmatrix} \dot{s}_1(t) \\ v_{1,1}(t) \\ v_{1,2}(t) \\ \dot{\boldsymbol{\sigma}}_1(t) \\ \mathbf{v}_2(t) \end{bmatrix}, \tag{17}$$

where $v_{1j} : [t_1, t_2] \to \mathbb{R}$ is the j-th component of $\mathbf{v}_1(t)$.

A more suitable choice of Lagrange coordinates is given by $\mathbf{q}(t) = \left[\mathbf{x}_1^T(t),\ \mathbf{r}_2^T(t)\right]^T$ since

$$\det\left(\frac{\partial\left[\tilde{\mathbf{f}}_{\text{ineq}}^T(\mathbf{s},\mathbf{x}_1,\mathbf{x}_2),\ \mathbf{f}_{\text{eq}}^T(t,\mathbf{s},\mathbf{x}_1,\mathbf{x}_2),\ \mathbf{q}^T(t,\mathbf{s},\mathbf{x}_1,\mathbf{x}_2)\right]^T}{\partial\left[\mathbf{s}^T\ \mathbf{x}_1^T,\ \mathbf{x}_2^T\right]^T}\right) = s_1(t)s_2(t)$$

for all $(t,\mathbf{s},\mathbf{x}_1,\mathbf{x}_2) \in (t_1,t_2) \times \mathbb{R}^2 \times \text{int}\,(D_{\text{rel}})$, and hence, the components of $\mathbf{q}$ are suitable Lagrange coordinates if $r_{\min} < \|\mathbf{r}_1(t) - \mathbf{r}_2(t)\|_2^2 < r_{\max}$. In this case, (9) gives

$$\mathbf{q}_{\text{dot}}(t) = \begin{bmatrix} \mathbf{v}_1(t) \\ \boldsymbol{\omega}_1(t) \\ \mathbf{v}_2(t) \end{bmatrix} = \begin{bmatrix} \mathbf{I}_3 & \mathbf{0}_{3\times3} & \mathbf{0}_{3\times3} \\ \mathbf{0}_{3\times3} & \mathbf{R}_{\text{rod}}^{-1}(\boldsymbol{\sigma}_1(t)) & \mathbf{0}_{3\times3} \\ \mathbf{0}_{3\times3} & \mathbf{0}_{3\times3} & \mathbf{I}_3 \end{bmatrix} \begin{bmatrix} \mathbf{v}_1(t) \\ \dot{\boldsymbol{\sigma}}_1(t) \\ \mathbf{v}_2(t) \end{bmatrix}. \tag{18}$$

Since we use this example throughout the paper, we define $\mathbf{q}_{\text{dot},1} \triangleq [\mathbf{v}_1^T, \boldsymbol{\omega}_1^T]^T$, $\mathbf{q}_{\text{dot},2} \triangleq \mathbf{v}_2$, and

$$\boldsymbol{\Psi}_1(\mathbf{x}_1(t)) \triangleq \begin{bmatrix} \mathbf{I}_3 & \mathbf{0}_{3\times3} \\ \mathbf{0}_{3\times3} & \mathbf{R}_{\text{rod}}^{-1}(\boldsymbol{\sigma}_1(t)) \end{bmatrix}.$$

Finally, note that if $r_{\min} < \|\mathbf{r}_1(t) - \mathbf{r}_2(t)\|_2^2 < r_{\max}$ for $t \in (t_1^*, t_2^*) \supset [t_1, t_2]$, then (14) and (15) reduce to

$$\|\mathbf{r}_1(t_1) - \mathbf{r}_2(t_1)\|_2^2 - \left(\frac{r_{\max} + r_{\min}}{2}\right) = 0, \tag{19}$$

$$\|\mathbf{r}_1(t_2) - \mathbf{r}_2(t_2)\|_2^2 - \frac{2\,(r_{\max} - r_{\min})}{3} = 0. \tag{20}$$

### 4.3 Constrained formation dynamic equations

The formation's kinetic energy is given by *König's theorem* (Pars, 1965) and for our problem takes the form

$$
k\left(\mathbf{q}(t), \mathbf{q}_{dot}(t)\right) = \frac{1}{2}\sum_{i=1}^{n} m_i \mathbf{v}_i^{T}\left(\mathbf{q}(t), \mathbf{q}_{dot}(t)\right)\mathbf{v}_i\left(\mathbf{q}(t), \mathbf{q}_{dot}(t)\right)
$$

$$
+ \frac{1}{2}\sum_{i=1}^{n} \boldsymbol{\omega}_i^{T}\left(\mathbf{q}(t), \mathbf{q}_{dot}(t)\right)\mathbf{I}_{in,i}\boldsymbol{\omega}_i\left(\mathbf{q}(t), \mathbf{q}_{dot}(t)\right), \tag{21}
$$

where $m_i$ is the mass of the i-th vehicle, which is assumed to be constant. The dynamic equations of the constrained formation can be written in terms of Lagrange coordinates by applying the *Boltzmann-Hammel equation* (Greenwood, 2003) to give

$$
\frac{d}{dt}\left(\frac{\partial k\left(\mathbf{q}, \mathbf{q}_{dot}\right)}{\partial \mathbf{q}_{dot}}\right) = \sum_{i=1}^{n} m_i \mathbf{v}_i^{T}\left(\mathbf{q}(t), \mathbf{q}_{dot}(t)\right)\frac{d}{dt}\frac{\partial \mathbf{v}_i\left(\mathbf{q}, \mathbf{q}_{dot}\right)}{\partial \mathbf{q}_{dot}}
$$

$$
+ \sum_{i=1}^{n} \boldsymbol{\omega}_i^{T}\left(\mathbf{q}(t), \mathbf{q}_{dot}(t)\right)\mathbf{I}_{in,i}\frac{d}{dt}\frac{\partial \boldsymbol{\omega}_i\left(\mathbf{q}, \mathbf{q}_{dot}\right)}{\partial \mathbf{q}_{dot}}
$$

$$
+ \sum_{i=1}^{n}\left(\mathbf{a}\left(\tilde{\mathbf{x}}_i\left(\mathbf{q}(t), \mathbf{q}_{dot}(t)\right)\right) + \mathbf{u}_{i,tran}(t)\right)\frac{\partial \mathbf{v}_i\left(\mathbf{q}, \mathbf{q}_{dot}\right)}{\partial \mathbf{q}_{dot}}
$$

$$
+ \sum_{i=1}^{n}\left(\mathbf{m}\left(\tilde{\mathbf{x}}_i\left(\mathbf{q}(t), \mathbf{q}_{dot}(t)\right)\right) + \mathbf{u}_{i,rot}(t)\right)\frac{\partial \boldsymbol{\omega}_i\left(\mathbf{q}, \mathbf{q}_{dot}\right)}{\partial \mathbf{q}_{dot}}. \tag{22}
$$

Equations (10) and (11) are the boundary conditions for (22). It is important to note that the dynamic equation (22) is written in terms of Lagrange coordinates, and hence, accounts for (5) and (6).

Analytical optimization techniques such as Pontryagin's minimum principle, Bellman's theorem, and calculus of variations require the dynamic equations to be written as a first-order ordinary differential equation in explicit form. Therefore, using the hypothesis on $\mathbf{q}_{dot}$, the second-order ordinary differential equation (22) needs to be written in a first-order form

$$
\dot{\mathbf{q}}_{dot}(t) = \mathbf{f}_{dyn}(\mathbf{q}(t), \mathbf{q}_{dot}(t), \tilde{\mathbf{u}}(t)), \tag{23}
$$

where $\tilde{\mathbf{u}}(t) \triangleq [\mathbf{u}_1^{T}(t), \ldots, \mathbf{u}_n^{T}(t)]^{T}$ and $\mathbf{f}_{dyn} : D_q \times \mathbb{R}^{6n-n_4} \times \mathbb{R}^{12n} \rightarrow \mathbb{R}^{6n-n_4}$. In order to isolate the contribution of $\tilde{\mathbf{u}}$ in (24), we define $\hat{\mathbf{f}}_{dyn}(\mathbf{q}(t), \mathbf{q}_{dot}(t)) \triangleq \mathbf{f}_{dyn}(\mathbf{q}(t), \mathbf{q}_{dot}(t), \tilde{\mathbf{u}}(t)) - \mathbf{u}_{i,tran}(t)\frac{\partial \mathbf{v}_i(\mathbf{q}, \mathbf{q}_{dot})}{\partial \mathbf{q}_{dot}} - \mathbf{u}_{i,rot}(t)\frac{\partial \boldsymbol{\omega}_i(\mathbf{q}, \mathbf{q}_{dot})}{\partial \mathbf{q}_{dot}}$.

Equation (22) or, equivalently, (23) gives a set of $6n - n_4$ equations in $2(6n - n_4)$ unknowns, which are $\mathbf{q}$ and $\mathbf{q}_{dot}$. Thus, (22) needs to be solved together with (9) (Greenwood, 2003) to give

$$
\begin{bmatrix} \mathbf{q}_{dot}(t) \\ \dot{\mathbf{q}}_{dot}(t) \end{bmatrix} = \begin{bmatrix} \boldsymbol{\Psi}\left(\mathbf{q}(t)\right)\dot{\mathbf{q}}(t) + \boldsymbol{\psi}\left(\mathbf{q}(t)\right) \\ \mathbf{f}_{dyn}(\mathbf{q}(t), \mathbf{q}_{dot}(t), \tilde{\mathbf{u}}(t)) \end{bmatrix}. \tag{24}
$$

From (21) it follows that the formation's kinetic energy k is not an explicit function of $\mathbf{s}$, and hence, if $\mathbf{q}$ is chosen as an explicit function of p components of $\mathbf{s}(t) \in \mathbb{R}^{n_3}$, then p of the $6n - n_4$ equations in (22) cannot be straightforwardly recast in the explicit form given by (23). In this case, assume, without loss of generality, that $\mathbf{q}$ explicitly depends on the first p components of $\mathbf{s}$ and substitute the corresponding p equations in (24) with

$$s_j(t)\ddot{s}_j(t) = -\dot{s}_j^2(t) - \frac{d^2}{dt^2}f_{\text{ineq},j}(x_1(\mathbf{q}(t), \mathbf{q}_{\text{dot}}(t)), \ldots, x_n(\mathbf{q}(t), \mathbf{q}_{\text{dot}}(t))), \qquad (25)$$

which is obtained by differentiating (7). In this case, the boundary conditions are given by

$$\tilde{f}_{\text{ineq}}(\mathbf{s}(\mathbf{q}(t_1), \mathbf{q}_{\text{dot}}(\mathbf{q}(t_1))), x_1(\mathbf{q}(t_1), \mathbf{q}_{\text{dot}}(\mathbf{q}(t_1))), \ldots, x_n(\mathbf{q}(t_1), \mathbf{q}_{\text{dot}}(\mathbf{q}(t_1)))) = 0_{r_3}, \qquad (26)$$

$$\tilde{f}_{\text{ineq}}(\mathbf{s}(\mathbf{q}(t_2), \mathbf{q}_{\text{dot}}(\mathbf{q}(t_2))), x_1(\mathbf{q}(t_2), \mathbf{q}_{\text{dot}}(\mathbf{q}(t_2))), \ldots, x_n(\mathbf{q}(t_2), \mathbf{q}_{\text{dot}}(\mathbf{q}(t_2)))) = 0_{r_3}, \qquad (27)$$

where $f_{\text{ineq},j} : \mathbb{R}^{n_3} \times D_{\text{rel}} \to \mathbb{R}$ is the j-th component of $f_{\text{ineq}}(\mathbf{s}(t), x_1(t), \ldots, x_n(t))$ (Jacobson & Lele, 1969), for $j = 1, \ldots, p$. If $s_j(t^*) = 0$ for some $t^* \in [t_1, t_2]$, then (25) can be replaced by

$$3\dot{s}_j(t)\ddot{s}_j(t) + s_j(t)\frac{d^3 s_j(t)}{dt^3} = -\frac{d^3}{dt^3}f_{\text{ineq},j}(x_1(\mathbf{q}(t), \mathbf{q}_{\text{dot}}(t)), \ldots, x_n(\mathbf{q}(t), \mathbf{q}_{\text{dot}}(t))), \qquad (28)$$

where $\mathbf{s} \in C^3(t_1, t_2)$. In general, (7) must be differentiated so that $\ddot{s}_j(t)$, or one of its higher-order derivatives, explicitly appears and is multiplied by a term that is non-zero for all $t \in [t_1, t_2]$. In this case, the differentiability assumptions on $\mathbf{s}$ and $f_{\text{ineq}}$ must be modified accordingly.

**Example 4.2.** Consider Example 4.1 with $\mathbf{q}(t) = [s_1(t), r_{1,1}(t), r_{1,2}(t), \sigma_1^T(t), r_2^T(t)]^T$. In this case, the formation's kinetic energy is given by

$$k(\mathbf{q}(t), \mathbf{q}_{\text{dot}}(t)) = \frac{1}{2}m_1 v_1^T(\mathbf{q}(t), \mathbf{q}_{\text{dot}}(t)) v_1(\mathbf{q}(t), \mathbf{q}_{\text{dot}}(t))$$

$$+ \frac{1}{2}m_2 v_2^T(t)v_2(t) + \frac{1}{2}\sum_{i=1}^{2}\omega_1^T(t)I_{\text{in},i}\omega_1(t).$$

The dynamic equations can now be found by applying (22) and accounting for (17) giving

$$v_{1,j}(t) = \frac{dr_{1,j}(t)}{dt}, \quad \omega_1(t) = R_{\text{rod}}^{-1}(\sigma_1(t))\dot{\sigma}_1(t), \quad v_2(t) = \frac{dr_2(t)}{dt}, \qquad (29)$$

$$m_1\frac{dv_{1,1}(t)}{dt} = m_1a_1(\tilde{x}_1(\mathbf{q}(t), \mathbf{q}_{\text{dot}}(t))) + m_1u_{1,\text{tran},1}(t), \qquad (30)$$

$$m_1\frac{dv_{1,2}(t)}{dt} = m_1a_2(\tilde{x}_1(\mathbf{q}(t), \mathbf{q}_{\text{dot}}(t))) + m_1u_{1,\text{tran},2}(t), \qquad (31)$$

$$I_{\text{in},1}\frac{d\omega_1(t)}{dt} = -\omega_1^\times(\omega_1(t))I_{\text{in},1}\omega_1(t) + m(\tilde{x}_1(\mathbf{q}(t), \mathbf{q}_{\text{dot}}(t, \mathbf{q}(t)))) + I_{\text{in},1}u_{1,\text{rot}}(t), \qquad (32)$$

$$m_2\frac{dv_2(t)}{dt} = m_2a(\tilde{x}_2(\mathbf{q}(t), \mathbf{q}_{\text{dot}}(t))) + m_2u_{2,\text{tran}}(t), \qquad (33)$$

where, for $j = 1,2$, $u_{1,\text{tran},j} : [t_1, t_2] \to \mathbb{R}$ (respectively, $a_j : \mathbb{R}^{12} \to \mathbb{R}$) is the $j$-th component of $\mathbf{u}_{1,\text{tran}}(t)$ (respectively, $\mathbf{a}\,(\tilde{\mathbf{x}}_1\,(t, \mathbf{q}(t), \mathbf{q}_{\text{dot}}(t, \mathbf{q}(t)))))$. Instead of deducing the dynamics of $s_1(t)$ from

$$m_1 \frac{dv_{1,3}(t)}{dt} = m_1 a_3\,(\tilde{\mathbf{x}}_1\,(\mathbf{q}(t), \mathbf{q}_{\text{dot}}(t))) + m_1 u_{1,\text{tran},3}(t),$$

we use (16) and (25) to obtain

$$\dot{s}_1^2(t) + s_1(t)\ddot{s}_1(t) = -2||\mathbf{v}_1(t) - \mathbf{v}_2(t)||_2^2$$

$$+ (\mathbf{r}_1(t) - \mathbf{r}_2(t))^{\mathsf{T}}(\mathbf{a}\,(\tilde{\mathbf{x}}_1\,(\mathbf{q}(t), \mathbf{q}_{\text{dot}}(t))) + \mathbf{u}_{1,\text{tran}}(t))$$

$$- (\mathbf{r}_1(t) - \mathbf{r}_2(t))^{\mathsf{T}}(\mathbf{a}\,(\tilde{\mathbf{x}}_2\,(\mathbf{q}(t), \mathbf{q}_{\text{dot}}(t))) + \mathbf{u}_{2,\text{tran}}(t)), \qquad (34)$$

which can be solved for $s_1(t)$ if $||\mathbf{r}_1(t) - \mathbf{r}_2(t)||_2^2 < r_{\max}$, $t \in [t_1, t_2]$. In this case, the boundary conditions to (34) are given by

$$\begin{bmatrix} ||\mathbf{r}_1(t_1) - \mathbf{r}_2(t_1)||_2^2 - r_{\max} + \frac{1}{2}s_1^2(t_1) \\ r_{\min} - ||\mathbf{r}_1(t_1) - \mathbf{r}_2(t_1)||_2^2 + \frac{1}{2}s_2^2(t_1) \end{bmatrix} = \mathbf{0}_2,$$

$$\begin{bmatrix} ||\mathbf{r}_1(t_2) - \mathbf{r}_2(t_2)||_2^2 - r_{\max} + \frac{1}{2}s_1^2(t_2) \\ r_{\min} - ||\mathbf{r}_1(t_2) - \mathbf{r}_2(t_2)||_2^2 + \frac{1}{2}s_2^2(t_2) \end{bmatrix} = \mathbf{0}_2.$$

If, alternatively, $\mathbf{q}(t) = [\mathbf{x}_1^{\mathsf{T}}(t), \mathbf{r}_2^{\mathsf{T}}(t)]^{\mathsf{T}}$, then the formation's kinetic energy is given by

$$k\,(\mathbf{q}(t), \mathbf{q}_{\text{dot}}(t)) = \frac{1}{2}\sum_{i=1}^{2} m_i \mathbf{v}_i^{\mathsf{T}}(t)\mathbf{v}_i(t) + \frac{1}{2}\sum_{i=1}^{2} \boldsymbol{\omega}_1^{\mathsf{T}}(t)\mathbf{I}_{\text{in},i}\boldsymbol{\omega}_1(t) \qquad (35)$$

and the dynamic equations, obtained by applying (22) and (18), are given by

$$\mathbf{v}_1(t) = \frac{d\mathbf{r}_1(t)}{dt}, \quad \boldsymbol{\omega}_1(t) = \mathbf{R}_{\text{rod}}^{-1}(\sigma_1(t))\dot{\sigma}_1(t), \quad \mathbf{v}_2(t) = \frac{d\mathbf{r}_2(t)}{dt}, \qquad (36)$$

$$m_1 \frac{d}{dt}\mathbf{v}_1(t) = m_1\mathbf{a}\,(\tilde{\mathbf{x}}_1(t)) + m_1\mathbf{u}_{1,\text{tran}}(t), \qquad (37)$$

$$\mathbf{I}_{\text{in},1}\frac{d}{dt}\boldsymbol{\omega}_1(t) = -\boldsymbol{\omega}_1^{\times}\,(\boldsymbol{\omega}_1(t))\,\mathbf{I}_{\text{in},1}\boldsymbol{\omega}_1(t) + \mathbf{m}\,(\tilde{\mathbf{x}}_1(t)) + \mathbf{I}_{\text{in},1}\mathbf{u}_{1,\text{rot}}(t), \qquad (38)$$

$$m_2 \frac{d}{dt}\mathbf{v}_2(t) = m_2\mathbf{a}\,\left(\left[\mathbf{r}_2^{\mathsf{T}}(t), \mathbf{v}_2^{\mathsf{T}}(t), \sigma_1^{\mathsf{T}}(t), \boldsymbol{\omega}_1^{\mathsf{T}}(t)\right]^{\mathsf{T}}\right) + m_2\mathbf{u}_{2,\text{tran}}(t). \qquad (39)$$

The Lagrange coordinates chosen imply that the first vehicle can be considered as unconstrained, that is, subject to (3), (14), and (15) only, and therefore, the dynamic equations (36) – (38) can be directly deduced from (3). Similarly, the translational dynamics of the second vehicle can be considered as unconstrained. Thus, (39) can be directly obtained from (3). Recall from Example 4.1 that the components of $\mathbf{q}$ are suitable Lagrange coordinates if $r_{\min} < ||\mathbf{r}_1(t) - \mathbf{r}_2(t)||_2^2 < r_{\max}$, whereas (29) – (33) hold if $r_{\min} < ||\mathbf{r}_1(t) - \mathbf{r}_2(t)||_2^2 < r_{\max}$ and $r_{1,3}(t) \neq r_{2,3}(t)$. Thus, $\mathbf{q}(t) = [\mathbf{x}_1^{\mathsf{T}}(t), \mathbf{r}_2^{\mathsf{T}}(t)]^{\mathsf{T}}$ is a more convenient choice of Lagrange coordinates than $\mathbf{q}(t) = [s_1(t), r_{1,1}(t), r_{1,2}(t), \sigma_1^{\mathsf{T}}(t), \mathbf{r}_2^{\mathsf{T}}(t)]^{\mathsf{T}}$.

This example will be further elaborated on in Section 6 for $\mathbf{q}(t) = \left[\mathbf{x}_1^\mathsf{T}(t), \mathbf{r}_2^\mathsf{T}(t)\right]^\mathsf{T}$, and hence, for notational convenience define $\mathbf{f}_{dyn,2}\left(\tilde{\mathbf{x}}_2(t), \mathbf{u}_{2,tran}(t)\right) \triangleq \mathbf{a}\left(\tilde{\mathbf{x}}_2(t)\right) + \mathbf{u}_{2,tran}(t)$ and

$$
\mathbf{f}_{dyn,1}(\mathbf{x}_1, \mathbf{q}_{dot,1}(\mathbf{x}_1), \mathbf{u}_1) \triangleq
\begin{bmatrix}
\mathbf{a}\left(\tilde{\mathbf{x}}_1(t)\right) + \mathbf{u}_{1,tran}(t) \\
-\mathbf{I}_{in,1}^{-1}\boldsymbol{\omega}_1^\times\left(\boldsymbol{\omega}_1(t)\right)\mathbf{I}_{in,1}\boldsymbol{\omega}_1(t) + \tilde{\boldsymbol{\omega}}_i\left(\tilde{\mathbf{x}}_1(t)\right) + \mathbf{u}_{1,rot}(t)
\end{bmatrix}.
$$

## 4.4 Path planning optimization problem revisited

The trajectory optimization problem defined in Section 3.4 can be reformulated as follows. For all $i = 1, ..., n$ and $t \in [t_1, t_2]$, find $\mathbf{u}_{i,tran}(t)$ (respectively, $\mathbf{u}_{i,rot}(t)$) among all admissible controls in $\Gamma_{i,tran}$ (respectively, $\Gamma_{i,rot}$) such that the performance measure (2) is minimized and $\mathbf{q}(t)$ satisfies (24), (10), and (11).

By comparing this problem statement to the problem statement given in Section 3.4, it is clear that (5) and (6) are not explicitly accounted for in the above reformulation of the optimization problem. Hence, the constrained optimization problem has been reduced to an unconstrained optimization problem by the introduction of slack variables and Lagrange coordinates.

## 4.5 Transversality condition

Let $\mathbf{S} : D_1 \to D_2$, where $D_1 \subset \mathbb{R}^p$ and $D_2 \subset \mathbb{R}^m$, be a a continuously differentiable manifold and let the *manifold tangent* to $\mathbf{S}$ at $\mathbf{y}_0$ be given by

$$
\frac{\partial \mathbf{S}(\mathbf{y})}{\partial \mathbf{y}}\bigg|_{\mathbf{y}=\mathbf{y}_0} (\mathbf{y} - \mathbf{y}_0) = \mathbf{0}_m. \tag{40}
$$

Every vector $\mathbf{v} \in \mathbb{R}^p$ that is normal to the manifold tangent to $\mathbf{S}$ at $\mathbf{y}_0$, that is, $\mathbf{v}^\mathsf{T}\mathbf{y} = 0$ for all $\mathbf{y} \in \mathbb{R}^p$ such that (40) holds, is said to verify the *transversality condition* for $\mathbf{S}$ at $\mathbf{y}_0$.

## 4.6 Pontryagin's minimum principle

Assume that a set of Lagrange coordinates has been found and that the formation's dynamic equations can be written in the form given by (24). Define the *costate vectors* $\boldsymbol{\lambda}_{dot} : [t_1, t_2] \to \mathbb{R}^{6n-n_4}$ and $\boldsymbol{\lambda}_{dyn} : [t_1, t_2] \to \mathbb{R}^{6n-n_4}$ so that the *costate equation*

$$
\frac{d}{dt}\begin{bmatrix} \boldsymbol{\lambda}_{dot}(t) \\ \boldsymbol{\lambda}_{dyn}(t) \end{bmatrix} = -\left(\frac{\partial}{\partial[\mathbf{q}^\mathsf{T}, \mathbf{q}_{dot}^\mathsf{T}]^\mathsf{T}}\begin{bmatrix} \boldsymbol{\Psi}\left(\mathbf{q}(t)\right)\dot{\mathbf{q}}(t) + \psi\left(\mathbf{q}(t)\right) \\ \mathbf{f}_{dyn}(\mathbf{q}(t), \mathbf{q}_{dot}(t), \tilde{\mathbf{u}}(t)) \end{bmatrix}\right)^\mathsf{T}\begin{bmatrix} \boldsymbol{\lambda}_{dot}(t) \\ \boldsymbol{\lambda}_{dyn}(t) \end{bmatrix} \tag{41}
$$

holds. The boundary conditions for (41) are given in Theorem 4.2 below. Given $\lambda_0 \in \mathbb{R}$, define the *Hamiltonian function*

$$
\mathfrak{h}\left(\mathbf{q}(t), \mathbf{q}_{dot}(t), \tilde{\mathbf{u}}(t), \boldsymbol{\lambda}_{dyn}(t), \boldsymbol{\lambda}_{dot}(t)\right) \triangleq \lambda_0 \sum_{i=1}^{n} \mu_i \|\mathbf{u}_i(t)\|_2 + \boldsymbol{\lambda}_{dot}^\mathsf{T}(t)\mathbf{q}_{dot}(t)
$$
$$
+ \boldsymbol{\lambda}_{dyn}^\mathsf{T}(t)\mathbf{f}_{dyn}(\mathbf{q}(t), \mathbf{q}_{dot}(t), \tilde{\mathbf{u}}(t)). \tag{42}
$$

Finally, define

$$\mathfrak{m}\left(\mathbf{q}(t), \mathbf{q}_{\text{dot}}(t), \boldsymbol{\lambda}_{\text{dyn}}(t), \boldsymbol{\lambda}_{\text{dot}}(t)\right) \triangleq \min_{\tilde{\mathbf{u}} \in \prod_{i=1}^{n}(\Gamma_{i,\text{tran}} \times \Gamma_{i,\text{rot}})} \mathfrak{h}\left(\mathbf{q}(t), \mathbf{q}_{\text{dot}}(t), \tilde{\mathbf{u}}(t), \boldsymbol{\lambda}_{\text{dyn}}(t), \boldsymbol{\lambda}_{\text{dot}}(t)\right).$$
(43)

The following theorem is known as the *Pontryagin minimum principle*. For details on this theorem and its numerous applications to optimal control, see Pontryagin et al. (1962).

**Theorem 4.2.** *(Pontryagin et al., 1962) For all i = 1, ..., n, let $\mathbf{u}^*_{i,\text{tran}}(t)$ and $\mathbf{u}^*_{i,\text{rot}}(t)$, $t \in [t_1, t_2]$, be admissible controls in $\Gamma_{i,\text{tran}}$ and $\Gamma_{i,\text{rot}}$, respectively, such that $\mathbf{q}^*(t)$ satisfies (24), (10), and (11). If $\mathbf{u}^*_{i,\text{tran}}(t)$ and $\mathbf{u}^*_{i,\text{rot}}(t)$ solve the trajectory optimization problem stated in Section 4.4, then there exist $\lambda^*_0 \in \overline{\mathbb{R}}_+$, $\boldsymbol{\lambda}^*_{\text{dyn}}(t)$, and $\boldsymbol{\lambda}^*_{\text{dot}}(t)$ such that i) $|\lambda^*_0| + ||\boldsymbol{\lambda}^*_{\text{dyn}}(t)||_2 + ||\boldsymbol{\lambda}^*_{\text{dot}}(t)||_2 \neq 0$, $t \in [t_1, t_2]$, ii) (41) holds, iii) $\mathfrak{h}\left(\mathbf{q}^*(t), \tilde{\mathbf{u}}^*(t), \boldsymbol{\lambda}^*_{\text{dyn}}(t), \boldsymbol{\lambda}^*_{\text{dot}}(t)\right)$ attains its minimum almost everywhere on $[t_1, t_2]$ except on a finite number of points, and iv) $\boldsymbol{\lambda}^*_{\text{dyn}}(t_1)$ and $\boldsymbol{\lambda}^*_{\text{dot}}(t_1)$ (respectively, $\boldsymbol{\lambda}^*_{\text{dyn}}(t_2)$ and $\boldsymbol{\lambda}^*_{\text{dot}}(t_2)$) satisfy the transversality condition for $\mathbf{S}_1$ (respectively, $\mathbf{S}_2$) at $\mathbf{q}^*(t_1)$ (respectively, $\mathbf{q}^*(t_2)$).*

Pontryagin minimum principle is a necessary condition for optimality, and hence, it provides *candidate* optimal control vectors. Sufficient conditions for optimality that are currently available in the literature do not apply to the optimization problem discussed herein.

It is worth noting that, instead of introducing the Lagrange coordinates, the equality constraints (7) and (5) can be accounted for by introducing Lagrange multipliers. This approach requires modifying the assigned performance measure and introducing additional costate vectors (Giaquinta & Hildebrandt, 1996; Lee & Markus, 1968). The dynamics of the costate vectors are characterized by ordinary differential equations known as costate equations, which need to be integrated numerically together with the dynamic equations of the state vector. Therefore, the computational complexity of finding optimal trajectories for large formations increases drastically when Lagrange multipliers are employed (L'Afflitto & Sultan, 2010). Alternatively, finding a suitable set of Lagrange coordinates can be a demanding task and in some cases the Lagrange coordinates may not have physical meaning (Pars, 1965); however, this reduces the dimension of the costate equation and consequently reduces computational complexity.

Finally, we say the optimization problem is *normal* if $\lambda_0 \neq 0$, otherwise the optimization problem is *abnormal*. Normality can be shown by using the *Euler necessary condition*

$$\frac{\partial \mathfrak{h}\left(\mathbf{q}(t), \mathbf{q}_{\text{dot}}(t), \tilde{\mathbf{u}}(t), \boldsymbol{\lambda}_{\text{dyn}}(t), \boldsymbol{\lambda}_{\text{dot}}(t)\right)}{\partial \tilde{\mathbf{u}}}\Bigg|_{\tilde{\mathbf{u}}=\tilde{\mathbf{u}}^*} = \mathbf{0}_{6n}^\mathsf{T},$$
(44)

where $\tilde{\mathbf{u}}^*(t) \triangleq \left[[\mathbf{u}^{*T}_{1,\text{tran}}(t), \mathbf{u}^{*T}_{1,\text{rot}}(t)]^\mathsf{T}, ..., [\mathbf{u}^{*T}_{n,\text{tran}}(t), \mathbf{u}^{*T}_{n,\text{rot}}(t)]^\mathsf{T}\right]^\mathsf{T} \in \text{int}\left(\prod_{i=1}^{n}(\Gamma_{i,\text{tran}} \times \Gamma_{i,\text{rot}})\right)$. In particular, assume, *ad absurdum*, that $\lambda_0 = 0$. Now, if (41) and (44) imply that $\boldsymbol{\lambda}_{\text{dot}}(t) = \mathbf{0}_{6n-n_4}$ and $\boldsymbol{\lambda}_{\text{dyn}}(t) = \mathbf{0}_{6n-n_4}$ for some $t \in [t_1, t_2]$, then assertion i) of Theorem 4.2 is contradicted. Therefore, $\lambda_0 \neq 0$, and hence, the optimization problem is normal. In this case, we assume without loss of generality that $\lambda_0 = 1$.

## 5. Analytical and numerical approaches to the optimal path planning problem

Finding minimizers to (2) subject to the constraints (3) – (6) can be formulated as a Lagrange optimization problem (Ewing, 1969), which has been extensively studied both analytically and numerically in the literature. Analytical methods rely on either Lagrange's variational approach using calculus of variations or on the direct approach. In the classical variational approach, candidate minimizers for a given performance functional can be found by applying the Euler necessary condition. In order to find the minimizers, candidate optimal solutions need to be further tested by applying the Clebsh necessary condition, Jacobi necessary condition, Weierstrass necessary condition, as well as the associated sufficient conditions (Ewing, 1969; Giaquinta & Hildebrandt, 1996).

This classical analytical approach is not practical since applying the Euler necessary condition involves solving a differential-algebraic boundary value problem, whose analytical solutions are impossible to find for many practical problems of interest. Moreover, numerical solutions to this boundary value problem are affected by a strong sensitivity to the boundary conditions (Bryson, 1975). Furthermore, verifying the Jacobi necessary condition or the Weierstrass necessary condition can be a dauting task (L'Afflitto & Sultan, 2010).

A variational approach to the optimal path planning problem for a single vehicle, known as *primer vector theory*, was addressed by Lawden (1963). Lawden's problem was formulated using the assumptions that the acceleration vector **a** induced by external forces due to the environment is function of only the position vector, the vehicle is a 3 DoF point mass, and the state and control are only subject to equality constraints (Lawden, 1963). Primer vector theory is successfully employed in spacecraft trajectory optimization (Jamison & Coverstone, 2010), orbit transfers (Petropoulos & Russell, 2008), and optimal rendezvous problems (Zaitri et al., 2010), however, vehicles are often assumed to be point masses subject to only gravitational acceleration. Among the few studies on primer vector theory applied to vehicle formations, it is worth noting the work of Mailhe & Guzman (2004), where the formation initialization problem is addressed. Applications of primer vector theory to 6 DoF single vehicles have been employed to optimize the descent on Mars (Topcu et al., 2007). These studies, however, assume that the spacecraft is subject to a constant gravity acceleration, the control variables are the translational acceleration and the angular rates, and the translational acceleration can be pointed in any direction by rotating the vehicle.

Pontryagin's minimum principle is a variational method that is equivalent to the Weierstrass necessary condition with the advantage of addressing constraints on the control more effectively than applying the classical variational approach. State constraints need to be addressed by applying an optimal switching condition on the costate equation (Pontryagin et al., 1962), which generally increases the complexity of the problem. In the present formulation, the constraints on the formation are addressed by employing Lagrange coordinates, which does not introduce further conditions on the costate vector dynamics.

The direct approach in the calculus of variations, which is more recent than the variational approach, is based on defining a minimizing sequence of control functions $u_n(t)$ in some set $\Gamma$ such that $\lim_{n \to +\infty} u_n(t) = u(t)$ is a minimizer of the performance measure $J[u(\cdot)]$. To this end, the following conditions should be met. *i*) Compactness of $\Gamma$, so that a minimizing sequence contains a convergent subsequence, *ii*) closedness of $\Gamma$, so that the limit of such a subsequence is contained in $\Gamma$, and *iii*) lower semicontinuity of the sequence

$\{\mathbf{u}_n(\cdot)\}_{n=0}^{\infty}$, that is, if $\lim_{n \to +\infty} \mathbf{u}_n(t) = \mathbf{u}(t)$, then $J[\mathbf{u}(t)] \leq \liminf_{n \to +\infty} J[\mathbf{u}_n(t)]$, $\mathbf{u}_n \in \Gamma$. Finally, it is also worth noting that approximate analytical methods can be used to solve the optimal path planning problem such as shape-based approximation methods (Petropoulos & Longuski, 2004), which are generally less effective due to the arbitrary parameterization of the minimizers (Wall, 2008).

Most of the results on the fuel consumption optimization employ numerical methods (Betts, 1998), which can be categorized as indirect or direct. Indirect numerical methods, which mimic the variational approach, suffer from high computational complexity since adjoint variables must be introduced. Alternatively, direct numerical methods are computationally more efficient, however, they require casting the given problem into a parameter optimization problem (Herman & Conway, 1987). Among the numerical methods commonly in use, it is worth mentioning genetic algorithms (Seereram et al., 2000) and particle swarm optimizers (Hassan et al., 2005).

One of the contributions of the present paper is that it extends Lawden's results on primer vector theory to formations of vehicles modeled as 6 DoF rigid bodies subject to generic environmental forces and moments by applying Pontryagin's minimum principle. As in all classical variational methods, Pontryagin's minimum principle is not suitable for numerically computing the optimal trajectory of a formation. However, Pontryagin's minimum principle allows us to draw analytical conclusions since it provides a generalization of the necessary conditions used by Lawden (1963), allows us to formally implement bounded integrable functions as admissible controls, and allows us to account for control constraints. Prussing (2010) and Marec (1979) have used Pontryagin's minimum principle to address primer vector theory using the same assumptions as Lawden (1963). In contrast, the present work provides additional analytical results for generic mission scenarios and complex environmental conditions for which numerical results can be verified. Furthermore, this paper exploits some properties of the costate space and consequently provides further insight into the formation system dynamics problem.

## 6. Necessary conditions for optimality of UAV formation trajectories

The following propositions are needed to develop the necessary conditions for optimality of the UAV formation problem.

**Proposition 6.1.** *Consider the performance measure* $J_{formation}[\tilde{\mathbf{u}}(\cdot)]$ *given by (2). Then, there exists at least one* $\tilde{\mathbf{u}}^*$ *such that* $J_{formation}[\tilde{\mathbf{u}}^*(\cdot)] \leq J_{formation}[\tilde{\mathbf{u}}(\cdot)]$ *for all* $\tilde{\mathbf{u}} \in \prod_{i=1}^{n}(\Gamma_{i,tran} \times \Gamma_{i,rot})$.

*Proof.* Since the integrand of the performance measure (1) is a continuous function defined on the compact set $\Gamma_{i,tran} \times \Gamma_{i,rot}$, it follows from Weierstrass' theorem that (1) has a global minimizer on $\Gamma_{i,tran} \times \Gamma_{i,rot}$. Now, since $\mu_i \in [0,1]$ with $\sum_{i=1}^{n}\mu_i = 1$, the result is immediate. $\square$

**Proposition 6.2.** *Assume that the hypothesis of Theorem 4.1 hold. If* $\boldsymbol{\lambda}_{dot}^*(t) \in \mathcal{N}\left(\left.\dfrac{\partial \boldsymbol{\Psi}(\mathbf{q})}{\partial \mathbf{q}}\right|_{\mathbf{q}=\mathbf{q}^*} \dot{\mathbf{q}}^*(t)\right.$

$\left.+\left.\dfrac{\partial \boldsymbol{\psi}(\mathbf{q})}{\partial \mathbf{q}}\right|_{\mathbf{q}=\mathbf{q}^*}\right)$, *then the path planning problem is normal.*

*Proof.* First, note that the Hamiltonian function (42) can be rewritten as

$$\mathfrak{h}\left(\mathbf{q}(t), \mathbf{q}_{\text{dot}}(t), \tilde{\mathbf{u}}(t), \lambda_{\text{dyn}}(t), \lambda_{\text{dot}}(t)\right) = \lambda_0 \sum_{i=1}^{n} \mu_i \|\mathbf{u}_i(t)\|_2$$

$$+ \sum_{i=1}^{n} \mathbf{u}_{i,\text{tran}}(t) \frac{\partial \mathbf{v}_i\left(\mathbf{q}, \mathbf{q}_{\text{dot}}\right)}{\partial \mathbf{q}_{\text{dot}}} \lambda_{\text{dyn}}(t)$$

$$+ \sum_{i=1}^{n} \mathbf{u}_{i,\text{rot}}(t) \frac{\partial \boldsymbol{\omega}_i\left(\mathbf{q}, \mathbf{q}_{\text{dot}}\right)}{\partial \mathbf{q}_{\text{dot}}} \lambda_{\text{dyn}}(t)$$

$$+ \lambda_{\text{dyn}}^{\text{T}}(t) \hat{\mathbf{f}}_{\text{dyn}}(\mathbf{q}(t), \mathbf{q}_{\text{dot}}(t)) + \lambda_{\text{dot}}^{\text{T}}(t) \mathbf{q}_{\text{dot}}(t). \quad (45)$$

Furthemore, note that (44) implies that

$$\lambda_0^* \sum_{i=1}^{n} \mu_i \frac{\mathbf{u}_i^{*T}(t)}{\|\mathbf{u}_i^*(t)\|_2} = -\sum_{i=1}^{n} \left[ \left.\frac{\partial \mathbf{v}_i\left(\mathbf{q}, \mathbf{q}_{\text{dot}}\right)}{\partial \mathbf{q}_{\text{dot}}}\right|_{(\mathbf{q}^*, \mathbf{q}_{\text{dot}}^*)} \lambda_{\text{dyn}}^*(t), \left.\frac{\partial \boldsymbol{\omega}_i\left(\mathbf{q}, \mathbf{q}_{\text{dot}}\right)}{\partial \mathbf{q}_{\text{dot}}}\right|_{(\mathbf{q}^*, \mathbf{q}_{\text{dot}}^*)} \lambda_{\text{dyn}}^*(t) \right],$$

where $\tilde{\mathbf{u}}^* \in int\left(\Pi_{i=1}^n(\Gamma_{i,\text{tran}} \times \Gamma_{i,\text{rot}})\right)$ and where we use the subscript $(\mathbf{q}^*, \mathbf{q}_{\text{dot}}^*)$ for $(\mathbf{q}, \mathbf{q}_{\text{dot}}) = (\mathbf{q}^*, \mathbf{q}_{\text{dot}}^*)$. Now, assume, *ad absurdum*, that $\lambda_0^* = 0$ and note that $\frac{\partial \mathbf{v}_i(\mathbf{q}, \mathbf{q}_{\text{dot}})}{\partial \mathbf{q}_{\text{dot}}} = \frac{\partial \mathbf{v}_i(\mathbf{q}, \mathbf{q}_{\text{dot}})}{\partial \mathbf{q}} \frac{\partial \mathbf{q}}{\partial \mathbf{q}_{\text{dot}}}$ and $\frac{\partial \boldsymbol{\omega}_i(\mathbf{q}, \mathbf{q}_{\text{dot}})}{\partial \mathbf{q}_{\text{dot}}} = \frac{\partial \boldsymbol{\omega}_i(\mathbf{q}, \mathbf{q}_{\text{dot}})}{\partial \mathbf{q}} \frac{\partial \mathbf{q}}{\partial \mathbf{q}_{\text{dot}}}$. Since $\boldsymbol{\Psi}(\mathbf{q})$ is diffeomorphic and Theorem 4.1 holds, it follows that $\lambda_{\text{dyn}}^*(t) = \mathbf{0}_{6n-n_4}$. In this case, (41) can be explicitly written as

$$\frac{d}{dt}\begin{bmatrix} \lambda_{\text{dot}}^*(t) \\ \lambda_{\text{dyn}}^*(t) \end{bmatrix} = - \begin{bmatrix} \frac{\partial \boldsymbol{\Psi}(\mathbf{q})}{\partial \mathbf{q}}\dot{\mathbf{q}}(t) + \frac{\partial \boldsymbol{\psi}(\mathbf{q})}{\partial \mathbf{q}} & \mathbf{0}_{(6n-n_4)\times(6n-n_4)} \\ \frac{\partial \mathbf{f}_{\text{dyn}}(\mathbf{q}, \mathbf{q}_{\text{dot}}, \tilde{\mathbf{u}})}{\partial \mathbf{q}} & \frac{\partial \mathbf{f}_{\text{dyn}}(\mathbf{q}, \mathbf{q}_{\text{dot}}, \tilde{\mathbf{u}})}{\partial \mathbf{q}_{\text{dot}}} \end{bmatrix}_{(\mathbf{q}^*, \mathbf{q}_{\text{dot}}^*)}^{\text{T}} \begin{bmatrix} \lambda_{\text{dot}}^*(t) \\ \lambda_{\text{dyn}}^*(t) \end{bmatrix}, \quad (46)$$

and hence, $\lambda_{\text{dot}}^*(t) = \mathbf{0}_{6n-n_4}$, which contradicts *i*) of Theorem 4.2. $\square$

If follows from Proposition 6.2 that the path planning optimization problem for a constrained formation is abnormal. Example 6.1 below, however, shows that this problem is normal for unconstrained 3 DoF vehicles, which is a well known result in the literature (Lawden, 1963).

**Theorem 6.1.** *Consider the path planning optimization problem. If* $\sum_{i=1}^{n} \mathbf{u}_{i,\text{tran}}^*(t) \left.\frac{\partial \mathbf{v}_i(\mathbf{q}, \mathbf{q}_{\text{dot}})}{\partial \mathbf{q}_{\text{dot}}}\right|_{(\mathbf{q}^*, \mathbf{q}_{\text{dot}}^*)} +$ $\sum_{i=1}^{n} \mathbf{u}_{i,\text{rot}}^*(t) \left.\frac{\partial \boldsymbol{\omega}_i(\mathbf{q}, \mathbf{q}_{\text{dot}})}{\partial \mathbf{q}_{\text{dot}}}\right|_{(\mathbf{q}^*, \mathbf{q}_{\text{dot}}^*)}$ *and* $-\lambda_{\text{dyn}}^*(t)$ *are parallel, then the performance measure (2) is minimized. Moreover, for all* $i = 1, \ldots, n$, *the following conditions hold.*

i) *If* $\lambda_0^* \mu_i > \left\| \left.\frac{\partial \mathbf{v}_i(\mathbf{q}, \mathbf{q}_{\text{dot}})}{\partial \mathbf{q}_{\text{dot}}}\right|_{(\mathbf{q}^*, \mathbf{q}_{\text{dot}}^*)} \lambda_{\text{dyn}}^*(t) \right\|_2$, *then* $\mathbf{u}_{i,\text{tran}}^*(t) = \mathbf{0}_3$.

ii) *If* $\lambda_0^* \mu_i > \left\| \left.\frac{\partial \boldsymbol{\omega}_i(\mathbf{q}, \mathbf{q}_{\text{dot}})}{\partial \mathbf{q}_{\text{dot}}}\right|_{(\mathbf{q}^*, \mathbf{q}_{\text{dot}}^*)} \lambda_{\text{dyn}}^*(t) \right\|_2$, *then* $\mathbf{u}_{i,\text{rot}}^*(t) = \mathbf{0}_3$.

iii) If $\lambda_0^* \mu_i < \left\| \dfrac{\partial \mathbf{v}_i(\mathbf{q}, \mathbf{q}_{dot})}{\partial \mathbf{q}_{dot}} \right|_{(\mathbf{q}^*, \mathbf{q}_{dot}^*)} \boldsymbol{\lambda}_{dyn}^*(t) \right\|_2$, then $\mathbf{u}_{i,tran}^*(t) = \rho_{i,2}$.

iv) If $\lambda_0^* \mu_i < \left\| \dfrac{\partial \boldsymbol{\omega}_i(\mathbf{q}, \mathbf{q}_{dot})}{\partial \mathbf{q}_{dot}} \right|_{(\mathbf{q}^*, \mathbf{q}_{dot}^*)} \boldsymbol{\lambda}_{dyn}^*(t) \right\|_2$, then $\mathbf{u}_{i,rot}^*(t) = \rho_{i,4}$.

v) If $\lambda_0^* \mu_i = \left\| \dfrac{\partial \mathbf{v}_i(\mathbf{q}, \mathbf{q}_{dot})}{\partial \mathbf{q}_{dot}} \right|_{(\mathbf{q}^*, \mathbf{q}_{dot}^*)} \boldsymbol{\lambda}_{dyn}^*(t) \right\|_2$, then $\mathbf{u}_{i,tran}^*(t)$ is unspecified.

vi) If $\lambda_0^* \mu_i = \left\| \dfrac{\partial \boldsymbol{\omega}_i(\mathbf{q}, \mathbf{q}_{dot})}{\partial \mathbf{q}_{dot}} \right|_{(\mathbf{q}^*, \mathbf{q}_{dot}^*)} \boldsymbol{\lambda}_{dyn}^*(t) \right\|_2$, then $\mathbf{u}_{i,rot}^*(t)$ is unspecified.

*Proof.* It follows from (45) that $\mathfrak{h}\left(\mathbf{q}(t), \tilde{\mathbf{u}}(t), \boldsymbol{\lambda}_{dyn}(t), \boldsymbol{\lambda}_{dot}(t)\right)$ is minimized if, for all $i = 1, \ldots, n$, $-\dfrac{\partial \mathbf{v}_i(\mathbf{q}, \mathbf{q}_{dot})}{\partial \mathbf{q}_{dot}}\bigg|_{(\mathbf{q}^*, \mathbf{q}_{dot}^*)} \boldsymbol{\lambda}_{dyn}^*(t)$ is parallel to $\mathbf{u}_{i,tran}^*(t)$ and if $-\dfrac{\partial \boldsymbol{\omega}_i(\mathbf{q}, \mathbf{q}_{dot})}{\partial \mathbf{q}_{dot}}\bigg|_{(\mathbf{q}^*, \mathbf{q}_{dot}^*)} \boldsymbol{\lambda}_{dyn}^*(t)$ is parallel to $\mathbf{u}_{i,rot}^*(t)$. Thus, using the triangular inequality, it follows that

$$\mathfrak{h}\left(\mathbf{q}^*(t), \tilde{\mathbf{u}}^*(t), \boldsymbol{\lambda}_{dyn}^*(t), \boldsymbol{\lambda}_{dot}^*(t)\right) - \boldsymbol{\lambda}_{dot}^{*T}(t)\mathbf{q}_{dot}(\mathbf{q}^*(t)) - \boldsymbol{\lambda}_{dyn}^T(t)\hat{\mathbf{f}}_{dyn}(\mathbf{q}^*(t), \mathbf{q}_{dot}(\mathbf{q}^*(t)))$$

$$\leq \sum_{i=1}^{n}\left[\left(\lambda_0^* \mu_i - \left\| \dfrac{\partial \mathbf{v}_i(\mathbf{q}, \mathbf{q}_{dot})}{\partial \mathbf{q}_{dot}} \right|_{(\mathbf{q}^*, \mathbf{q}_{dot}^*)} \boldsymbol{\lambda}_{dyn}^*(t) \right\|_2 \right) \|\mathbf{u}_{i,tran}^*(t)\|_2\right]$$

$$+ \sum_{i=1}^{n}\left[\left(\lambda_0^* \mu_i - \left\| \dfrac{\partial \boldsymbol{\omega}_i(\mathbf{q}, \mathbf{q}_{dot})}{\partial \mathbf{q}_{dot}} \right|_{(\mathbf{q}^*, \mathbf{q}_{dot}^*)} \boldsymbol{\lambda}_{dyn}^*(t) \right\|_2 \right) \|\mathbf{u}_{i,rot}^*(t)\|_2\right], \qquad (47)$$

which proves i) – iv). Next, if $\lambda_0^* \mu_i = \left\| \dfrac{\partial \mathbf{v}_i(\mathbf{q}, \mathbf{q}_{dot})}{\partial \mathbf{q}_{dot}} \right|_{(\mathbf{q}^*, \mathbf{q}_{dot}^*)} \boldsymbol{\lambda}_{dyn}^*(t) \right\|_2$ (respectively, $\lambda_0^* \mu_i = \left\| \dfrac{\partial \boldsymbol{\omega}_i(\mathbf{q}, \mathbf{q}_{dot})}{\partial \mathbf{q}_{dot}} \right|_{(\mathbf{q}^*, \mathbf{q}_{dot}^*)} \boldsymbol{\lambda}_{dyn}^*(t) \right\|_2$), then Pontryagin's minimum principle does not provide any information about the optimal control, and hence, v) and vi) hold. □

Analogous to Lawden's (Lawden, 1963) primer vector theory, $\dfrac{\partial \mathbf{v}_i(\mathbf{q}, \mathbf{q}_{dot})}{\partial \mathbf{q}_{dot}}\bigg|_{(\mathbf{q}^*, \mathbf{q}_{dot}^*)} \boldsymbol{\lambda}_{dyn}^*(t)$ and $\dfrac{\partial \boldsymbol{\omega}_i(\mathbf{q}, \mathbf{q}_{dot})}{\partial \mathbf{q}_{dot}}\bigg|_{(\mathbf{q}^*, \mathbf{q}_{dot}^*)} \boldsymbol{\lambda}_{dyn}^*(t)$ determine the magnitude and the direction of the control forces, and hence, we denote them as the *translational primer vector* and the *rotational primer vector*, respectively. Moreover, the trajectory given by each of the cases in Theorem 6.1 are called *arcs*. For each $i = 1, \ldots, n$, the arcs corresponding to i) (respectively, ii)) are called *maximum translational* (respectively, *rotational*) *thrust arcs*. Similarly, arcs corresponding to iii) (respectively, iv)) are called *null translational* (respectively, *rotational*) *thrust arcs*. Finally, arcs corresponding to v) (respectively, vi)) are called *singular translational* (respectively, *rotational*) *thrust arcs*. The optimal translational and rotational control vectors for v) and vi) in Theorem

6.1 need to be deduced by applying the generalized Legendre-Clebsch condition (Giaquinta & Hildebrandt, 1996).

**Theorem 6.2.** *Consider the path planning optimization problem. Then, there exists* $c^* \in \mathbb{R}$ *such that*

$$\mathfrak{m}\left(\mathbf{q}^*(t), \mathbf{q}^*_{\text{dot}}(t), \boldsymbol{\lambda}^*_{\text{dyn}}(t), \boldsymbol{\lambda}^*_{\text{dot}}(t)\right) = c^*. \tag{48}$$

*Proof.* It follows from the Weierstrass - Erdmann condition (Giaquinta & Hildebrandt, 1996) that on an optimal trajectory,

$$\frac{d}{dt}\mathfrak{h}\left(\mathbf{q}^*(t), \mathbf{q}^*_{\text{dot}}(t), \tilde{\mathbf{u}}^*(t), \boldsymbol{\lambda}^*_{\text{dyn}}(t), \boldsymbol{\lambda}^*_{\text{dot}}(t)\right) = \frac{\partial}{\partial t}\mathfrak{h}\left(\mathbf{q}(t), \mathbf{q}^*_{\text{dot}}(t), \tilde{\mathbf{u}}^*(t), \boldsymbol{\lambda}^*_{\text{dyn}}(t), \boldsymbol{\lambda}^*_{\text{dot}}(t)\right)$$

holds for all $t \in (t_1, t_2)$. Now, since $\mathfrak{h}$ does not explicitly depend on $t$, it follows that there exists $c^* \in \mathbb{R}$ such that $\mathfrak{h}\left(\mathbf{q}^*(t), \mathbf{q}^*_{\text{dot}}(t), \mathbf{0}_{6n}, \boldsymbol{\lambda}^*_{\text{dyn}}(t), \boldsymbol{\lambda}^*_{\text{dot}}(t)\right) = c^*$, which proves (48). $\square$

**Proposition 6.3.** *Consider the costate dynamics given by* (46). *Then, the dynamics of* $\boldsymbol{\lambda}^*_{\text{dyn}}(t)$ *are decoupled from the dynamics of* $\boldsymbol{\lambda}^*_{\text{dot}}(t)$.

*Proof.* The result is immediate from the form of (46). $\square$

It follows from Proposition 6.3 that the translational primer vector and the rotational primer vector dynamics are independent of the choice of $\mathbf{q}_{\text{dot}}$. Moreover, in solving for $\boldsymbol{\lambda}^*_{\text{dyn}}(t)$ we need not integrate a system of $2(6n - n_4)$ ordinary differential equations as in (41), but rather a system of $(6n - n_4)$ ordinary differential equations, which is very advantageous for large formations.

**Proposition 6.4.** *The translational primer vector and the rotational primer vector are continuously differentiable functions.*

*Proof.* First, note that $\boldsymbol{\lambda}^*_{\text{dyn}}(\cdot)$ and $\boldsymbol{\lambda}^*_{\text{dot}}(\cdot)$ are continuous with continuous derivatives almost everywhere on $t \in (t_1, t_2)$ except for a finite number of points (Pontryagin et al., 1962). Next, the differentiability assumption on the environmental model for $\mathbf{a}(\cdot)$ and $\mathbf{m}(\cdot)$ implies that the matrix on the right-hand side of (41) is of class $C^1(\mathbb{R}^{6n-n_4} \times \mathbb{R}^{6n-n_4} \times \mathbb{R}^{12n})$. Hence, $\frac{d}{dt}\boldsymbol{\lambda}^*_{\text{dyn}}(\cdot)$ and $\frac{d}{dt}\boldsymbol{\lambda}^*_{\text{dot}}(\cdot)$ are continuous on $(t_1, t_2)$. $\square$

In order to elucidate the translational primer vector and rotational primer vector dynamics for a vehicle formation problem, we focus on specific formation configurations and on a specific environmental model. Hence, in the reminder of the paper we concentrate on the case where $n_v$ components of $\mathbf{v}_i$ and $n_\omega$ components of $\boldsymbol{\omega}_i$ are also components of $\mathbf{q}_{\text{dot}}$. A justification for this model is as follows. Assume that the i-th formation vehicle behaves as unconstrained, e.g., the first vehicle in Examples 4.1 and 4.2, or the dynamics of the i-th vehicle can be addressed as partly unconstrained, e.g., the second formation vehicle in the aforementioned examples. In either of these cases, it is natural to choose the unconstrained components of $\mathbf{v}_i$ and $\boldsymbol{\omega}_i$ as some of the components of $\mathbf{q}_{\text{dot}}$. This model includes the classical formation configuration known as the *leader-follower* model, whose trajectories are computed as a function of the leader's path (Wang, 1991).

To simplify the environmental model assume that

$$\mathbf{a}\left(\widetilde{\mathbf{x}}_i\left(\mathbf{q}(t), \mathbf{q}_{\mathrm{dot}}(t)\right)\right) = \mathbf{a}\left(\left[0_3^{\mathrm{T}}, \mathbf{v}_i^{\mathrm{T}}(\mathbf{q}(t), \mathbf{q}_{\mathrm{dot}}(t)), 0_3^{\mathrm{T}}, 0_3^{\mathrm{T}}\right]^{\mathrm{T}}\right), \tag{49}$$

$$\widetilde{\boldsymbol{\omega}}_i\left(\widetilde{\mathbf{x}}_i\left(\mathbf{q}(t), \mathbf{q}_{\mathrm{dot}}(t)\right)\right) = \widetilde{\boldsymbol{\omega}}_i\left(\left[0_3^{\mathrm{T}}, \mathbf{v}_i^{\mathrm{T}}(\mathbf{q}(t), \mathbf{q}_{\mathrm{dot}}(t)), 0_3^{\mathrm{T}}, \boldsymbol{\omega}_i^{\mathrm{T}}(\mathbf{q}(t), \mathbf{q}_{\mathrm{dot}}(t))\right]^{\mathrm{T}}\right). \tag{50}$$

For notational convenience, we will refer to (49) and (50) as $\mathbf{a}(\mathbf{v}_i(t))$ and $\widetilde{\boldsymbol{\omega}}_i(\mathbf{v}_i(t), \boldsymbol{\omega}_i(t))$, respectively. This assumption on the accelerations induced by external forces and external moments is justified by a common environmental model given by (Anderson, 2001)

$$\mathbf{a}\left(\widetilde{\mathbf{x}}_i\left(\mathbf{q}(t), \mathbf{q}_{\mathrm{dot}}(t)\right)\right) = \mathbf{g} + \|\mathbf{v}_i(t)\|_2^2\left(-k_{i,\mathrm{D}}\widehat{\mathbf{v}}_i(t) + k_{i,\mathrm{L}}\widehat{\mathbf{v}}_i^{\mathrm{L}}(t) - k_{i,\mathrm{S}}\widehat{\mathbf{v}}_i^{\mathrm{S}}(t)\right), \tag{51}$$

$$\mathbf{m}\left(\widetilde{\mathbf{x}}_i\left(\mathbf{q}(t), \mathbf{q}_{\mathrm{dot}}(t)\right)\right) = \|\mathbf{v}_i(t)\|_2^2\left(k_{i,\mathrm{R}}\widehat{\boldsymbol{\omega}}_i^{\mathrm{R}}(t) + k_{i,\mathrm{P}}\widehat{\boldsymbol{\omega}}_i^{\mathrm{P}}(t) + k_{i,\mathrm{Y}}\widehat{\boldsymbol{\omega}}_i^{\mathrm{Y}}(t)\right), \tag{52}$$

where $\mathbf{g}$ is the constant gravitational acceleration, $\widehat{\mathbf{v}}_i \triangleq \mathbf{v}_i/\|\mathbf{v}_i\|_2$, $\widehat{\mathbf{v}}_i^{\mathrm{L}} : [t_1, t_2] \rightarrow \mathbb{R}^3$ (respectively, $\widehat{\mathbf{v}}_i^{\mathrm{S}} : [t_1, t_2] \rightarrow \mathbb{R}^3$) is the unit vector in the direction of the aerodynamic lift (respectively, in the direction opposite to the aerodynamic side force), $\widehat{\boldsymbol{\omega}}_i^{\mathrm{R}} : [t_1, t_2] \rightarrow \mathbb{R}^3$ (respectively, $\widehat{\boldsymbol{\omega}}_i^{\mathrm{P}} : [t_1, t_2] \rightarrow \mathbb{R}^3$ and $\widehat{\boldsymbol{\omega}}_i^{\mathrm{Y}} : [t_1, t_2] \rightarrow \mathbb{R}^3$) is the unit vector in the direction of roll (respectively, pitch and yaw), and $k_{i,\mathrm{D}}$, $k_{i,\mathrm{L}}$, $k_{i,\mathrm{S}}$, $k_{i,\mathrm{R}}$, $k_{i,\mathrm{P}}$, and $k_{i,\mathrm{Y}}$, are the drag, lift, side force, roll, pitch, and yaw coefficients, respectively.

Using the above assumptions, it follows from (22) that

$$\dot{\mathbf{v}}_i(t) = \widehat{\mathbf{a}}(\mathbf{v}_i(t)) + \widehat{\mathbf{u}}_{i,\mathrm{tran}}(t), \tag{53}$$

$$\dot{\boldsymbol{\omega}}_i(t) = \widehat{\widetilde{\boldsymbol{\omega}}}_i(\mathbf{v}_i(t), \boldsymbol{\omega}_i(t)) + \widehat{\mathbf{u}}_{i,\mathrm{rot}}(t), \tag{54}$$

where $\widehat{\mathbf{v}}_i : [t_1, t_2] \rightarrow \mathbb{R}^{n_v}$ (respectively, $\widehat{\boldsymbol{\omega}}_i : [t_1, t_2] \rightarrow \mathbb{R}^{n_\omega}$) represents the components of $\mathbf{v}_i(\mathbf{q}(t), \mathbf{q}_{\mathrm{dot}}(t))$ (respectively, $\boldsymbol{\omega}_i(\mathbf{q}(t), \mathbf{q}_{\mathrm{dot}}(t))$) that are also components of $\mathbf{q}_{\mathrm{dot}}(t)$, and $\widehat{\mathbf{a}} : \mathbb{R}^3 \rightarrow \mathbb{R}^{n_v}$ and $\widehat{\mathbf{u}}_{i,\mathrm{tran}} : [t_1, t_2] \rightarrow \mathbb{R}^{n_v}$ (respectively, $\widehat{\widetilde{\boldsymbol{\omega}}}_i : \mathbb{R}^3 \times \mathbb{R}^3 \rightarrow \mathbb{R}^{n_\omega}$ and $\widehat{\mathbf{u}}_{i,\mathrm{rot}} : [t_1, t_2] \rightarrow \mathbb{R}^{n_\omega}$) are the corresponding components of $\mathbf{a}(\mathbf{v}_i(t))$ and $\mathbf{u}_{i,\mathrm{tran}}(t)$ (respectively, $\widetilde{\boldsymbol{\omega}}_i(\mathbf{v}_i(t), \boldsymbol{\omega}_i(t))$ and $\mathbf{u}_{i,\mathrm{rot}}(t)$).

Next, it follows from (46), (53), and (54) that

$$\frac{d}{dt}\begin{bmatrix} \boldsymbol{\lambda}_{\mathrm{dyn},i,\widehat{v}}^*(t) \\ \boldsymbol{\lambda}_{\mathrm{dyn},i,\widehat{\omega}}^*(t) \end{bmatrix} = -\begin{bmatrix} \left(\dfrac{\partial \widehat{\mathbf{a}}(\mathbf{v}_i)}{\partial \widehat{\mathbf{v}}_i}\right)^{\mathrm{T}} & \left(\dfrac{\partial \widehat{\widetilde{\boldsymbol{\omega}}}_i(\mathbf{v}_i, \boldsymbol{\omega}_i)}{\partial \widehat{\mathbf{v}}_i}\right)^{\mathrm{T}} \\ 0_{n_\omega \times n_v} & \left(\dfrac{\partial \widehat{\widetilde{\boldsymbol{\omega}}}_i(\mathbf{v}_i, \boldsymbol{\omega}_i)}{\partial \widehat{\boldsymbol{\omega}}_i}\right)^{\mathrm{T}} \end{bmatrix}_{(\widehat{\mathbf{v}}_i^*, \widehat{\boldsymbol{\omega}}_i^*)} \begin{bmatrix} \boldsymbol{\lambda}_{\mathrm{dyn},i,\widehat{v}}^*(t) \\ \boldsymbol{\lambda}_{\mathrm{dyn},i,\widehat{\omega}}^*(t) \end{bmatrix}, \tag{55}$$

where $\boldsymbol{\lambda}_{\mathrm{dyn},i,\widehat{v}} : [t_1, t_2] \rightarrow \mathbb{R}^{n_v}$ and $\boldsymbol{\lambda}_{\mathrm{dyn},i,\widehat{\omega}(t)} : [t_1, t_2] \rightarrow \mathbb{R}^{n_\omega}$ are the $n_v$ and $n_\omega$ components of $\boldsymbol{\lambda}_{\mathrm{dyn},i}^*(t)$ corresponding to the $n_v$ and $n_\omega$ components of $\dot{\mathbf{v}}_i(t)$ and $\dot{\boldsymbol{\omega}}_i(t)$, respectively.

**Theorem 6.3.** *Assume that* $\|\hat{\mathbf{u}}_{i,\text{tran}}^*(t)\|_2 = \hat{\rho}_{i,\text{tran}}$, $\|\hat{\mathbf{u}}_{i,\text{rot}}^*(t)\|_2 = \hat{\rho}_{i,\text{rot}}$, $\|\mathbf{u}_{i,\text{tran}}^*(t)\|_2 = \rho_{i,\text{tran}}$, $\|\mathbf{u}_{i,\text{rot}}^*(t)\|_2 = \rho_{i,\text{rot}}$, where $\hat{\rho}_{i,\text{tran}}$ and $\rho_{i,\text{tran}} \in (\rho_{i,1}, \rho_{i,2})$, $\hat{\rho}_{i,\text{rot}}$ and $\rho_{i,\text{rot}} \in (\rho_{i,3}, \rho_{i,4})$, and $\left[\frac{\partial \hat{\omega}_i(\mathbf{v}_i, \omega_i)}{\partial \hat{\omega}_i}\right]_{(\mathbf{v}_i^*, \omega_i^*)}$ *is invertible. Then,*

$$\left\| \left[\frac{\partial \hat{\omega}_i(\mathbf{v}_i, \omega_i)}{\partial \hat{\omega}_i}\right]_{(\mathbf{v}_i^*, \omega_i^*)}^{-T} \left(\mathbf{I}_3 + \left[\frac{\partial \hat{\mathbf{a}}_i(\mathbf{v}_i)}{\partial \hat{\mathbf{v}}_i}\right]_{\mathbf{v}_i^*}^T\right) \mathbf{u}_{i,\text{tran}}^*(t) \right\|_2 \leq \sqrt{\rho_{i,\text{tran}}^2 + \rho_{i,\text{rot}}^2}, \tag{56}$$

$$\left\| \left[\frac{\partial \hat{\omega}_i(\mathbf{v}_i, \omega_i)}{\partial \hat{\omega}_i}\right]_{(\mathbf{v}_i^*, \omega_i^*)}^{-T} \dot{\mathbf{u}}_{i,\text{rot}}^*(t) \right\|_2 \leq \sqrt{\rho_{i,\text{tran}}^2 + \rho_{i,\text{rot}}^2}. \tag{57}$$

*Proof.* It follows from (44), (42), (53), and (54) that

$$\lambda_0^* \mu_i \frac{\hat{\mathbf{u}}_{i,\text{tran}}^*(t)}{\|\tilde{\mathbf{u}}^*(t)\|_2} = -\lambda_{\text{dyn},i,\hat{v}}^*(t),$$

$$\lambda_0^* \mu_i \frac{\hat{\mathbf{u}}_{i,\text{rot}}^*(t)}{\|\tilde{\mathbf{u}}^*(t)\|_2} = -\lambda_{\text{rot},i,\hat{\omega}(t)}^*(t). \tag{58}$$

Recalling that $\left\| \frac{\hat{\mathbf{u}}_{i,\text{rot}}^*(t)}{\|\tilde{\mathbf{u}}^*(t)\|_2} \right\|_2 \leq 1$ and using (55) and (58) we obtain

$$\left\| \lambda_0^* \mu_i \left[\frac{\partial \hat{\omega}_i(\mathbf{v}_i, \omega_i)}{\partial \hat{\omega}_i}\right]_{(\mathbf{v}_i^*, \omega_i^*)}^{-T} \frac{\|\tilde{\mathbf{u}}_i^*(t)\|_2 \dot{\hat{\mathbf{u}}}_{i,\text{tran}}^*(t) + \dot{\tilde{\mathbf{u}}}_i^{*T}(t) \tilde{\mathbf{u}}_i^*(t) \hat{\mathbf{u}}_{i,\text{tran}}^*(t)}{\|\tilde{\mathbf{u}}_i^*(t)\|_2^2} \right.$$

$$\left. + \lambda_0^* \mu_i \left[\frac{\partial \hat{\omega}_i(\mathbf{v}_i, \omega_i)}{\partial \hat{\omega}_i}\right]_{(\mathbf{v}_i^*, \omega_i^*)}^{-T} \left[\frac{\partial \hat{\mathbf{a}}_i(\mathbf{v}_i)}{\partial \hat{\mathbf{v}}_i}\right]_{\mathbf{v}_i^*}^T \frac{\hat{\mathbf{u}}_{i,\text{tran}}^*(t)}{\|\tilde{\mathbf{u}}_i^*(t)\|_2} \right\|_2 \leq \lambda_0^* \mu_i,$$

$$\left\| \lambda_0^* \mu_i \left[\frac{\partial \hat{\omega}_i(\mathbf{v}_i, \omega_i)}{\partial \hat{\omega}_i}\right]_{(\mathbf{v}_i^*, \omega_i^*)}^{-T} \frac{\|\tilde{\mathbf{u}}_i^*(t)\|_2 \dot{\hat{\mathbf{u}}}_{i,\text{rot}}^*(t) + \dot{\tilde{\mathbf{u}}}_i^{*T}(t) \tilde{\mathbf{u}}_i^*(t) \hat{\mathbf{u}}_{i,\text{rot}}^*(t)}{\|\tilde{\mathbf{u}}_i^*(t)\|_2^2} \right\|_2 \leq \lambda_0^* \mu_i.$$

Now, noting that $\dot{\tilde{\mathbf{u}}}_i^{*T}(t) \tilde{\mathbf{u}}_i^*(t) = 0$, the result follows. $\square$

Since Theorem 6.3 is proven using the Euler necessary condition, it follows that $(\mathbf{u}_{i,\text{tran}}^*, \mathbf{u}_{i,\text{rot}}^*) \in \text{int}(\Gamma_{i,\text{tran}} \times \text{int}(\Gamma_{i,\text{rot}}))$. However, the parameter bounds $\rho_{i,j}$, $j = 1, 2, 3, 4$, are imposed by physical and not mathematical considerations, and hence, for practical applications we can assume that there exists $\epsilon > 0$ such that Theorem 6.3 holds for $\rho_{i,\text{tran}} \in (\rho_{i,1} - \epsilon, \rho_{i,2} + \epsilon)$ and $\rho_{i,\text{rot}} \in (\rho_{i,3} - \epsilon, \rho_{i,4} + \epsilon)$. Consequently, for engineering applications we can assume that Theorem 6.3 also holds on arcs of maximum translational and rotational thrust.

**Corollary 6.1.** *Assume that the hypothesis of Theorem 6.3 hold. If* $n_\omega = 0$, *then*

$$\left\| \left[ \frac{\partial \hat{a}_i(v_i)}{\partial \hat{v}_i} \right]_{v_i^*}^{-T} \hat{u}_{i,\text{tran}}^*(t) \right\|_2 \leq \sqrt{\rho_{i,\text{tran}}^2 + \rho_{i,\text{rot}}^2}. \tag{59}$$

*Alternatively, if* $n_v = 0$, *then*

$$\left\| \left[ \frac{\partial \hat{\omega}_i(v_i, \omega_i)}{\partial \hat{\omega}_i} \right]_{(v_i^*, \omega_i^*)}^{-T} \hat{u}_{i,\text{rot}}^*(t) \right\|_2 \leq \sqrt{\rho_{i,\text{tran}}^2 + \rho_{i,\text{rot}}^2}. \tag{60}$$

*Proof.* The proof is a direct consequence of Theorem 6.3.                □

**Example 6.1.** Consider the formation of the two vehicles addressed in Examples 4.1 and 4.2, and assume that $q(t) = [x_1^T(t), r_2^T(t)]^T$. As shown in Example 4.2, if $r_{\min} < ||r_1(t) - r_2(t)||_2^2 < r_{\max}$, then the first vehicle and the translational dynamics of the second vehicle can be considered unconstrained. Thus, the costate equation (41) can be rewritten as two decoupled ordinary differential equations given by

$$\frac{d}{dt} \begin{bmatrix} \lambda_{\text{dot},1}(t) \\ \lambda_{\text{dyn},1}(t) \end{bmatrix} = - \begin{bmatrix} \begin{bmatrix} 0_{3\times3} & 0_{3\times3} \\ 0_{3\times3} & \frac{\partial R_{\text{rod}}^{-1}(\sigma_1)}{\partial \sigma_1} \dot{\sigma}_1 \end{bmatrix}^T & 0_{6\times6} \\ \left( \frac{\partial f_{\text{dyn},1}(x_1, q_{\text{dot},1}(x_1), u_1)}{\partial x_1} \right)^T & \left( \frac{\partial f_{\text{dyn},1}(x_1, q_{\text{dot},1}(q), u_1)}{\partial q_{\text{dot},1}} \right)^T \end{bmatrix} \begin{bmatrix} \lambda_{\text{dot},1}(t) \\ \lambda_{\text{dyn},1}(t) \end{bmatrix}, \tag{61}$$

$$\frac{d}{dt} \begin{bmatrix} \lambda_{\text{dot},2}(t) \\ \lambda_{\text{dyn},2}(t) \end{bmatrix} = - \begin{bmatrix} 0_{3\times3} & \frac{\partial f_{\text{dyn},2}(\tilde{x}_2(t), u_{2,\text{tran}}(t))}{\partial r_2} \\ 0_{3\times3} & \frac{\partial f_{\text{dyn},2}(\tilde{x}_2(t), u_{2,\text{tran}}(t))}{\partial v_2} \end{bmatrix}^T \begin{bmatrix} \lambda_{\text{dot},2}(t) \\ \lambda_{\text{dyn},2}(t) \end{bmatrix}, \tag{62}$$

where $\lambda_{\text{dyn}}(t) \triangleq [\lambda_{\text{dyn},1}^T(t) \ \lambda_{\text{dyn},2}^T(t)]^T$, $\lambda_{\text{dyn},1} : [t_1, t_2] \to \mathbb{R}^6$, $\lambda_{\text{dyn},2} : [t_1, t_2] \to \mathbb{R}^3$, $\lambda_{\text{dot}}(t) \triangleq [\lambda_{\text{dot},1}^T(t), \lambda_{\text{dot},2}^T(t)]^T$, $\lambda_{\text{dot},1} : [t_1, t_2] \to \mathbb{R}^6$, and $\lambda_{\text{dot},2} : [t_1, t_2] \to \mathbb{R}^3$.

From (61) and (62) it follows that the path planning optimization problem for the first vehicle is possibly abnormal since we cannot verify a priori whether or not

$$\lambda_{\text{dot},1}^*(t) \in \mathcal{N} \left( \begin{bmatrix} 0_{3\times3} & 0_{3\times3} \\ 0_{3\times3} & \frac{\partial R_{\text{rod}}^{-1}(\sigma_1)}{\partial \sigma_1} \end{bmatrix} \dot{\sigma}_1(q(t)) \right]_{q=q^*} \right),$$

whereas the path planning optimization problem for the second vehicle is normal since its rotational dynamics are not expressed by (62). Normality for the second formation vehicle can also be proven by rewriting the unconstrained dynamic equations (3) for a 3 DoF vehicle. For details, see L'Afflitto & Sultan (2008).

Using (18) it follows that (45) can be written as

$$\mathfrak{h}\left( q(t), q_{\text{dot}}(t), \tilde{u}(t), \lambda_{\text{dyn}}(t), \lambda_{\text{dot}}(t) \right) = \mathfrak{h}_1 \left( x_1(t), u_1(t), \lambda_{\text{dyn},1}(t), \lambda_{\text{dot},1}(t) \right)$$

$$+ \mathfrak{h}_2 \left( x_2(t), u_{2,\text{tran}}(t), \lambda_{\text{dyn},2}(t) \right), \tag{63}$$

where

$$\mathfrak{h}_1\left(\mathbf{x}_1(t), \mathbf{u}_1(t), \lambda_{\text{dyn},1}(t), \lambda_{\text{dot},1}(t)\right) = \lambda_0 \mu_1 ||\mathbf{u}_1(t)||_2 + \lambda_{\text{dyn},1,1}^{\text{T}}(t)\mathbf{u}_{1,\text{tran}}(t)$$
$$+ \lambda_{\text{dyn},1,2}^{\text{T}}(t)\mathbf{u}_{1,\text{rot}}(t) + \lambda_{\text{dyn},1,1}^{\text{T}}(t)\mathbf{a}\left(\tilde{\mathbf{x}}_1(t)\right)$$
$$+ \lambda_{\text{dyn},1,2}^{\text{T}}(t)\left(\tilde{\omega}_1\left(\tilde{\mathbf{x}}_1(t)\right) - \mathbf{I}_{\text{in},1}^{-1}\boldsymbol{\omega}_1^{\times}\left(\boldsymbol{\omega}_1(t)\right)\mathbf{I}_{\text{in},1}\boldsymbol{\omega}_1(t)\right)$$
$$+ \lambda_{\text{dot},1,1}^{\text{T}}\mathbf{v}_1(t) - \lambda_{\text{dot},1,2}^{\text{T}}\mathbf{R}_{\text{rod}}^{-1}(\sigma_1(t))\dot{\sigma}_1(t), \tag{64}$$

$$\mathfrak{h}_2\left(\mathbf{x}_2(t), \mathbf{u}_{2,\text{tran}}(t), \lambda_{\text{dyn},2}(t)\right) = \mu_2 ||\mathbf{u}_{2,\text{tran}}(t)||_2 + \lambda_{\text{dyn},2,1}^{\text{T}}(t)\mathbf{u}_{2,\text{tran}}(t)$$
$$+ \lambda_{\text{dyn},2,1}^{\text{T}}(t)\mathbf{a}\left(\tilde{\mathbf{x}}_2(t)\right) + \lambda_{\text{dot},2,1}^{\text{T}}\mathbf{v}_2(t), \tag{65}$$

where $\lambda_{\text{dyn},1}(t) \triangleq [\lambda_{\text{dyn},1,1}^{\text{T}}(t), \lambda_{\text{dyn},1,2}^{\text{T}}(t)]^{\text{T}}$, $\lambda_{\text{dyn},2}(t) \triangleq [\lambda_{\text{dyn},2,1}^{\text{T}}(t), \lambda_{\text{dyn},2,2}^{\text{T}}(t)]^{\text{T}}$, and $\lambda_{\text{dyn},j,k}$ : $[t_1, t_2] \to \mathbb{R}^3$, j, k = 1, 2. Now, using Theorem 6.3 we can construct a candidate optimal control law. Remarkably, the same candidate optimal control law can be obtained by applying Theorem 6.3 to (64) and (65) independently. The fact that the candidate optimal control law for the the first vehicle can be found independently from the second vehicle is another advantage in employing Lagrange coordinates. The minimization of $\mathfrak{h}_2$ leads to the same candidate optimal control law as given by primer vector theory with the only difference being that the arcs of maximum, null, and singular thrust are not characterized by the sign of $||\lambda_{\text{dyn},2,1}^{*}(t)||_2 - 1$ as in Lawden's work (Lawden, 1963) but rather by the sign of $||\lambda_{\text{dyn},2,1}^{*}(t)||_2 - \mu_2$.

Singular translational thrust arcs for the first vehicle occur when

$$(\lambda_0 \mu_1)^2 = \lambda_{\text{dyn},1,1}^{\text{T}}(t)\lambda_{\text{dyn},1,1}(t) \tag{66}$$

and, as shown in Theorem 6.3, $\mathbf{u}_{2,\text{tran}}^{*}$ cannot be found on singular arcs by applying Pontryagin's minimum principle. However, from (44) and (64), we note that $\lambda_0 \mu_1 \frac{\mathbf{u}_{1,\text{tran}}^{*}(t)}{||\mathbf{u}_1^{*}(t)||_2} = -\lambda_{\text{dyn},1,1}^{*}(t)$, and hence, (66) yields

$$||\mathbf{u}_1^{*}(t)||_2^2 = \mathbf{u}_{1,\text{tran}}^{*\text{T}}(t)\mathbf{u}_{1,\text{tran}}^{*}(t). \tag{67}$$

Thus, on singular translational thrust arcs for the first vehicle $\mathbf{u}_{1,\text{rot}}^{*}(t) = \mathbf{0}_3$. Similarly, it can be shown that $\mathbf{u}_{1,\text{tran}}^{*}(t) = \mathbf{0}_3$ on singular rotational thrust arcs for the first vehicle. Finally, singular arcs for the second vehicle occur when

$$\mu_2^2 = \lambda_{\text{dyn},2,1}^{*\text{T}}(t)\lambda_{\text{dyn},2,1}^{*}(t). \tag{68}$$

From (44) and (65) it follows that $\mu_2 \frac{\mathbf{u}_{2,\text{tran}}^{*}(t)}{||\mathbf{u}_{2,\text{tran}}^{*}(t)||_2} = -\lambda_{\text{dyn},2,1}^{*}(t)$, which satisfies (68). Hence, any admissible $\mathbf{u}_{2,\text{tran}}$ can be applied on singular arcs. This was first noted by Lawden (1963).

## 7. Illustrative numerical example

In this section, we present a numerical example to highlight the efficacy of the framework presented in the paper. In particular, we consider the two vehicles presented in Examples

4.1, 4.2, and 6.1 with masses 0.1kg and inertia matrices $0.40 I_3$ kgm$^4$ flying in an environment modeled by (51) and (52), where $\mathbf{g} = [0, 0, -9.81]^T \frac{m}{s^2}$, $k_{i,D} = 0.20$, $k_{i,L} = 1.20$, $k_{i,S} = 0.50$, $k_{i,R} = 0.30$, $k_{i,P} = 0.30$, and $k_{i,Y} = 0.30$, for i $= 1, 2$. Furthermore, we assume that $t_1 = 0.00$ s, $t_2 = 60.00$ s, $\mathbf{r}_1(t_1) = [0.00, 0.00, 0.00]^T$ m, $\mathbf{r}_1(t_2) = [0.90, -10.00, -1.80]^T$ m, $\sigma_1(t_1) = [0.00, 0.00, 0.00]^T$, and $\sigma_1(t_2) = [0.00, 0.00, 120.00 \frac{\pi}{180.00}]^T$. For our simulation we take $\rho_{i,1} = 10.00 \frac{m}{s^2}$, $\rho_{i,2} = 45.00 \frac{m}{s^2}$, $\rho_{i,3} = 10.00 \frac{1}{s^2}$, and $\rho_{i,1} = 20.00 \frac{1}{s^2}$, for i $= 1, 2$. The boundary conditions for the second vehicle are deduced from (14) and (15) by assuming that $r_{max} = \frac{21}{25}$m and $r_{min} = \frac{33}{50}$m. It can be easily verified that the constraints given by (12) and (13) hold for all $t \in [t_1, t_2]$. Letting $\mu_1 = \mu_2 = \frac{1}{2}$ and applying Theorem 6.1, we obtain the optimal trajectory shown in Figure 1. Figures 2 and 3 show the optimal control as a function of the norm of the translational primer vector and the rotational primer vector, as well as time, respectively. For this example $J[\mathbf{u}_1(\cdot)] = 10.00 \frac{m}{s}$ and $J[\mathbf{u}_2(\cdot)] = 11.60 \frac{m}{s}$. Since $m\left(\mathbf{q}^*(t), \mathbf{q}^*_{dot}(t), \lambda^*_{dyn}(t), \lambda^*_{dot}(t)\right) = 22.30 \frac{m}{s^2}$, Theorem 6.2 holds. Finally, Figure 4 shows the translational primer vector and the rotational primer vector of the first vehicle as a function of time.

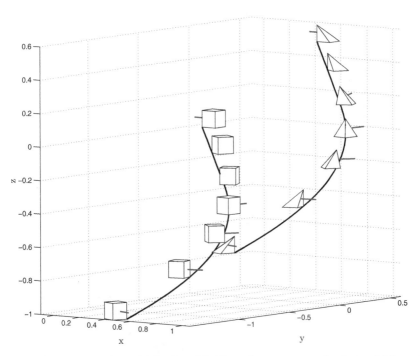

Fig. 1. Optimal trajectories for vehicles 1 and 2. The cube represents the first vehicle and the prism represents the second vehicle.

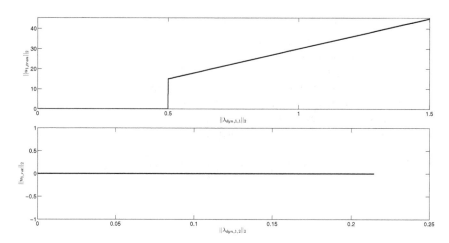

Fig. 2. Optimal control for the first vehicle as function of the norm of the translational primer vector and the rotational primer vector.

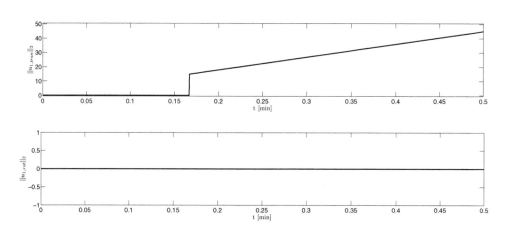

Fig. 3. Optimal control for the first vehicle as function of time.

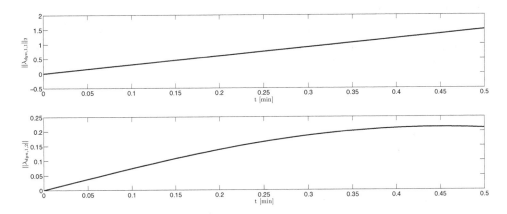

Fig. 4. Translational and rotational primer vector norms as functions of time for the first vehicle.

## 8. Conclusion and recommendations for future research

In this paper, we addressed the problem of minimizing the control effort needed to operate a formation of n UAVs. Specifically, a candidate optimal control law as well as necessary conditions for optimality that characterize the resulting optimal trajectories are derived and discussed assuming that the formation vehicles are 6 DoF rigid bodies flying in generic environmental conditions and subject to equality and inequality constraints. The results presented extend Lawden's seminal work (Lawden, 1963) and several papers predicated on his work.

An illustrative numerical example involving a formation of two vehicles is provided to illustrate the mathematical path planning optimization framework presented in the paper. Furthermore, we show that our framework is not restricted to UAV formations and can be applied to formations of robots, spacecraft, and underwater vehicles.

The results of the present paper can be further extended in several directions. Specifically, an analytical study of the translational primer vector and the rotational primer vector can be useful in identifying numerous properties of the formation's optimal path. In particular, the translational primer vector and the rotational primer vector can be used to measure the sensitivity of the candidate optimal control law to uncertainties in the dynamical model. In this paper, we provide a generic formulation to the optimal path planning problem in order to address a large number of formation problems. However, specializing our results to a particular formation and a particular environmental model can lead to analytical tools that can be amenable to efficient numerical methods. Additionally, nonholonomic constraints have not been accounted in our framework and can be addressed by modifying Theorem 4.1. Finally, in this paper, we penalize vehicle control effort by tuning the constants $\mu_1,...,\mu_n$ in (2). In many practical applications, however, it is preferable to trade-off the control effort in a formation of vehicles by optimizing over the free parameters $\mu_1,...,\mu_n$.

## 9. Acknowledgments

The first-named author would like to thank Drs. C. Sultan and E. Cliff at Virginia Polytechnic Institute and State University for several helpful discussions. This research was supported in part by the Domenica Rea d'Onofrio Fellowship Foundation and the Air Force Office of Scientific Research under Grant FA9550-09-1-0429.

## 10. References

Ambrosia, V. & Hinkley, E. (2008). NASA science serving society: Improving capabilities for fire characterization to effect reduction in disaster losses, *IEEE International Geoscience and Remote Sensing Symposium, 2008. IGARSS 2008.*, Vol. 4, pp. IV –628 –IV –631.

Anderson, J. D. (2001). *Fundamentals of Aerodynamics*, McGraw Hill, New York, NY.

Bataillé, B., Moschetta, J. M., Poinsot, D., Bérard, C. & Piquereau, A. (2009). Development of a VTOL mini UAV for multi-tasking missions, *The Aeronautical Journal* 13: 87–98.

Betts, J. T. (1998). Survey of numerical methods for trajectory optimization, *AIAA Journal of Guidance, Control, and Dynamics* 21: 193–207.

Blackmore, L. (2008). Robust path planning and feedback design under stochastic uncertainty, *Proceedings of the AIAA Guidance, Navigation, and Control Conference*, AIAA, Honolulu, HI.

Bryson, A. E. (1975). *Applied Optimal Control*, Hemisphere, New York, NY.

Ewing, E. G. (1969). *Calculus of Variations with Applications*, Dover Edition, New York, NY.

Giaquinta, M. & Hildebrandt, S. (1996). *Calculus of Variations I*, Springer-Verlag, Berlin, Germany.

Greenwood, T. D. (2003). *Advanced Dynamics*, Cambridge University Press, New York, NY.

Haddal, C. C. & Gertler, J. (2010). Homeland security: Unmanned aerial vehicles and border surveillance, *Technical Report RS21698*, Congressional Research Service, Washington, D.C.

Hassan, R., Cohanim, B. & de Weck, O. (2005). A comparison of particle swarm optimization and the genetic algorithm, *Proceedings of the 46th AIAA Structures, Structural Dynamics and Materials Conference*, AIAA, Breckenridge, CO.

Herman, A. L. & Conway, B. A. (1987). Direct optimization using nonlinear programming and collocation, *AIAA Journal of Guidance, Control, and Dynamics* 10: 338–342.

Jacobson, D. & Lele, M. (1969). A transformation technique for optimal control problems with a state variable inequality constraint, *IEEE Transactions on Automatic Control* 14(5): 457–464.

Jamison, B. R. & Coverstone, V. (2010). Analytical study of the primer vector and orbit transfer switching function, *AIAA Journal of Guidance, Control, and Dynamics* 33: 235–245.

Jang, J. S. & Tomlin, C. J. (2005). Control strategies in multi-player pursuit and evasion game, *Proceeding AIAA Guidance, Navigation, and Control Conference*, AIAA, San Francisco, CA.

L'Afflitto, A. & Sultan, C. (2008). Applications of calculus of variations to aircraft and spacecraft path planning, *Proceedings of the AIAA Guidance, Navigation, and Control Conference*, AIAA, Chicago, IL.

L'Afflitto, A. & Sultan, C. (2010). On calculus of variations in aircraft and spacecraft formation flying path planning, *Proceedings of the AIAA Guidance, Navigation, and Control Conference*, AIAA, Toronto, Canada.

Lawden, D. F. (1963). *Optimal Trajectories for Space Navigation*, Butterworths, London, UK.

Lee, E. B. & Markus, L. (1968). *Foundations of Optimal Control Theory*, Wiley, New York, NY.

Lillesand, T., Kiefer, R. W. & Chipman, J. (2007). *Remote Sensing and Image Interpretation*, Wiley, New York, NY.

Mailhe, L. & Guzman, J. (2004). Initialization and resizing of formation flying using global and local optimization methods, *Proceedings IEEE Aerospace Conference*, Vol. 1, pp. 547–556.

Majewski, S. E. (1999). Naval command and control for future UAVs. MS Thesis, Naval Postgraduate School, Monterey, CA.

Marec, J. P. (1979). *Optimal Space Trajectories*, Elsevier, New York, NY.

Neimark, J. I. & Fufaev, N. A. (1972). *Dynamics of Nonholonimic Systems*, American Mathematical Society, New York, NY.

Oyekan, J. & Huosheng, H. (2009). Toward bacterial swarm for environmental monitoring, *IEEE International Conference on Automation and Logistics*, pp. 399 –404.

Pars, L. A. (1965). *A Treatise on Analytical Dynamics*, Wiley, New York, NY.

Petropoulos, A. E. & Longuski, J. M. (2004). Shape-based algorithm for automated design of low-thrust, gravity-assist trajectories, *AIAA Journal of Guidance, Control, and Dynamics* 32: 95–101.

Petropoulos, A. E. & Russell, R. P. (2008). Low-thrust transfers using primer vector theory and a second-order penalty method, *Proceedings of the AIAA Astrodynamics Specialist Conference*, AIAA, Honolulu, HI.

Plnes, D. & Bohorquez, F. (2006). Challenges facing future micro-air-vehicle development, *AIAA Journal of Aircraft* 43: 290–305.

Pontryagin, L. S., Boltyanskii, V. G., Gamkrelidze, R. V. & Mishchenko, E. F. (1962). *The Mathematical Theory of Optimal Processes*, Interscience Publishers, New York, NY.

Prussing, J. E. (2010). Primer vector theory and applications, *in* B. A. Conway (ed.), *Spacecraft Trajectory Optimization*, Cambridge University Press, Chicago, IL, pp. 155–188.

Ramage, J., Avalle, M., Berglund, E., Crovella, L., Frampton, R., Krogmann, U., Ravat, C., Robinson, M., Shulte, A. & Wood, S. (2009). Automation technologies and application considerations for highly integrated mission systems, *Technical Report TR-SCI-118*, North Atlantic Treaty Organisation.

Scharf, D., Hadaegh, F. & Ploen, S. (2003a). A survey of spacecraft formation flying guidance and control (part 1): Guidance, *Proceedings of the American Control Conference*, pp. 1733 – 1739.

Scharf, D., Hadaegh, F. & Ploen, S. (2003b). A survey of spacecraft formation flying guidance and control (part 2): Control, *Proceedings of the American Control Conference*, pp. 1740 – 1748.

Schouwenaars, T., Feron, E. & How, J. (2006). Multi-vehicle path planning for non-line of sight communication, *Proceedings of the American Control Conference*, pp. 5758–5762.

Seereram, S., Li, E., Ravichandran, B., Mehra, R. K., Smith, R. & Beard, R. (2000). Multispacecraft formation initialization using genetic algorithm techniques,

*Proceedings of the 23rd Annual AAS Guidance and Control Conference*, AAS, Breckenridge, CO.

Shanmugavel, M., Tsourdos, A. & White, B. (2010). Collision avoidance and path planning of multiple UAVs using flyable paths in 3D, *15th International Conference on Methods and Models in Automation and Robotics*, pp. 218–222.

Shuster, M. D. (1993). Survey of attitude representations, *Journal of the Astronautical Sciences* 11: 439–517.

Topcu, U., Casoliva, J. & Mease, K. D. (2007). Minimum-fuel powered descent for Mars pinpoint landing, *AIAA Journal of Spacecraft and Rockets* 44(2): 324–331.

Valentine, F. A. (1937). The problem of Lagrange with differential inequalities as added side conditions, *in* G. A. Bliss (ed.), *Contributions to the Calculus of Variations*, Chicago University Press, Chicago, IL, pp. 407–448.

Wall, B. J. (2008). Shape-based approximation method for low-thrust trajectory optimization, *Proceedings of the AIAA Astrodynamics Specialist Conference*, AIAA, Honolulu, HI.

Wang, P. K. C. (1991). Navigation strategies for multiple autonomous mobile robots moving in formation, *Journal of Robotic Systems* 8: 177 – 195.

Zaitri, M. K., Arzelier, D. & Louembert, C. (2010). Mixed iterative algorithm for solving optimal impulsive time-fixed rendezvous problem, *Proceedings of the AIAA Guidance, Navigation, and Control Conference*, AIAA, Toronto, Canada.

# 11

# Subjective Factors in Flight Safety

Jozsef Rohacs

*Budapest University of Technology and Economics*
*Hungary*

## 1. Introduction

The central deterministic element of the aircraft conventional control systems is the pilot – operator. Such systems are called as active endogenous subjective systems, because (i) the actively used control inputs (ii) origin from inside elements (pilots) of the system as (iii) results of operators' subjective decisions. The decisions depend on situation awareness, knowledge, practice and skills of pilot-operators. They may make decisions in situations characterized by a lack of information, human robust behaviors and their individual possibilities. These attributes as subjective factors have direct influences on the system characteristics, system quality and safety.

Aircraft control containing human operator in loop can be characterized by subjective analysis and vehicle motion models. The general model of solving the control problems includes the passive (information, energy - like vehicle control system in its physical form) and active (physical, intellectual, psychophysiology, etc. behaviors of subjects - operators) resources. The decision-making is the appropriate selection of the required results leading to the best (effective, safety, etc.) solutions.

This chapter defines the flight safety and investigates aircraft stochastic motion. It shows the disadvantages of the stochastic approximation and discusses, how, the methods of subjective analysis can be applied for the evaluation of flight safety.

The applicability of the developed method of investigation will be demonstrated by analysis of the aircraft controlled landing. The applied equation of motions describes the motion of aircraft in vertical plane, only. The boundary constraints are defined for velocity, trajectory angle and altitude. The subjective factor is the ratio of required and available time to decision on the go-around. The decision depends on the available information and psycho-physiological condition of operator pilots and can be determined by the theory of statistical hypotheses. The endogenous dynamics of the given active system is modeled by a modified Lorenz attractor.

## 2. Flight safety

### 2.1 Definitions

Safety is the condition of being safe; freedom from danger, risk, or injury. From the technical point of view, safety is a set of methods, rules, technologies applied to avoid the emergency situation caused by unwanted system uncertainties, errors or failures appearing randomly.

Safety and security are the twin brothers. The difference between them could be defined such as follows:

- *Safety:* avoid emergency situation caused by unwanted system uncertainties, errors or failures appearing randomly.
- *Security:* avoid emergency situations caused by unlawful acts (of unauthorized persons) – threats.

Safety related investigations start as early as the development of the given system. At the definition and preliminary phase of a new system, one should also concentrate some efforts on the (i) potential safety problems, (ii) critical situations, (iii) critical system failures, (iv) and their possible classification, identification. After the risk assessment, the next step is the development of a set of policies and strategies to mitigate those risks. Generally, the safety policies and strategies are based on the synergy of the

- physical safety (characteristics of the applied materials, structural solutions, system architecture that help to overcome safety critical – emergency situations),
- technical safety (dedicated active or passive safety systems including e.g. sensors to enhance situation awareness),
- non-technical safety (such as policy manuals, traffic rules, awareness and mitigation programs).

The safety of any systems can be evaluated by using the risk analysis methods. Risk is the probability that an emergency situation occurs in the future, and which could also be avoided or mitigated, rather than present problems that must be immediately addressed.

## 2.2 Flight safety metrics

The evaluation of the flight safety is not a simple task. There is no uniformly applicable metrics for the evaluation. Some governments have already published (CASA, 2005; FAA, 2006; Transport, 2007) their opinion and possible methodologies for flight safety measures that are applied by evaluators (Ropp & Dillmann, 2008). The problem is associated with the very complex character of flight safety depending on the developed and applied

- safety plan with management commitment,
- documentation management,
- risk monitoring,
- education and training,
- safety assurance (quality management on safety),
- emergency response plan.

Risk analyses methods defining the probability of emergency situations or risks are very widely used for flight safety evaluation. Metrics of risk is the probability of the given risk as an unwanted danger event. This probability has at least four slightly different interpretations:

- classic - the unwanted event,
- logic - the necessary evil,
- objective - relative frequency,
- subjective - individual explanation of the events.

In practice, the analysis of accident statistics could characterize the flight risks. Such statistics give the evidences for the well-known facts (Rohacs, 1995, 2000; Statistical 2008): (i) the longest part of the flight (with about 50 - 80 % of flight time) is the cruise phase, which only accounts for 5 - 8 % of the total accidents and 6 - 10 % of the total fatal accidents, (ii) the most dangerous phases of flight are the take-off and landing, because during this about 2 % of flight time the 25 - 28 % of fatal accidents are occurring, and (iii) generally nearly 80 % of the accidents are caused by human factors and about 50 % of them are initiated by the pilots.

A good example of using accident statistics is shown in Figure 1. Beside showing the effects of technological development on the reduction of flight risks, it also shows that since 2003, the European fatal accident rate - as fatalities per 10 million flights - has increased, without knowing – so far – the reason causing it.

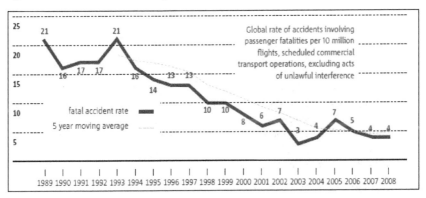

Fig. 1. Characterization of the European accident statistics (EASA, 2008).

The accident statistics could be also used for flight safety analysis in original, or unusual method. While accident statistics demonstrate a considerable higher risk, accident rates for small aircraft, according to the Figure 2., the ratio of all and fatal accidents are nearly the same for airlines and general aviation. This means that the small and larger civilian aircraft are developed, designed, and produced with the same philosophy, at least the same safety approach and 'structural damping of damage processes'. The flight performances, flight dynamics, load conditions, structural solutions are different for small and larger aircraft, and therefore the accidents rates are also different. However, the risk of hard aftermath, appearing the fatal accident following the accidents are the same.

## 2.3 Human factors

In 1908, 80 % of licensed pilots were killed in flight accident (Flight, 2000). Since that, the World and the aviation have changed a lot. After 1945, the role of technical factors in causing the accidents (and generally in safe piloting) is continuously decreasing while the role of human factors is increasing.

As it was outlined already, nearly 80 % of accidents are caused by human factors. (Rohacs, 1995, 2000; Statistical, 2008). While, only 4 -7 % of accidents are defined by the "independent investigators" as accident caused by unknown factors. According to Ponomerenko (2000) this figure might be changed when one tries to establish the truth in fatal accidents,

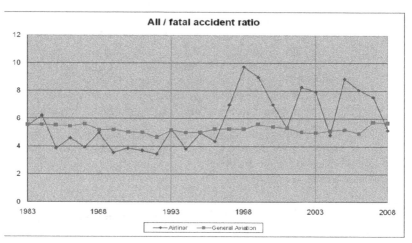

Fig. 2. An original way to compare airliner and GA accident statistics.

especially by taking into account the socio-psychological aspects and use of " 'guilt' and 'guilty' as the 'master key' to unlock the true cause of the accident. Hence, the bias of the investigators often does not represent the interest of the victims, but that of the administrative superstructure. It side steps the legal and socio-psychological estimation of aircrew behavior, and replaces it by formal logic analysis of known rules: permitted/forbidden, man or machine, chance/relationship, violated/not violated, etc."

Accident investigations show that human factors could be divided into three groups depending on their origins.

- Technical factors: disharmony in human - machine interface. Most known cases from this group are called as PIDs (pilot induced oscillations). Some of these factors, like limitations of the control stick forces are included even into the airworthiness requirements.
- Ergonomic factors: a lack of ergonomic information display, guidance control, out-of-cockpit visibility, design of instrument panel, as well as of adequate training [Ponomarenko 2000].
- Subjective factors: un-predictable and non-uniform man's behavior. Making wrong decisions because the lack of knowledge and practice of operators.

The different groups have nearly the same role in accident casualty, equal to 25, 35, 40 %, respectively. Others (Lee, 2003) call the same type of factors as system data problems, human limitation and time related problems.

The first group of human factors, harmonization of the man-machine interface from the technical side of view is well investigated and such type of human factors are taken into account in aircraft development and design processes. Generally, the handling quality or (nowadays) the car free characteristics are the merits and used as main philosophical approaches to solve these types of problems.

The ergonomic factors have been investigated a lot for last 40 - 50 years. The third generation of the fighters had been developed with the use of ergonomics, especially in development of the cockpit, that were radically redesigned for that period. However, the ergonomic investigations had used the governing idea, how to make better for operator. A new approach has developed for last 20 years that investigates the 'ergatic' systems (see for example Pavlov & Chepijenko, 2009) in which the operator (pilot) one of the important (might be most important) element of the systems, and the psycho-physiological behaviors of the operator may play determining roles in operation of system.

The third group of human factors has not investigated on the required level yet. Generally, the key element of human reaction on the situation, especially on the emergency situation is the time. *However, the speed and time of reaction is "... not determined by the amount of processed information, but by the choice of the signal's importance, which is always subjective and affected by individual personality traits" (Ponomarenko 2000).* In an emergency situation, flight safety does not depend as much on the detailed information on the emergency situation and the size of pilot supporting information, as on the whole picture including space and time, knowledge and practice of pilots and the actual determination of the ethical limits of man's struggle with the arisen situation.

Flight safety could also be analyzed with the prediction of the future air transport characteristics. For example, the NASA initiated zero accident project, (Commercial, 2000; Shin, 2000; White, 2009) leads to the following general conclusion: before introducing the wide-body aircraft, the risk of flight was decreased by a factor of 10, but this cannot be further reduced with the present technical and technological methods (Rohacs, 1998; Shin, 2000). Even so, the number of aircraft and the number of yearly, daily flights are continuously increasing (Fig. 3.); Seeing this, the absolute number of accidents is expected to increase in the future, which might even lead to the vision made by Boeing, in which by 2016/17, each week one large-body aircraft is envisioned to have an accident. "Given the very visible, damaging, and tragic effects of even a single major accident, this number of accidents would clearly have an unacceptable impact upon the public's confidence in the aviation system and impede the anticipated growth of the commercial air-travel market" (Shin, 2000). Therefore, new methods like emergency management might need to be developed and applied to keep the absolute number of accidents on the present level.

Seeing the envisioned rapid development of the future aviation, especially the small aircraft transportation system, the conclusion derived from the zero accident program and use of the subjective analysis in flight safety investigation might be relevant to be kept in mind.

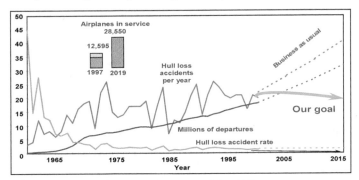

Fig. 3. The NASA zero accident program (Commercial, 2000).

## 3. Flight safety evaluation

### 3.1 Technical approach to flight safety evaluation

Technically, flight risks are always initiated by the deviations in the system parameters. Therefore, the investigation of the system parameter uncertainties and anomalies might be applied as a basis to evaluate flight safety. Flight safety is the risk that an emergency situation occurs, when the system parameters (at least one of them) are out of the tolerance zones. In the view of this, flight safety might be characterized by the probability of the deviations (in the structural and operational characteristics) being larger than those predetermined by the airworthiness (safety) requirements (Bezapasnostj 1988, Rohacs & Németh, 1997).

Mathematically, flight operation quality, $Q_r(t)$, could be given in the following simple form:

$$Q \equiv \{a_i\} \quad , \quad i = 1, n \tag{1}$$

where $a_i$ are the parameters defining the attributes of the given aircraft or system. In a more general form, it could be given as:

$$a_i = f\left(a_1, a_2, \ldots a_{i-1}, a_{i+1}, \ldots a_n\right) . \tag{2}$$

In real flight situations, the real quality of operation $Q_r(t)$ is deviated from the design (nominal) quality $Q_r^n(t)$:

$$\delta\, Q_r(t) = Q_r(t) - Q_r^n(t) \quad . \tag{3}$$

For each case, the acceptable level of deviation is maximized by the flight safety threshold ($\delta_{fs}$),

$$|\delta\, Q_r(t)| \geq \delta_{fs} \quad , \tag{4}$$

where $P\left(|\delta\, Q_r(t)| \geq \delta_{fs}\right)$ describes the probability of a flight event (flight out of prescribed operational modes).

By summing all the potential flight events, flight safety $(P_{fs})$ could be given with the following probability:

$$P_{fs} = 1 - \sum_{i=1}^{n} R_i(t) P_i(t) \tag{5}$$

where $R_i(t)$ - is the risk of flight accident.

For time period [0, T] the following integral risk can be applied:

$$\tilde{P}_{fs} = 1 - \frac{1}{T} \int_0^T \left(1 - P_{fs}(t)\right) dt \quad . \tag{6}$$

Because $\delta \mathbf{Q}_r(t)$ is the random value with probability density, $\rho(\delta \mathbf{Q}_r(t))$, the flight safety level can be given as:

$$P_{fs} \equiv P\left(\left|\delta \mathbf{Q}_r(t)\right| \leq \delta_{fs}\right) = \int_{-\delta_{fs}}^{\delta_{fs}} \rho(\delta \mathbf{Q}_r) d\delta \mathbf{Q}_r \cdot \tag{7}$$

According to the Tchebyshev inequality

$$P\left(\left|\delta \mathbf{Q}_r(t)\right| \succ \delta_{fs}\right) \leq D(\delta \mathbf{Q}_r) / \delta_{fs}^2 \tag{8}$$

the flight safety level takes the form:

$$P_{fs} \equiv P\left(\left|\delta \mathbf{Q}_r(t)\right| \leq \delta_{fs}\right) \geq 1 - D(\delta \mathbf{Q}_r) / \delta_{fs}^2 , \tag{9}$$

where $D(\delta \mathbf{Q}_r)$ is the dispersion of $\delta \mathbf{Q}_r$.

Such type of system approach was developed, applied and improved. Generally, once the aircraft is investigated as a dynamic system, the effects of the system anomalies could be given by the following type of probabilities (Rohacs 1986; Rohacs & Nemeth, 1997):

$$P_1\left\{ \mathbf{y}(t)\big|_{t_0 \leq t \leq t+\tau, \, \mathbf{x}\in\Omega_x, \, \mathbf{u}\in\Omega_u, \, \mathbf{z}\in\Omega_z, \, \mathbf{p}\in\Omega_p} \right\} , \tag{10.a}$$

$$P_2\left\{ \mathbf{u}(t)\big|_{t_0 \leq t \leq t+\tau, \, \mathbf{x}\in\Omega_x, \, \mathbf{y}\in\Omega_y, \, \mathbf{z}\in\Omega_z, \, \mathbf{p}\in\Omega_p} \right\} , \tag{10.b}$$

where $\mathbf{y} \in R_r$ defines the output (measurable) signal vector (measured vector of operational characteristics) $\mathbf{x} \in R_n$ is the state vector, $\mathbf{u} \in R_m$ gives the input (control) vector, $\mathbf{z} \in R_i$ stands for the vector of environmental characteristics (vector of service conditions), $\mathbf{p} \in R_k$ is the parameter vector characterizing the state of the aircraft, $t$ defines the time, $\tau$ provides the elementary time, $\Omega_x$, $\Omega_y$, $\Omega_z$, $\Omega_u$, $\Omega_p$ are the allowed ranges for the given characteristics.

If the joint density function,

$$f_\Sigma = f\left[\mathbf{x}(t), \mathbf{u}(t), \mathbf{z}(t), \mathbf{p}(t), \mathbf{y}(t)\right] \tag{11}$$

is known, then the recommended characteristics can be calculated as:

$$P_1\left\{\mathbf{y}(t) \in \mathbf{\Omega}_y |...\right\} = \frac{\int_{\Omega_i} f_\Sigma dxdudzdpdy}{\int_{-\infty}^{+\infty} dy \int_{\Omega_j} f_\Sigma dxdudzdp} \quad \begin{array}{l} (i \in \mathbf{x}, \mathbf{u}, \mathbf{z}, \mathbf{p}, \mathbf{y}) \\ (j \in \mathbf{x}, \mathbf{u}, \mathbf{z}, \mathbf{p}) \end{array} , \tag{12.a}$$

$$P_2\left\{\mathbf{u}(t) \in \mathbf{\Omega}_u |...\right\} \cdot = \frac{\int_{\Omega_i} f_\Sigma dxdudzdpdy}{\int_{-\infty}^{+\infty} du \int_{\Omega_j} f_\Sigma dxdzdpdy} \quad \begin{array}{l} (i \in \mathbf{x}, \mathbf{u}, \mathbf{z}, \mathbf{p}, \mathbf{y}) \\ (j \in \mathbf{x}, \mathbf{z}, \mathbf{p}, \mathbf{y}) \end{array} . \tag{12.b}$$

Unfortunately, this method of determining the effects of the system anomalies on the flight safety is often considered to be too complex, while it is found to be reasonable, since the formulas given above could be supported with statistical data collected during aircraft operation. The method of determining the flight risk on the probability approach (as given in (Gudkov & Lesakov, 1968; Howard, 1980)) is envisioned to be too complicated, once it is also desirable to consider the so-called common (failures appearing at the same time due to different reasons) and depending failures or errors. The Figures 4 and 5 show a nice example of using the described method is the investigation changes in geometrical and operational characteristics of aircraft investigated by (Rohacs 1986) and published in several articles, like (Rohacs, 1990).

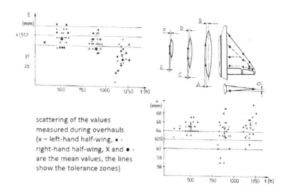

Fig. 4. The level book and examples of the measuring data for Mig-21.

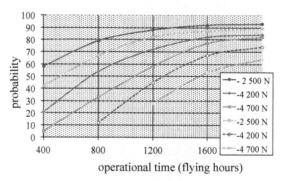

operational time (flying hours)

Fig. 5. Probability of lack of generated lift at fighters Míg-21 due the changes in wing geometry during the operation (line - single seat, dot line - double seats aircraft)

## 3.2 Stochastic model of flight risk

The aircraft's motion is the result of the deterministic control and the stochastic disturbance processes. Such motion might be mathematically given by the following stochastic (random) differential equation, called as diffusion process (Gardiner, 2004):

$$\dot{x} = f(x,t) + \sigma(x,t)\eta(t) \ , \tag{13}$$

Naturally, this equation might be also given in vector form. The first part of the right side of the equation describes the drift (direction of the changes) of the stochastic process passing through $x(t) = X$ at the moment $t$, while the second part shows the scattering (variance) of the random process. Here $\eta(t)$ is the random disturbance (e.g. air turbulence, or cumulative effects of random load processes, including even extreme loads as hard touchdown, etc.).

Seeing that the future states depend only on the present sate, the equation (13) is in fact a Markov process (Ibe, 2008; Rohacs & Simon, 1989; Tihonov, 1977). Such process can be fully described by its transition probability density function:

$$p(x_2, t_2 | X_1, t_1), \qquad (t_2 > t_1) , \tag{14}$$

which characterizes the distribution probability of the continuous random process $(x(t))$ at the moment $t_2$, once it's passing through the $x(t) = X$ at time $t_1$.

The transition probability density function can be described by the following Fokker - Planck - Kolmogorov equations (Gardiner, 2004):

$$\frac{\partial p(x_2, t_2 | X_1, t_1)}{\partial t_2} = -\frac{\partial}{\partial x_2}\left[f(x_2, t_2)p(x_2, t_2 | X_1, t_1)\right] +$$

$$+\frac{1}{2}\frac{\partial^2}{\partial x_2^2}\left[\sigma^2(x_2, t_2)p(x_2, t_2 | X_1, t_1)\right] , \tag{15.a}$$

or

$$\frac{\partial(x, t)}{\partial t} = -\frac{\partial}{\partial x}\left[f(x, t)p(x, t)\right] + \frac{1}{2}\frac{\partial^2}{\partial x^2}\left[\sigma^2(x, t)p(x, t)\right]. \tag{15.b}$$

Statistic flight mechanics has already worked out several methods for the application of such models. For example, the statistical linearization through the proof of the sensitivity function matrix to the flight mechanic models and generating out the set of equations for the moments of the investigated stochastic process could be used to study the scattering of the process.

Using the equations (15.a), (15.b), which define the Markov process, the following definition could be made:

$$p(X_2, t_2 | X_1, t_1) = \sum_{X(t)} p(X_2, t_2 | x, t)p(x, t | X_1, t_1), \qquad (t_2 \geq t \geq t_1) , \tag{16}$$

This is called Chapman - Kolmogorov – Smoluchovski equation. It gives the possibility to approximate the investigated non-linear stochastic process with continuous time and state space with a Markov chain with continuous time and discrete state space. This leads us back to the situation chain process.

The space of the motion variables can be divided into several subspaces, called as situations. The motion of the aircraft is in fact a time invariant series of situations. This is the situation dynamics.

Accidents are the results of the situation process, which is assumed to be similar to the one given in the Figure 6. Here, $N$ marks the normal, conventional flight, $S_1$, $S_2$, $S_3$ are different states related to the case when the aircraft has one (F1), two (F2), or three serious system failures ($F_3$), while $A$ shows the accident situation (Rohacs & Nemeth, 1997; Rohacs, 2000).

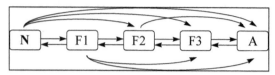

Fig. 6. Simple graph model of aircraft pre-accident process

The Markov chain can be described by the transition probabilities, $\beta_{i,j}$. These variables give the probability of moving the aircraft from a state (situation) $S_i$ to a state $S_j$. As it is known, this type of process can be approximated by Markov process, under the following conditions:

- the transition from one state into another occurs in a significantly short time,
- the probability of a transfer from one state into another through one or more other states is a limited, and
- the time spent in the states could be approximated by an exponential distribution.

Under the conditions mentioned above, the process could be described with the following model:

$$\dot{P}(t) = P_t(t)P(t) , \tag{17}$$

where $P(t)=[Pi(t)]$ is a vector of probabilities defining the states $S_i$ ($i=N$, F1, F2, F3, A).

At this stage, one should give the applicable graph model and estimate the transition probability matrix.

In this simple case, the aircraft's operational process – as a stochastic process with continuous time and discrete states shown in the Fig. 6. – could be approximated by the following Markov model:

$$\dot{P}(t) = \beta(t)P(t) \tag{18}$$

where $P(t)=[P\,i(t)]$ is a vector of probabilities that the aircraft is in the states $S_i$ ($i=N$, F1, F2, $F_3$, A), and

$$\beta(t)=[\beta\,i,j] \tag{19}$$

is a time depending transition matrix:

$$\beta(t) = \begin{bmatrix} -\beta_{N,F1}-\beta_{N,F2}-\beta_{N,F3}-\beta_{N,A} & -\beta_{F1,N} & 0 & 0 & 0 \\ \beta_{N,F1} & -\beta_{F1,N}-\beta_{F1,F2}-\beta_{F1,F3}-\beta_{F1,A} & -\beta_{F2,F3} & 0 & 0 \\ \beta_{N,F2} & \beta_{F1,F2} & -\beta_{F2,F1}-\beta_{F2,F3}-\beta_{F2,A} & -\beta_{F3,F2} & 0 \\ \beta_{N,F3} & \beta_{F1,F3} & \beta_{F2,F3} & -\beta_{F3,F2}-\beta_{F2,A} & -\beta_{A,F3} \\ \beta_{N,A} & \beta_{F1,A} & \beta_{F2,A} & \beta_{F3,A} & -\beta_{A,F3} \end{bmatrix}$$

Our theoretical and practical investigations on flight safety showed that the aircraft's operational process is a complicated process. For example, if a pilot reports an in-operating engine, than ATCOs are often to make 40 - 100 times more mistakes relative to normal circumstances. The simplified graph model of flight situations - taking into account such effects - is given in the Figure 7. The advantage of this representation method over the others, could be summarized in the followings. Firstly, this model includes a new state, called state of anomalies (*An*), in which the aircraft does not have any failures or errors, but still, its characteristics are essentially deviating from their nominal values. Secondly, the total amount of states are decomposed or grouped into four subparts (structure, pilot, air traffic control, surroundings).

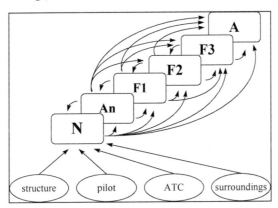

Fig. 7. The suggested general graph model of aircraft.

To simplify the representation of this method, the Figure 7. shows only the nominal state decomposition (Rohacs & Nemeth, 1997; Rohacs, 2000). Even so, the different numbers of failures are further decomposed. States *N* is a prescribed nominal state. States *An* and *F1* might only be initiated by the anomalies or failures in one of the aircraft's flight operation subsystems (e.g. aircraft structure, pilot, ATC, surroundings). On the other hand, the states *F2*, *F3* might be initiated by two or three failures appearing in any combination of the subsystems. For example *F2* may contain mistake of the pilot and ATCO, or two aircraft structural (system) failures.

According to these specific features of the model, the general Markov model should have 43 states. For example in our model, the state number 21, is the state with two failures generated in the structure and one is initiated by the mistake of the pilot. As a consequence, the transfer matrix is composed of 43 x 43 elements, while the elements of the matrix are the linear functions of $P(t)$:

$$\beta_{i,j} = \beta_{i,j,0} + K_{i,j} P(t) ; \qquad (20)$$

where, $\beta_{i,j,0}$ is the initial transfer matrix element, $K_{i,j}$ is the vector of coefficients. The vector $K_{i,j}$ may contain zero elements, too, if the given state has no influence on transfer process.

The determination of the vector elements $K_{i,j}$, is based on the theory of anomalies, dealing with the calculation of the real deviations, characteristics, and distributions. For example, human error depends on weather, traffic situations, or possible system failures. Naturally, if the aircraft is piloted by pilot with limited skills, then the coefficients would be higher than it is for the conventional small aircraft operations. After the evaluation of different models based on the above discussed Markov and semi-Markov processes, we found that the inadequate initial data and the relatively large number of states makes the semi-Markov process irrelevant for our purposes.

Due to the large number of states, the developed model might be seen too complex. On the other hand, by the analysis of the potential methods to simplify the model, it was found that the suggested approach can be transferred to the model shown in the Figure 7. This is reasonable, since from a flight safety point of view, the most important is the transfer of one state into another, and not the detail how that transfer could be made. Therefore, the transition matrix element, $\beta_{F1}$, $\beta_{F2}$, describing the transfer from one failure state ($F1$) into the state with two failures ($F2$) can be given in the following form:

$$\beta_{F1,F2} = \frac{\sum_{i,j} \beta_{F1_i,F2_j} P_{F1_i}}{\sum_{k,i} \beta_{An_k,F1_i} P_{An_k}} , \qquad (21)$$

where $An$ indicates the state with anomalies, and $k, i, j$ are indexes defining the states.

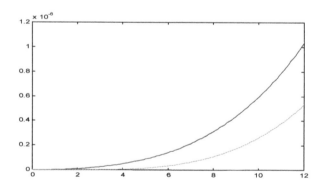

Fig. 8. Flight risk by considering (state An included - solid blue line ) or neglecting the effects of anomalies (green dashed line).

As a result, the general model – describing the real interactions between different types of failures, distinguishing common and depending failures – could be reduced to a simple model.

The developed model was used for the analysis of the aircraft control. Some results are shown in Figures 8. and 9.

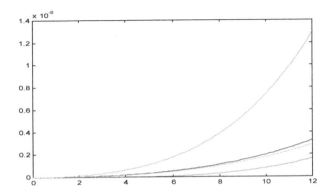

Fig. 9. Probability of appearance of first failures (solid blue line - pilot error (failure), green dashed line - pilot error in case of system anomalies red dashed line - structure failure, blue dashed dot line - structure failure calculated considering the influences of the anomalies).

## 4. Subjective analysis and flight safety

### 4.1 Theoretical background

The major determinative element of the aircraft's conventional control systems is the pilot. Such systems are called as ergatic active endogenous systems [Kasyanov 2007], since the systems are actively controlled by solutions initiated by ergates (Greek ἐργάτης ergatēs - worker), human organism (e.g. nervous cells). So the control solution becomes from inside the system, from the operator. Such effects are called often as endogenous feedback or endogenous dynamics (Banos, Lamnabhi-Lagarrigue & Montoya, 2001; Fliens et all, 1999, Nieuwstadt 1997]. Because pilots make their decision upon their situation awareness, knowledge, practice and skills, e.g. on the subjective way, the system would be also subjective. Beside human robust behaviors and individual possibilities, pilots – in certain circumstances – should also make decisions, even if the information for an appropriate reaction is limited.

Safety of active systems is determined by risks initiated by subjects being the central elements of the given system. For example, flight safety is the probability that a flight happens without an accident. Aircraft are moving in the three dimensional space, in function of their aerodynamic characteristics, flight dynamics, environmental stochastic disturbances (e.g. wind, air turbulence) and applied control. Pilots make decision upon their situation awareness. They must define the problem and choose the solution from their resources, which makes human controlled active systems endogenous. Resources are methods or technologies that can be applied to solve the problems (Kasyanov, 2007). These could be classified into the so-called (i) passive (finance, materials, information, energy - like aircraft control system in its physical form) and (ii) active (physical, intellectual, psycho-physiological behaviors, possibilities of subjects) resources. The passive resources are therefore the resources of the system (e.g. air transportation system, ATM, services provided), while the active resources are related to the pilot itself. Based on these, decision making is in fact the process of choosing the right resources that leads to an optimal solution.

Subjects (like pilots) could develop their active resources (or competences) with theoretical studies and practical lessons. However, the ability of choosing and using the right resources is highly depending on (i) the information support, (ii) the available time, (iii) the real knowledge, (iv) the way of thinking, and (v) the skills of the subject. Such decisions are the results of the subjective analysis.

There is insufficient information on the physical, systematic, intellectual, physiological characteristics of the subjective analysis, as well as on the way of thinking, and making decision of subjects-operators like pilots. Only limited information is available on the time effects, possible damping the non-linear oscillations, the long-term memory, which makes the decision system chaotic.

Flight safety can be evaluated by the combination of subjective analysis and aircraft motion models.

At first, the pilot as subject ($\Sigma$) must identify and understand the problem or the situation ($S_i$), then from the set of accessible or possible devices, methods and factors ($S_p$) must choose the disposable resources ($R^{\text{disp}}$) available to solve the identified problems, to finally decide and apply the required resources ($R^{\text{req}}$) (Kasyanov 2007) (Fig.10.). For this task, the pilot applies its active and passive resources. The active resources will define how the passive resources are used:

$$R_{\text{a}}^{\text{req}} = f\left(R_{\text{p}}^{\text{req}}\right) \tag{22}$$

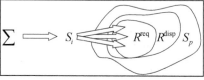

Fig. 10. Pilot decision – action process (endogenous dynamics) in aircraft operation (control) system.

Instead of the function between the resources (22), the literature often uses the velocity of transferring the passive resources into the actives:

$$v_{\text{a}}^{\text{req}} = f_v\left(v_{\text{a}}^{\text{req}}\right)v_{\text{a}}^{\text{req}}, \tag{23}$$

where

$$v_{\text{a}}^{\text{req}} = \frac{dR_{\text{a}}^{\text{req}}}{dt}, \qquad v_{\text{p}}^{\text{req}} = \frac{dR_{\text{p}}^{\text{req}}}{dt}, \tag{24}$$

and in simple cases

$$f_v = \frac{\partial R_{\text{a}}^{\text{req}}}{\partial R_{\text{p}}^{\text{req}}}. \tag{25}$$

It is clear that the operational processes can be given by a series of situations: pilot identifies the situation ($S_i$,), makes decision, controls ($R_a^{req}$), which transits the aircraft into the next situation ($S_j$,). (The situation $S_j$, is one of the set of possible situations). This is a repeating process (Fig. 11.), in which the transition from one situation into another depends on (i) the evaluation (identification) of the given situation, (ii) the available resources, (iii) the appropriate decision of the pilot, (iv) the correct application of the active resources, (v) the limitation of the resources and (vi) the affecting disturbances.

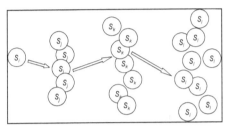

Fig. 11. Situation chain process of aircraft operational process as a result of an active subjective endogenous control.

The situation chain process can be given by the following mathematical formula:

$$c(t): \quad \left(x_0, t_0, \omega\left(t_f \in [t_0, t_0 + \tau]\right); R^{disp}(t_0), R^{req}(t_0), ...\right), \tag{26}$$

or in a more general approach:

$$c(t): \quad \left(P : \sigma_0(t_0) \rightarrow \sigma_j\left(t_f \in [t_0, t_0 + \tau]\right) \in S_f \subset S_a, R^{disp}(t_0), R^{req}(t_0), ...\right); \tag{27}$$

where $x_0$ is the vector of parameters at the initial (actually starting) state at $t_0$ time; $\sigma$ gives the state of the system in the given time; $\tau$ defines the available time for the transition of the state vector into the set of $\omega$ not later than $[t_0, t_0 + \tau]$; $P$ are the problems how to transit the system from the initial state into the one of the possible state $S_f \subset S_a$ not later than $\tau$.

During a flight, one flight situation is followed by another. Therefore, the aircraft flight operational process with continuous state space and time can be approximated by the stochastic process with continuous time and discrete state space, flight situations. This means that a flight is a typical situation chain process. (This is a basis for using the stochastic model of flight risk - see 2.2. point.)

### 4.2 Using the developed model to investigation of the aircraft landing

Final approach and landing are the most dangerous phases of flights. It is even a more significant problem for personal flights, controlled by less-skilled pilots.

The developed method using the subjective analysis to the flight safety evaluation was applied to investigate a landing procedure of a small aircraft.

In this investigation, no side wing, and no lateral motion were considered. By using the trajectory reference system – in which the $x$ axis shows the direction of the wind, $z$ axis is perpendicular to $x$ in the local vertical plane, while centre of the coordinate system is located in the aircraft's centre of gravity – the motion of the aircraft could be given by the motion and the rotation of its center of gravity (Kasyanov 2004):

$$m\frac{dV}{dt} = T(V,z,t) - W\sin\theta - D(V,z,t) \ , \tag{28.a}$$

$$mV\frac{d\theta}{dt} = L(V,z,t) - W\cos\theta \ , \tag{28.b}$$

$$I_y\frac{dq}{dt} = M(\alpha,q,V,z,t) \ . \tag{28.c}$$

Due to the applied control, the trust ($T$), the lift ($L$), the drag ($D$) and the aerodynamic moment ($M$) are all clearly depending on time. The altitude ($z$) has also an influence on the variable above, through the ground effect. Mass ($m$) and therefore the weight ($W$) of the aircraft are assumed to be constant. The aircraft's velocity ($V$) and pitch rate ($q$) describes the motion, while the flight path angle (or descent angle $\theta$) gives the position of the aircraft. The angle of attack ($\alpha$) is the difference between the pitch attitude, $\vartheta$ and flight path angles:

$$\alpha = \upsilon - \theta \ . \tag{29}$$

The pitch rate and the modification of the altitude could be easily given by:

$$q = \frac{d\upsilon}{dt} \ , \tag{30}$$

$$\frac{dH}{dt} = V\sin\theta \ . \tag{31}$$

According to the flight operational manuals and airworthiness requirements, limitations (mi - minimum and ma - maximum) should be applied on the velocity, the descent angle and the decision altitude:

$$V \in \left[V_{mi}^*, V_{ma}^*\right], \tag{32.a}$$

$$\theta \in \left[\theta_{mi}^*, \theta_{ma}^*\right], \tag{32.b}$$

$$H \geq H_{Dmi}^* \ . \tag{32.c}$$

A simple assumption could be applied: during an approach, pilots should decide whether to land or to make a go-around. For this decision they need time, which is the sum of (i) the time to understand and evaluate the given situation, $\sigma_k$, (ii) the time for decision making and (iii) the time to react (covering also the reaction time of the aircraft for the applied decision) (Kasyanov 2007):

$$t^{req} = t^{req}_{ue}(\sigma_k) + t^{req}_{dec}(S_a) + t^{req}_{react}(\sigma_k, S_a) .$$ (33)

Here $\sigma_k$ defines all possible situations (e.g. $\sigma_1$ might be the situation of landing at first approach without any problems, $\sigma_2$ could be related to the situation when the under carriage system could not be opened, $\sigma_3$ might stand for a landing on the fuselage, $\sigma_5$ for go-around, or $\sigma_5$ for a successful landing after second approach).

$S_a$ is the chosen solution from the set of possible solutions. It is clear that all solutions have a limited drawback, such as extra cost, or extra fuel.

The subjective factor of pilots might be introduced with the use of the ratio of the required and disposable resources (Kasyanov 2007):

$$\overline{r}_k = \frac{R^{req}(\sigma_k)}{R^{disp}(\sigma_k)} = \overline{t}_k = \frac{t^{req}(\sigma_k)}{t^{disp}(\sigma_k)} .$$ (34)

In this case, an endogenous index can be defined as

$$\varepsilon_k(\sigma_k) = \frac{\overline{r}_k}{1-\overline{r}_k} = \frac{t^{req}(\sigma_k)}{t^{disp}(\sigma_k) - t^{req}(\sigma_k)} \quad or \quad \varepsilon_k(\sigma_k) = \frac{t^{req}(\sigma_k) + t^{dec}(S_a)}{t^{disp}(\sigma_k) + t^{dec} - t^{req}(\sigma_k)} ,$$ (35)

where $t^{dec}(S_a)$ is a time required to recognize the set of alternative strategies.

Naturally, we can assume that pilots are able to evaluate the consequences of their decisions, and therefore they can evaluate the risk of the applied solutions. Such evaluation can be defined as the subjective probability of situations: $P(\sigma_k)$, canonic distribution of which as the distribution of canonic assemble of the preferences is assumed to hold the following form:

$$p(\sigma_k) = \frac{P^{-\alpha}(\sigma_k)e^{-\beta\varepsilon_k(\sigma_k)}}{\sum\limits_{q=1}^{2} P^{-\alpha}(\sigma_q)e^{-\beta\varepsilon_k(\sigma_q)}} ,$$ (36)

where $p(\sigma_k)$ describes the distribution of the best alternatives from a negative point of view.

The time-depending coefficients $\alpha$ and $\beta$ should be chosen in a way to model the endogenous dynamics, model the subjective psycho physiological personalities of pilots. The qualities of the pilots are depending on different factors including "periodical" incapacity to make decisions that increases while getting closer to the decision time (altitude) of go-around.

The (36) has special features: in case of $\overline{t}_k = \frac{t^{req}(\sigma_k)}{t^{disp}(\sigma_k)} \to 0$ preferences are determined by the

subjective probability, $P(\sigma_k)$, only, and in case $\overline{t}_k \to 1$, the preference turn into zero. The (36 ) comes from the solution of the following function:

$$\Phi_p = -\sum_{k=1}^{N} p(\sigma_k)\ln p(\sigma_k) - \beta \sum_{k=1}^{N} p(\sigma_k)\varepsilon_k(\sigma_k) - \alpha \sum_{k=1}^{N} p(\sigma_k)\ln P(\sigma_k) + \gamma \sum_{k=1}^{N} p(\sigma_k) \ . \tag{37}$$

A special feature of this function is that the structure of the efficiency function includes the logarithm of the subjective probability:

$$\eta_p = -\sum_{k=1}^{N} \left(\alpha \ln P(\sigma_k) + \beta\varepsilon(\sigma_k)\right) p(\sigma_k). \tag{38}$$

The complexity of decision making could be characterized by the uncertainties or the pilots' incapacity to make decisions, which is increasing while getting closer to the minimum decision altitude, $H^*_{Dmi}$. To make decisions, the pilots must overcome their "entropic barrier", $H_p$. The rate of incapacity could be defined with the norm of entropy:

$$\bar{H}_p = \frac{H_p}{\ln N} \ . \tag{39}$$

Figure 12. shows a simplified decision making situation at an approach about the go-around [Kasyanov 2004, 2007]. At $t_0, x_0$, $S_a : (\sigma_1, \sigma_2)$ indicates the set of alternative situations with the distribution of preferences $p(\sigma_1)$ and $p(\sigma_2)$ (where $\sigma_1$ indicates the landing and $\sigma_2$ defines the go-around).

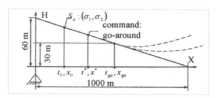

Fig. 12. Final phase of aircraft approach.

The preferences are oscillating, because of the exogenous fluctuation (while decision altitude is getting closer) and the endogenous processes (depending on the uncertainties in the situation awareness and operators (pilots) incapacity to make decisions). If pilots are able to overcome their entropy barrier up to command for go-around (reaching the decision minimum altitude), $t^*, x^*$, then they could make a decision. Due to this decision, the set of situations, $S_a$, can be given with the followings:

$$S_{a1} : (\sigma_2); p(\sigma_2) = 1; p(\sigma_1) = 0$$

$$S_a : (\sigma_2); p(\sigma_2)$$
$$t \prec t^*$$
$$p(\sigma_1) + p(\sigma_2) = 1$$

$$S_{a2} : (\sigma_1); p(\sigma_1) = 1; p(\sigma_2) = 0 \tag{40}$$
$$t \geq t^*$$

If pilots are not able to overcome their entropy barrier before reaching $t^*, x^*$, the flight situation would become more complex, and therefore the possibility to perform a go-around (case $\sigma_2$) might be even out of the possible set of situations.

## 4.3 Modeling the human way of thinking and decision making

A human as "biomotoric system" uses the information provided by sense organs (sight, hearing, balance, etc.) to determine the motoric actions (Zamora, 2004). From a piloting point of view, balance is the most important from the human sense organs. (As known, pilots are flying upon their "botty" for sensing the aircraft's real spatial position, orientation and motion dynamics (Rohacs, 2006).) The sense of balance (Zamora, 2004) is maintained by a complex interaction of visual inputs (the proprioceptive sensors being affected by gravity and stretch sensors found in muscles, skin, and joints), the inner ear vestibular system, and the central nervous system. Disturbances occurring in any part of the balance system, or even within the brain's integration of inputs, could cause dizziness or unsteadiness.

In addition to this, human has another sensing, kinesthesia (Zamora, 2004) that is the precise awareness of muscle and joint movement that allows us to coordinate our muscles when we walk, talk, and use our hands. It is the sense of kinesthesia that enables us to touch the tip of our nose with our eyes closed or to know which part of the body we should scratch when we itch. This type of sensing is very important in controlling an aircraft and moving in 3D space. (Some scientists believe that future aircraft control system must be operated by thumbs, as the new generation is trained on video-games such as "Game Boy" (Rohacs, 2006).)

The main element of the "human biomotoric system" is the human brain that is the anteriormost part of the central nervous system in humans as well as the primary control center for the peripheral nervous system.

The human brain (Russel, 1979; Davidmann, 1998). is a very complex system based on the net of brain cells called as neurons that specialize in communication. The brain contains circuits of interconnected neurons that pass information between themselves.

The neurons contain the dendrites, cell body and axon. In neurons, information passes from dendrites through the cell body and down the axon (Russel, 1979; Davidmann, 1998).

Principally, transmission of information through the neuron is an electrical process. The passage of a nerve impulse starts at a dendrite, it then travels through the cell body, down the axon to an axon terminal. Axon terminals lie close to the dendrites of neighboring neurons.

From control theory point of view, the most important behavior of human brain is the memory, namely learning, memorizing and remembering (Receiving, Storing and Recalling). Generally, human beings are learning all the time, storing information and then recalling it when it is required (Davidmann, 1998). After the investigation of human thinking, including recognition, information analysis, reasoning, decision support (Rohacs, 2006; 2007) the human way of thinking is found to be have the following behaviors:

- syntactic and semantic processing of the sensed information,
- working on the basis of large net of small and simplified articles (neurons),
- using the complex system oriented approach,
- making parallel thinking and activity,
- learning (synthesis of the new knowledge),
- model-formation and using the models (including verbal models applied in learning processes and complex mathematical representation),
- long-term memory,

- tacit knowledge (took in practice),
- intentional thinking (goal and wish),
- intuition (subconscious thinking),
- creativity (finding the contexts),
- innovativity (making originally new minds, things),
- unexpected values can be appeared,
- jumping from quantity to quality.

Seeing all the features listed above, it is clear that human thinking and decision making is a very complex process, containing some chaotic effects.

There is not enough information on the physical, systematic, intellectual, psychophysiology, etc. characteristics of the subjective analysis, about the way of thinking and making decision of subjects-operators like pilots. Only limited information is available on the time effects, possible damping the non-linear oscillations, long term memory, etc. making the decision system chaotic.

Professor Kasyanov introduced a special chaotic model (Kasyanov, 2007) based on the modified Lorenz attractor (Stogatz, 1994) for modeling the endogenous dynamics of the described process.

$$\frac{dX}{dt} = aY - bZ - hX^2 + f(t);$$

$$\frac{dY}{dt} = -Y - XZ + cX - mY^2; \tag{41}$$

$$\frac{dZ}{dt} = XY - dZ - nZ^2.$$

where $a$, $b$, $c$, $d$, $h$, $m$, $n$ are the constants while $f$ takes into account the disturbance. (In case of $h=m=n=0$ and $f(t)=0$ the model turns into the classic form of Lorenz attractor.)

Principally, there are no strong arguments explaining the use of Lorenz attractor to model the human way of decision making (human thinking) (Dartnell, 2010; Krakovska, 2009), but the results of application are close to real situations.

## 4.4 Results of investigations

Professor Kasyanov investigated various model types, and evaluated the model parameters (Kasyanov, 2007). For a medium sized aircraft (weight of aircraft, W = $10^6$ N; wing area, S = 100 m²; wing aspect ratio A = 7; thrust T = 9.4 x $10^4$ N; and velocity V = 70 m/sec) with commercial pilots, he recommended to use the following values: a=8; b=8; c=20; d.43; f=0.8; h= 0.065; m=0.065; n=0.065.

Using these parameters, the subjective probabilities might be chosen as $P(\sigma_1) = 0.53$, $P(\sigma_2) = 0.6$ and $\varepsilon_1 = 5.5 + 0.01t$, $\varepsilon_2 = 5.4 + 0.04t$ take into account the decreasing difference in the required and the available time for the decision. The typical results of using the described model are shown in the Figure 13., demonstrating the chaotic character of decision making.

In this example, the figures demonstrate that pilots are unfixed for a period of about 10 sec, during which their preferences (A, B) are changing by sudden oscillations and the H entropy at the beginning is rather high. If the limit for the entropy would be 0.7 (that is still quit high) then decisions could be made in about 10 sec. This means that pilots will not able to do that according to the Figure 12.

If the parameters are set to a=10; b=10; c=35; d=1; f=0; h= 0.065; m=0.065; n=0.065 and $P(\sigma_1)=0.53$, $P(\sigma_2)=0.6$, then (see Figure 14) the entropy would quickly decrease and the decision could be made in about 3 sec. According to the ICAO requirements, time $t = t_{ga} - t^*$ (see Figure 12.) should not be less than 3.16 sec. Therefore, if the situation presented in the Figure 12. appears before $t_0, x_0$, then the right decision could be made.

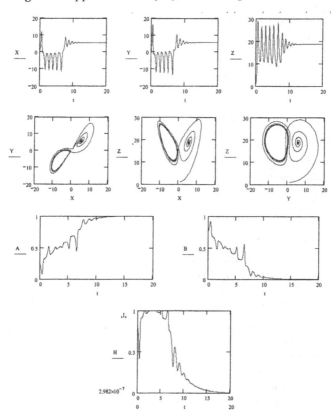

Fig. 13. Results of using the developed model to landing of a medium sized aircraft.

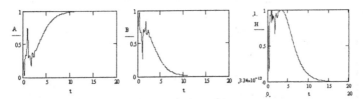

Fig. 14. Results, when the parameters are chosen for well-skilled pilots.

From the results of using the developed model to the landing phase of a small aircraft (such as analyzed in the Hungarian national projects SafeFly: development of the innovative safety technologies for a 4 seats composite aircraft and EU FP7 project PPlane: Personal Plane: Assessment and Validation of Pioneering Concepts for Personal Air Transport Systems, Grant agreement no.233805) several important conclusions had been made (Rohacs et all, 2011; Rohacs & Kasyanov, 2011; Rohacs, 2010).

During the final approach, the common airliner pilots require about three times more time for making decision on go-around than the well practiced colleagues.

Using the developed model and condition defined by Figure 12, the descent velocity of a small aircraft could be determined to about 100 km/h for airliner common pilots, and 75 km/h for those of less-skilled.

In this case, the airport can be designed with a landing distance of less than 600 m (runway about 250 - 300 m) and a protected zone under the approach (to overfly the altitude of 100 m) of about 1500 m. These characteristics enable to place small airports close / closer to the city center.

## 5. Conclusions

This chapter introduced the subjective analysis methodology into the investigation of the real flight situation, flight safety. The subject, as pilot operator generates his decision on the basis of his subjective situation analysis depending on the available information and his psycho-physiological condition. The subjective factor is the time available for the decision of the given tasks.

After the general discussion on flight safety, its metrics and accident statistics, an original approach was introduced to study the role of human factors in flight safety. The deterministic or stochastic models of flight safety are not included clearly the subjective behaviors of human operators. However, the subjective analysis may open a new vision on the flight safety and may result to improve the aircraft development methods and tools.

The subjective decision making of pilots was modeled by the modified Lorenz attractor that needs further investigation and explanation. The applicability of the developed methodology was applied to study the small aircraft final approach and landing. It demonstrates that the model is suitable to investigate the difference between the well trained and less-skilled pilots. The model helped in the definition of the aircraft and airport characteristics for the personal air transportation system.

This work is connected to the scientific program of the "Development of the innovative safety technologies for a 4 seats composite aircraft - SafeFly" (NKTH-MAG ZRt. OM-000167/2008) supported by Hungarian National Development Office and Personal Plane - PPLANE Projhect supported by EU FPO7 (Contract No - 233805) and the research is supported by the Hungarian National New Széchenyi Plan (TÁMOP-4.2.2/B-10/1-2010-0009)

## 6. References

Afrazeh A & Bartsch H (2007) Human reliability and flight safety. International Journal of Reliability, Quality and Safety Engineering 14(5): 501–516

Banos, A., Lamnabhi-Lagarrigue, F., Montoya, F. J. (2001) Advances in the Control of Nonlinear Systems, Lecture Notes in Control and Information Sciences 264, Spinger - Verlag London Berlin Heidelberg, 2001,

Bezopastnostj (1988) poletov letateljnüh apparatov (pod red. A.I. Starikova). Transport, Moscow, 1988.

CASA (2005) AC 139-16(0): Developing a Safety Management System at Your Aerodrome, Australian Government – Civil Aviation Safety Authority (CASA) Advisory Circular, 2005.

Commercial (2000) Aviation Safety Team (CAST), Process Overview,
http://www.icao.int/fsix/cast/CAST%20Process%20Overview%209-29-03.ppt

Dartnell, L. (2010) Chaos in the brain, (2010), http://plus.maths.org/content/chaos-brain

Davidmann, M. (1998) How the Human Brain Developed and How the Human Mind Works, 1998, 2006, http://www.solbaram.org/articles/humind.html

EASA (2008) Annual Safety Review, EASA 2008.

FAA (2006) Introduction to Safety Management Systems for Air Operators, Federal Aviation Administration Advisory Circular 120-92: Appendix 1, Jun. 22, 2006.

Fliens, M., Levine, J., Martin, P., Rouchen, P. (1999) A Lie-Bäcklund Approach to Equivalence and Flatness of Nonlinear Systems, IEEE Transactions on Automatic Control, Vol. 44, No. 5, MAY 1999, 922 - 937.

Flight (2000) Control design – Best Practice, NATO, RTO-TR-029, AC/323(SCI)TP/23, Neuilly-sur-Seine Cedex, France, 2000.

Gardiner, C. W. (2004) Handbook of Stochastic Methods for Physics, Chemistry and the natural Sciences, Springer Series in Synergetics, Springer-Verlag Berlin Heidenberg New York, 2004.

Gudkov A. I., Lesakov P. S. (1968) Vneisnie nagruzki i prochnostj letateljnih apparatov, Masinostroyeniye Moscow, 1968.

Howard R. W. (1980) Progress in the Use of Automatic Flight Controls in Safety Critical Applications, The Aeronautical Journals, 1980 v. 84. X. No. 837. pp.316-326.

Ibe, O. C. (2008) Markov Process for Stochastic Modeling, Academic press. 2008

Kasyanov, V. A. (2004) Flight modelling (in Russian), ), National Aviation University, Kiev, 2004, 400 p.

Kasyanov, V. A. (2007) Subjective analysis (in Russian), National Aviation University, Kiev, 2007, 512. p.

Krakovska A. (2009) Two Decades of Search for Chaos in Brain, MEASUREMENT 2009, Proceedings of the 7th International Conference, Smolenice, Slovakia, pp. 90 - 94.

Lee, C. A. (2003) Human error in aviation (2003)
http://www.carrielee.net/pdfs/HumanError.pdf

Pavlov, V.V., Chepijenko, V. I. (2009) Ergaticheskie sistemii upravleniya (Ergatic control systems), gasudarstvennij nauchno-Isledovatjelskij Institute Avitacii, Kiev, http://194.44.242.245:8080/dspace/bitstream/handle/123456789/7645/01-Pavlov.pdf?sequence=1

Ponomarenko, V. (2000) Kingdom in the Sky – Earthly Fetters and Heavenly Freedoms. The Pilot's Approach to the Military Flight Environment, NATO RTO-AG-338 AC/323(HFM)TP/5, July, 2000

Rohacs, J. (1986) Deviation of Aerodynamic Characteristics and Performance Data of Aircraft in the Operational Process. (Ph.D. thesis) KIIGA, Kiev , 1986.

Rohacs, J. (1990) Analysis of Methods for Modeling Real Flight Situations. 17th Congress of the International Council of the Aeronautical Sciences, Stockholm, Sweden, Sept. 9-14. 1990, ICAS Proceedings 1990. pp. 2046-2054.

Rohács, J. (1995) Repülések biztonsága (Safety of Flights) Bólyai János Műszaki Katonai Főiskola (Military Technology High School Named János Bólyai), Budapest, 1995.

Rohács, J. (1998) Revolution in Safety Sciences -- Application of the Micro Devices „Progress in Safety Sciences and Technology" (Edited by Zeng Quingxuan, Wang Liqiong, Xie Xianping, Qian Xinming) Science Press Beijing / New York, 1998, pp. 969 - 973.

Rohacs, J. (2000) Risk Analysis of Systems with System Anomalies and Common Failures „Progress in Safety Sciences and Technology" Vol. II. Part. A. (edited by Li Shengcai, Jing Guoxun, Qian Xinming), Chemical Industry Press, Beijing, 2000, 550–560.

Rohacs, J. (2006) Development of the control based on the biologycal principles, ICAS Congress, Hamburg, 2006 Sept. CD-ROM, ICAS, 2006

Rohacs, J. (2007) Some thoughts about the biological principle based control, Sixth International Conference on Mathematical Probéems and Engineering and Aerospace Sciences (ed. By Sivasundaram, S.), Cambridge Scientific Publisherm 2007, pp. 627-638, ISBN 978-1-904868-56-9

Rohacs, J. (2010) Subjective Aspects of the less-skilled Pilots, Performance, Safety and Well-being in Aviation, Proceedings of the 29th Conference of the European Association for Aviation Psychology, 20-24 September 2010, Budapest, Hungary, (edited by A. Droog, M. Heese), ISBN: 978-90-815253-2-9 pp. 153-159

Rohacs, J., Kasyanov, V. A. (2011) Pilot subjective decisions in aircraft active control system, J. Theor. Appl. Mech., 49, 1, pp. 175-186, 2011

Rohács, J., Németh, M. (1997) Effects of Aircraft Anomalies on Flight Safety „Aviation Safety (Editor: Hans M. Soekkha) VSP, Ultrecht, The Netherland, Tokyo, Japan, 1997, pp. 203–211.

Rohacs, J., Rohacs, D., Jankovics, I., Rozental, S., Hlinka, J, Katrnak, T., Helena, T. (2011) Personal aircraft system improvements Internal report, PPLANE (EU FP 7 projects), Budapest, 2011.

Rohacs, J., Simon I. (1989) Repülőgépek és helikopterek üzemeltetési zsebkönyve (The handbook of airplane and helicopter operation) Müszaki Könyvkiadó, Budapest, 1989.

Ropp, T. D., Dillman, B. G. (2008) Standardized Measures of Safety: Finding Global Common Ground for Safety Metrics, IAJC -IJME Conference, International Conference on Engineering and Technology, 2008, Nashville, TN, US, ENT 203: Topics in Aviation Safety, Paper No. 29.

Russel, P. (1979) The Brain Book, Penguin Group, new York, 1979.

Shin, J. (2000) The NASA Aviation Safety Program: Overview, Nasa, 2000, NASA/TM−2000-209810, http://gltrs.grc.nasa.gov/reports/2000/TM-2000-209810.pdf

Statistical (2008) summary of commercial jet airplane accidents worldwide operations 1959 - 2008, Boeing, http://www.boeing.com/news/techissues/pdf/statsum.pdf

Strogatz, S. (1994) Nonlinear dynamics and chaos : with applications to physics, biology, chemistry, and engineering. Perseus Books, Massachusetts, US, 1994.

Tihonov, V.I., Mironov, M.A. (1977) Markovskie processi. Sovetskoe Radio, Moscow, 1977.

Transport (2007) Canada, TP 14343, Implementation Procedures guide for Air Operators and Approved Maintenance Organizations, April, 2007.

White, J.: (2009) Aviation safety program, NASA, (2009) http://www.docstoc.com/docs/798142/NASA-s-Aviation-Safety-Program

Zamora, A. (2004) "Human Sense Organs - The Five Senses." Anatomy and Structure of Human Sense Organs. Scientific Psychic, 2004, 2011, http://www.scientificpsychic.com/workbook/chapter2.htm

# 12

# Measuring and Managing Uncertainty Through Data Fusion for Application to Aircraft Identification System

Peter Pong[1] and Subhash Challa[2]

[1]*Jacobs Australia / University of Melbourne*
[2]*NICTA Victoria Research Laboratory / University of Melbourne*
*Australia*

## 1. Introduction

Despite the use of modern Identification Friend Foe (IFF) technology, aircraft recognition remains problematic even though a great deal of research effort has already been invested in this area. In the military context, IFF identification is supposed to be initiated when the interrogator transmits a signal to the aircraft and friendly aircraft are 'supposed' to reply to the signal by transmitting an identification code to the interrogator. Hostile aircraft often become unresponsive to the interrogator because it is either does not have the appropriate transponder or is trying to avoid being identified as an unfriendly aircraft. In the civilian air transport system, the Secondary Surveillance Radar (SSR) allows the location of the civilian aircraft being transmitted (through transponder) to the Air Traffic Controller (ATC). However, in extreme incidents, such as the attacks on the World Trade Center on 11th September 2001, the SSR transponders were manually disabled, which prevented the ATC detecting flight path alternation. To avoid the drawback of the transponder based aircraft identification system, the technique of Non-Cooperative Target Recognition (NCTR) has become a useful technology, because it does not require the participation of friendly aircraft. The NCTR technique relies primarily on the ground based target classification technology. In a typical classification problem, the goal is to develop a classifier that is capable to discriminate targets. This technology shares a great deal of similarity with the modern Electronics Support Measures (ESM) system that often employs as a Radar Warning Receiver (RWR) for modern military aircraft self-protection. Acknowledging the number of successful classifier technologies reported in this area, the goal of this work is not to propose any new algorithm to enhance the classification technology. Instead, a novel method, based on uncertainty measures, is introduced to improve the classification function by employing a data fusion technique. Data fusion applying evidential reasoning framework is a well established technique to fuse diverse sources of information. A number of fusion methods within this formalism were introduced including Dempster-Shafer Theory (DST) Fusion, Dezert Samarandche Fusion (DSmT), and Smets' Transferable Belief Model (TBM) based fusion. However, the impact of fusion on the level of uncertainty within these techniques was not studied in detail. While the use of Shannon entropy with the Bayesian fusion is well understood, the measures of uncertainty within the Dempster-Shafer formalism is not widely regarded. In this paper, an uncertainty based technique is proposed to quantify the evolution of DST fusion. This technique is then utilised to determine the optimal combination of sensor information to achieve the least uncertainty in the context of the aircraft identification problem using sensors operating the NCTR technique.

## 2. Background

Information fusion is often used as a data-processing technique to integrate uncertain information from multiple sensors. Information often contains uncertainties, which are usually related to physical constrains, detection algorithms and the transmitting channel of the sensors. Whilst the intuitive approaches, such as Dempster-Shafer Fusion (Shafer, 1976), Dezert Samarandche Fusion (DSmT)(Dezert & Smarandache, 2006) and Smets' Transferable Belief Model (TBM) (B.Ristic & P.Smets, 2005) aggregate all available information, these approaches do not always guarantee optimum results. Acknowledging that these techniques have associated measurement costs, the essence is to derive a fusion technique to minimise global uncertainties.

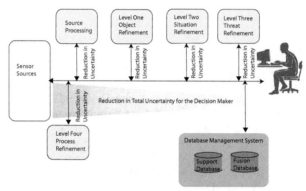

Fig. 1. JDL Model and Uncertainty

In the aerospace community, there is an increasing trend to automate decision processes based on information fusion techniques. As an example, fighter pilots may rely on various forms of data fusion models to assist in assessing the current situations, when uncertain information co-exists at all levels of fusion. Considering the many data fusion models, the Joint Defence Laboratory (JDL) model (Hall & Llinas, 2001) is one the most commonly referred frameworks, which consists of Level 1 Object Assessment, Level 2 Situation Assessment, Level 3 Impact Assessment and Level 4 Process Refinement. The decision maker is supposed to treat the JDL model at 4 independent levels of functions, however, each level of fusion often includes unavoidable uncertainties. That means any aircraft identification system employing real-time situation analysis technology is required to manage uncertainty in the most effective manner. The techniques based on statistical models employed in aircraft tracking were widely acknowledged, but the methods based on uncertainty measures for target identification are not well understood in the aviation community. In recognition of this deficiency, this paper explores a novel aircraft identification technique by leveraging a new uncertainty based fusion concept.

The new concept introduced in this work explores a number of uncertainty measures under the reasoning framework and attempts to introduce a methodology to manage uncertainty variation under the DST based fusion. An example derived from an Aircraft Identification (AI)

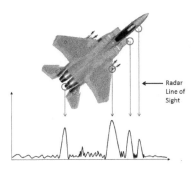

Fig. 2. Example of a radar range profile of a fighter aircraft

system is employed to demonstrate the characteristics of uncertainty variation. In terms of target tracking, significant advancements have been made in the past two decades to improve tracking technology by employing sophisticated data fusion techniques. Some of the earlier works went even further by incorporating Target Identification information, such as IFF data, to improve the overall track quality (Leung & Wu, 2000), (Carson & Peters, 26-30 Oct 1997), (Bastiere, 1997), and (Perlovsky & Schoendorf, 1995). When legitimate statistical information is presented, the techniques employed by tracking and identification using IFF information are relatively mature. However, when conflicting information is presented to the NCTR system, most techniques employed today may find it difficult to discriminate the contradicting information. In this work, we propose a technique based on uncertainty measures to resolve this problem. The employment of uncertainty in recent aviation research was reported in areas, such as air traffic control (Porretta & Ochieng, 2010), navigation (Deng & Liu, 2011) and airport surface movement management (Schuster & Ochieng, 2011), however, all these works essentially model uncertainty based on the target statistical characteristics, such as model based classified illustrated in Figure 2. Instead of treating uncertainty implicitly using their statistical values, the concept proposed in this work treats uncertainty measures directly as input parameters. In this way, we could explicitly quantify the fusion performance to make the best target identification.

## 3. Sensor selection and decision making

Information fusion is often perceived to produce improved decision. This assumption is generally true when sensor availability is limited, however, one has to question whether fusing all available data guarantee synergy. The focus of this work is on the reduction of uncertainties by expressing the relevant uncertainties in the reasoning system and utilise these measures to achieve the best information fusion strategy. In order to develop an uncertainty based information fusion in the aircraft identification context, the authors argue that the best fusion decision can only be observed when (i) the information fusion could provide the least ambiguous choice, (ii) the result produced by the fusion system induces the least vague answer under the reasoning framework, and (iii) the final recommendation provided by the fusion system has the fewest uncertainties. These three axioms underlying this paper are used to define the best fusion configuration. It is apparent that the goal of uncertainty based fusion is to choose the result with the least uncertainty. A fusion process based on uncertainties has the potential to lead to a biased result. However, it is difficult to neglect a decision based on information fusion when it is the least uncertain, least ambiguous and the most defined answer when compared with other potential solutions.

Figure 3 depicts an illustrative example where an aircraft identification scenario is considered. Assuming a model based classifier is employed to identify three kinds of aircraft types - Dual engines aircraft (D), Quadruple engines aircraft (Q) and Helicopter (H). Also assuming that the sensors produced an "unknown" state in the form of {D, Q, H}, where the decision of the aircraft type is not possible to be classified. Three sensors are utilised in this example to simplify the demonstration, where a classification value based on Basic Probability Assignment (BPA) are given to each of the classification reports with details also summarised in Figure 3.

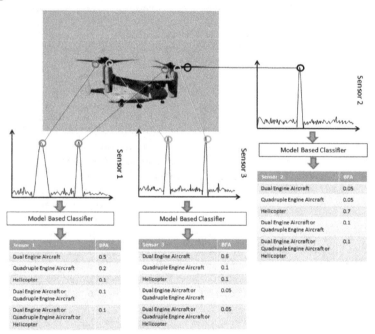

Fig. 3. Multi-sensor aircraft classification

If the identification process performed by each sensor is independent, information provided by Sensor 2 is clearly contradicting with Sensor 1 and Sensor 3. The errors can be induced by the incorrect scatter angle, or simply estimated by an inaccurate model. Based on the axioms discussed, it is observed that fusing Sensor 1 and Sensor 2, or Sensor 2 and Sensor 3 under DST (which be discussed in the next section) will not produce a pronounced result to identify the aircraft type. The result of the fusion is illustrated in Table 1, where only the combination of Sensor 1 and Sensor 3 could provide an unambiguous fusion result. This example highlights the criticality of uncertainty measures in relation to the standard DST fusion process. Section 5 and Section 6 of this paper provide an empirical uncertainty measures analysis in the

|        | Sensor 1&2 | Sensor 1&3 | Sensors 2&3 |
|--------|-----------|-----------|-------------|
| D,Q,H  | 0.026     | 0.022     | 0.0356      |
| D,Q    | 0.0779    | 0.037     | 0.0595      |
| H      | 0.3506    | 0.7704    | 0.3810      |
| Q      | 0.1558    | 0.1185    | 0.0833      |
| D      | 0.3896    | 0.0519    | 0.4405      |

Table 1. Sensor fusion example with contradicted information

reasoning framework, and provides an insight into how this method can be applied in an aircraft identification capability.

## 4. Evidential reasoning framework

The notion of Basic Probability Assignment (BPA) (Shafer, 1976) is defined with respect to a finite universe of propositions or frame of discernment, $\Omega$. The sum of the probabilities assigned to all subsets of $\Omega$ and all propositions which support $\Omega$ must be in unity, as such BPA is a function from the set of subsets, $2^{\Omega}$, of $\Omega$ to the unit interval $[0,1]$. In accordance with the convention proposed by Shafer (Shafer, 1976):

$$m(\emptyset) = 0 \tag{1}$$

and

$$\sum_{A \subseteq \Omega} m(A) = 1 \tag{2}$$

The *subset* $A$ of $\Omega$ such that $m(A) > 0$ is called a *focal element* of $m$, and $\emptyset$ is the empty set. Whilst the summation of BPA must be unity, it is not manditory for the BPA of a proposition $A$ and its negation $\overline{A}$ sum to unity.

### 4.1 Belief and plausibility measures

The idea of linking belief with evidential measures was first discussed by Shafer, and the idea of Belief function in reference to the BPA is defined as,

**Definition 1.** *Bel:* $2^{\Omega} \rightarrow [0,1]$ *is a belief function over* $\Omega$ *if it satisfies:*

- $Bel(\emptyset) = 0$
- $Bel(\Omega) = 1$
- *for every integer* $n > 0$ *and collection of subsets* $A_1, ...., A_n$ *of* $\Omega$

$$Bel(A_1 \cup ... \cup A_n) \geq \sum_i Bel(A_i) - \sum_{i<j} Bel(A_i \cap A_j) + ... + (-1)^{n+1} Bel(A_1 \cap ... \cap A_n)$$

BPA gives a measure of support that is assigned exactly to the focal elements of a given frame of discernment. In order to aggregate the total belief in a subset $A$, the extent to which all the available evidence supports $A$, one needs to sum together the BPAs of all the subsets of $A$ for a belief measurement.

$$Bel(A) = \sum_{B \subseteq A} m(B) \quad \forall A \subseteq \Omega \tag{3}$$

The remaining evidence may not necessarily support the negation $\overline{A}$. In fact some of them may be assigned to propositions which are not disjointed from $A$, and hence, could be plausibly transferred directly to $A$ for further information. Shafer called this the plausibility of A:

$$Pl(A) = \sum_{B \cap A \neq \emptyset} m(B) \quad \forall A \subseteq \Omega \tag{4}$$

### 4.2 Dempster-Shafer fusion under an iterative process

Dempster's rule of combination forms a new body of evidence with which the focal elements are all non-empty intersections $X \cap Y$. Given any $S \subseteq U$ there are many pairs $X, Y \subseteq U$ such that $X \cap Y = S$ and so the total weight of agreement assignable to the focal subset $X \cap Y$ is $\sum_{X \cap Y = S} m(X)m'(Y)$. Once normalising the agreement with the "non-conflicting values" $(1 - K)$, Dempster's rule of combination for imprecise evidence becomes,

$$(m * m')(S) = \frac{1}{1 - K} \sum_{X \cap Y = S} m(X)m'(Y) \tag{5}$$

for all $\emptyset \neq S \subseteq U$. The *conflict* between two bodies of evidence $m, m'$ is the total weight of contradiction between the events of $m$ and the events of $m'$:

$$K(m, m') = \sum_{X \cap Y = \emptyset} m(X)m'(Y) \tag{6}$$

The quantity $1 - K$ is the cumulative degree to which the two bodies of evidence do not contradict with each other and is called the *agreement* between $m$ and $m'$. In general evidential theory, Dampster-Shafer rules, belief functions, plausibility functions and BPA forms a suite of significant tools to construct probabilities through carefully modelled evidence. Through this combination process, two new measurement values - *non-specificity* and *conflict*, are also generated as a by-product. An empirical analysis is presented in Section 5 in conjunction with the theory of Aggregated Uncertainty (AU) and the recently proposed generalised Total Uncertainty (TU) measures.

## 5. Uncertainty measures within the evidential reasoning framework

While the classical uncertainties are often measured by the Hartley and Shannon functions, the two functions are tailored for different purposes. In order to cater for both uncertainties, evidential based uncertainty measures are adopted. Two types of classical evidential based uncertainties - non-specificity and conflict - are often measured as part of the DST fusion (Harmanec, 1996). In this section, an overview is introduced to the concept of Hartley Uncertainty measures, Aggregrated Uncertainty (AU) measures and Total Uncertainty (TU) measures which was proposed by Klir (Klir, 2006). This analysis covers the context of the DST fusion system and their subsequent implication. A practical example based on aircraft identification applying uncertainty measures as sensor discrimination matrices is discussed in Section 7 to verify our observations.

### 5.1 Hartley uncertainty

The technique of uncertainty measures was first addressed by Shannon. Under his proposal, the way to quantify uncertainty measures expressed by a probability distribution function $p$

on a singleton set is in the form of,

$$-c \sum p(x) \log_b p(x) \tag{7}$$

where $b$ and $c$ are positive constants, and $b \neq 1$. While this technique is useful to apply in sensor management system operating under the probabilistic framework, it cannot be used under a finite set condition. An alternative is to employ the legacy Hartley measures (Hartley, n.d.), where it seems to be the only meaningful way to measure uncertainty in the form of,

$$c \log_b \sum_{x \in \Omega} r_A(x) \tag{8}$$

or alternatively

$$c \log_b |A| \tag{9}$$

where $A$ is a finite set and $|A|$ is the cardinality of the finite set. $b$ and $c$ are positive constants, and $b \neq 1$. When uncertainty is measured in *bits*, $c \log_b 2 = 1$. Harley uncertainty measures, $H$, defined for any basic possibility functions, $r_A$,

$$H(r_A) = \log_2 |A| \tag{10}$$

On closer examination of (10), $H(r_A)$ is a measure directly related to the specificity of a finite set. In other words, the larger the size of a set, the less specific the measurement becomes. This type of measures was defined as *non-specificity* by Klir (Klir, 2006). In the reasoning framework, Hartley Measures are usually treated as a weighted average of all the focal subsets in the form of BPA function (Klir, 2006).The concept of generalised Harley measures in the context of DST framework is thus defined by the function,

$$GH(m) = \sum_{A \in \Omega} m(A) \log_2 |A| \tag{11}$$

where $\Omega$ is the superset of the focal elements.

### 5.2 Aggregated uncertainty measures

Suppose the goal of information fusion is to reduce global uncertainties, Harmanec (Harmanec, 1996) was the first to explore the concept of uncertainty measures in the DST framework. The idea of $AU$ uncertainty measures was proposed as the optimum uncertainty measures technique under the DST domain, because it is the only way to incorporate the value of non-specificity and conflict simultaneously, which often coexist in the DST framework.

**Definition 2.** *The measure of the Aggregated Uncertainty contained in Bel, denoted as $AU(Bel)$, is defined by*

$$AU(Bel) = \max\{-\sum_{x \in \Omega} p_x \log_2 p_x\} \tag{12}$$

*where the maximum is taken over all $\{p_x\}_{x \in \Omega}$ such that $p_x \in [0,1]$ for all $x \in \Omega$, $\sum_{x \in \Omega} p_x = 1$ and for all $A \subseteq \Omega$, $Bel(A) \leq \sum_{x \in A} p_x$.*

Although the $AU$ technique is not an efficient algorithm, it does satisfy all the properties defined as uncertainty measures (Harmanec, 1996), and specifically, the subadditivity/additivity characteristics.

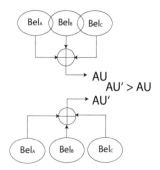

Fig. 4. Additivity and Subadditivity

**Subadditivity.** If $Bel$ is an arbitrary joint belief function on $X \times Y$ and the associated marginal belief functions are $Bel_X$ and $Bel_Y$, then

$$AU(Bel) \leq AU(Bel_X) + AU(Bel_Y) \tag{13}$$

**Additivity.** If $Bel$ is a joint belief function on $X \times Y$, and the marginal belief functions $Bel_X$ and $Bel_Y$ are noninteractive, then

$$AU(Bel) = AU(Bel_X) + AU(Bel_Y) \tag{14}$$

The property of additivity/subadditivity of AU call forth the assumption that uncertainties could be reduced if sensors share common interaction prior the information fusion process occurring. Assuming sensor dependency exists among $Bel_A$, $Bel_B$ and $Bel_C$, the characteristics of the resultant uncertainty under an evidential fusion system is illustrated pictorially in Figure 4. The algorithm to compute AU uncertainty was originated by Harmanec (Harmanec, 1996). Under the proposed algorithm, the input is treated in the form of a frame of discernment $X$, with a belief function $Bel$ on $X$. This algorithm's computation completes once a finite number of steps have been taken and the output is the correct value of the function $AU(Bel)$, since $\{p_x\}_{x \in X}$ maximises the Shannon entropy within the constraints induced by $Bel$.

## 5.3 Total uncertainty measures

The concept of generalised Total Uncertainty (TU) was proposed by Klir (Klir & Smith, 2001) not long after the introduction of AU uncertainty. This measure is defined as a combination of $AU$ uncertainty and Generalised Hartley Measures,

$$TU = \langle GH, GS \rangle \tag{15}$$

where $GH$ represent the Generalised Hartley measures which was discussed in (11). The factor $GS$ is called Generalised Shannon measurement (Klir, 2006), which is the conflicts measurement with the consideration of evident specificity. In other words, it is $GS = AU - GH$, the Aggregated Uncertainty with the reduction of specificity consideration. One advantage of the disaggregated TU, in comparison with AU, is that it expresses amounts of both types of uncertainty (non-specificity and conflict) explicitly, and consequently, it is highly sensitive to changes in evidence. These new features of uncertainty measures allow one to work with any set of recognised and well-developed theories of uncertainty as a whole, which are commonly seen in any evidential based fusion problem.

| Classification | 1 Sensor | 3 Sensors | 7 Sensors |
|:---:|:---:|:---:|:---:|
| A | 0.22 | 0.3485 | 0.4125 |
| B | 0.25 | 0.3309 | 0.3525 |
| C | 0.26 | 0.2845 | 0.2343 |
| D | 0,00 | 0.0015 | 0.0001 |
| A,B | 0.07 | 0.0163 | 0.0004 |
| A,C | 0.03 | 0.005 | 0.0001 |
| A,D | 0.03 | 0.005 | 0.0001 |
| B,C | 0.015 | 0.0022 | 0.0000 |
| B,D | 0.005 | 0.0007 | 0.0000 |
| C,D | 0.01 | 0.0014 | 0.0000 |
| A,B,C,D | 0.1 | 0.0042 | 0.0000 |

Table 2. Classification Results with DST Fusion

## 6. Analysis of uncertainty measures under the Dempster Shafer fusion framework

To appreciate the impact of uncertainty variation, an example with a set of arbitrary data is illustrated in Table 2. The data set is exactly the same measurement values, such that an iterative DST fusion can be performed. The results in Table 2 confirmed that sensor information can be refined and appears to have a reduction of ambiguity under an iterative DST fusion process. However, the merit of these results cannot be examined further, unless an acceptable matrices is used to quantify the fusion. To address this point, the results illustrated in Figure 5 a demonstrate how AU uncertainty reduction could quantify the DST fusion process. Whilst the AU uncertainty measure are a useful index to quantify the DST fusion process, it is suggested to be insensitive to small change in evidences (Klir, 2006). Acknowledging the inherited issues with the AU uncertainty measures, this work also examines the concept of employing Total Uncertainty Map (TUM) to evaluate a standard DST Fusion process. Considering TU is an amalgamation of GH and GS, the uncertainty variation becomes significant if it is illustrated in two dimensional space. Figure 5b is an illustration of how a TUM can be used to visualise the recursive DST fusion. To assist the interpretation, the results of $GS/GH$ are also provided in Figure 5a to enhance the illustration. In this case, $GS$ and $GH$ are treated as an unified parameters with the variation under the DST fusion process observed. Due to the equivalent sensor input for the DST fusion, the weighted average of

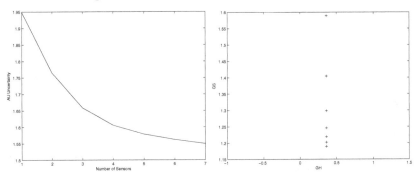

(a) AU Uncertainty Variation Under the DS Fusion   (b) TU Map Variation Under the DS Fusion

Fig. 5. Uncertainty Variation

| Sensor 1 | Sensor 2 | Sensor 3 | Sensor 4 |
|---|---|---|---|
| { A } = 0.26 | {B} = 0.2 | {B} = 0.1 | {A} = 0.05 |
| {B}= 0.26 | {A,B}= 0.1 | {C} = 0.1 | {B} = 0.05 |
| {C}= 0.26 | {A,C} = 0.1 | {A,B}=0.16 | {D} = 0.2 |
| {A,B}= 0.07 | {A,B,C}=0.1 | {B,C}=0.14 | {A,B} = 0.11 |
| {A,C}= 0.01 | {A,C,D}=0.1 | {B,D}=0.05 | {A,C} = 0.03 |
| {A,D}= 0.01 | {B,C,D}=0.3 | {A,C}=0.1 | {A,D} = 0.03 |
| {B,C}= 0.01 | | {A,B,C}=0.2 | {C,D} = 0.03 |
| {B,D}= 0.01 | | {B,C,D}=0.15 | {B,C,D} = 0.3 |
| {C,D}= 0.01 | | | {A,B,C,D} = 0.2 |
| {A,B,C,D}= 0.1 | | | |

Table 3. Random Sensor Input

each focal subset are virtually unchanged, which is why the GH values displayed in Figure 5b remain constant throughout the iterative DST fusion process. Further observation shows, however, that other uncertainty in the form of conflicts are gradually reduced as part of the DST fusion process. To further explore the characteristics of uncertainty variation, four arbitrary sensor data sets are outlined in Table 3. The TU uncertainty is displayed in Figure 6 b. These results are further broken down into four levels and each level represents the number of sensors fused by the DST fusion. Based on the sample results, it is difficult to provide a consolidated uncertainty variation within the DST fusion framework. However, a potential optimisation solution exists when the fusion goal is to present the most specific and least conflicted information to the decision maker. This concept will be covered in Section 7 by leveraging a NCTR based AI example.

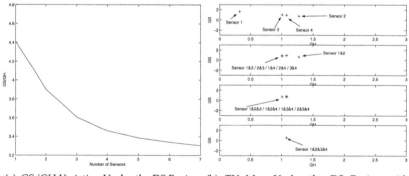

(a) GS/GH Variation Under the DS Fusion (b) TU Map Under the DS Fusion with Random Sensors Input Data

Fig. 6. Extended Uncertainty Variation Modelling

## 7. NCTR based Aircraft Identification (AI)

This case study utilises an example commonly encountered in a model based classification system. Assuming each NCTR sensor has a potential to produce feature detection of,

$$B = \{E0, E1, E2, ....., E36\}$$

where B is the frame of discernment of the aircraft's type attributes, and this example allows seven model based classifiers to report aircraft type identification. To reduce the

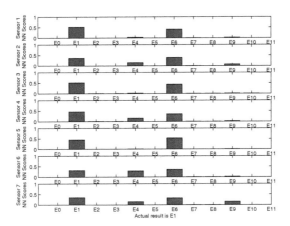

Fig. 7. Model based classifier for aircraft type detection

| Sensor1 | Sensor2 | Sensor3 | Sensor4 | Sensor5 | Sensor6 |
|---------|---------|---------|---------|---------|---------|
| $m\{e_1\} = 0.5188$ | $m\{e_1\} = 0.3617$ | $m\{e_1\} = 0.5126$ | $m\{e_1\} = 0.4565$ | $m\{e_1\} = 0.4414$ | $m\{e_1\} = 0.3480$ |
| $m\{e_6\} = 0.4124$ | $m\{e_4\} = 0.1540$ | $m\{e_6\} = 0.4387$ | $m\{e_4\} = 0.1551$ | $m\{e_6\} = 0.5342$ | $m\{e_4\} = 0.1533$ |
| $m\{\Theta\} = 0.0687$ | $m\{e_6\} = 0.3971$ | $m\{\Theta\} = 0.0487$ | $m\{e_6\} = 0.3546$ | $m\{\Theta\} = 0.0244$ | $m\{e_6\} = 0.3254$ |
| | $m\{e_9\} = 0.0733$ | | $m\{\Theta\} = 0.0337$ | | $m\{e_9\} = 0.1602$ |
| | $m\{\Theta\} = 0.0138$ | | | | $m\{\Theta\} = 0.0357$ |

Table 4. Normalised Aircraft Detection

computational workload this example only employs 12 of the target type signature instead of the potential 37 type of targets, where the results are depicted in Figure 7. The 12 aircraft type signatures selected for this simulation share similar characteristics, and often cause confusion to this particular NCTR platform. The remaining 25 emitter detections are not discarded, but are consolidated as detection CLUTTER. This method is similar to the strategy reported in (Yu & Sycara, 2006), instead this case study treats all aircraft signatures as the total frame of discernment $\Theta_E$, $\{e_0, e_1, e_2, e_3, e_4, e_5, e_6, e_7, e_8, e_9, e_{10}, e_{11}\}$. In terms of the simulation, each emitter signature is considered as $e_i \in E$, where $m(e_i)$ is the normalised confidence level assigned by the post threshold detection process. For instance, the normalised post-detection confidence level with Sensor 2 are $m\{e_1\} = 0.3617$, $m\{e_4\} = 0.1540$, $m\{e_6\} = 0.3971$ and $m\{e_9\} = 0.0733$. To include the non-mutually exclusive aircraft type as CLUTTER, $m\{\Theta_E\} = c(CLUTTER)$, where we assign the confidence of CLUTTER to the set of all possible aircraft types. In this case, the normalised $m\{\Theta_E\}$ based on the pre-detection process is 0.0138.

Upon completion with the BPA preparation, we performed a DST based fusion with a permutation space of $2^7$. Figure 8 shows the uncertainty in the form of AU as gradually reduced with the increment of DST fusion. However, the results become less effective when more sensors are fused together. In accordance with the discussions covered in Section 6, the authors believe the optimum approach when conducting an uncertainty based DST fusion cannot rely on one single parameter alone. Depending on the computational workload and the tolerance of conflicts, the uncertainty based fusion process ought to be determined by a TU map, where $GS$ and $GH$ are to be treated separately. The preliminary results based on this concept are illustrated in Figure 9.

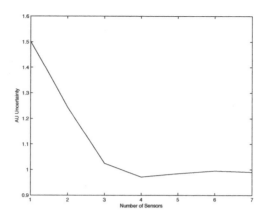

Fig. 8. AU Uncertainty Variation Under the model based classifier DST Fusion

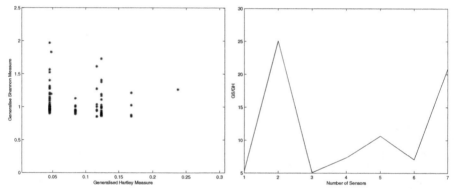

(a) TU Map Variation Under the model based    (b) GS/GH Variation Under the model based
classifier DST Fusion                          classifier DST Fusion

Fig. 9. Uncertainty Variation

Notwithstanding the treatment of uncertainty in the DST context, Figure 9a outlined a method when adopting the theory of AU uncertainty to search for the least uncertain post-fusion results. For comparison purposes, the results of $GS/GH$ measures are also displayed in Figure 9b. Under such a process, the final result is to be determined by the fusion that produces the minimum AU uncertainty. In this particular example, Sensors 1, 3, 4 and 7 can be selected to participate in the fusion process. Based on the least AU uncertainty, the final BPA for the detected emitters are given below:

$$m\{\Theta_E\} = 0$$
$$m\{e_1\} = 0.5604$$
$$m\{e_6\} = 0.4396$$

With a similar approach, and adopting the $GS/GH$ characteristics, Sensors 1, 2 and 3 are selected to join the fusion process. Based on the least $GS/GH$ uncertainty, the final BPA for

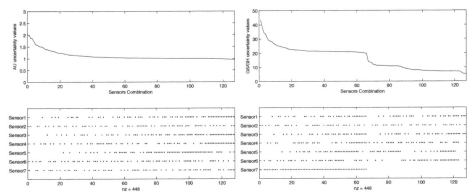

(a) AU Uncertainty measurements in conjunction with sensor fusion

(b) GS/GH Uncertainty measurements in conjunction with sensor fusion

Fig. 10. Uncertainty Variation

the detected emitters are given below, which is equivalent to sensor combination with the least AU uncertainty:

$$m\{\Theta_E\} = 0$$
$$m\{e_1\} = 0.5604$$
$$m\{e_6\} = 0.4396$$

Although the final results obtained from the uncertainty based DST fusion do not yield distinct decisions, the results justify that aircraft type $e_1$ or aircraft type $e_6$ are detected.

## 8. Conclusion

This paper reviews the role of uncertainty measures in the data fusion framework within the context of evidential reasoning. An empirical analysis of the AU and TU uncertainty variations is conducted under the DST fusion framework. A preliminary method to choose sensors based on the uncertainty level is proposed. This technique is illustrated with an aircraft identification problem, when the radar range profile classifier is employed to support an identification system such as NCTR. Since the amount of reflected radar energy is different for different parts of the aircraft, inconsistency often occurs even when the same target is being observed by a number of sensors despite using the same classifier model. It is this inconsistency which makes the uncertainty based fusion technique useful in resolving aircraft identification problems. While the proposed technique can be computationally intensive, the idea underwrites a conservative result with the least measurable uncertainty. This approach essentially yields the potential to evaluate all kinds of reasoning based fusion systems. We have certainly not reached the end of our research effort yet, as the proposed concept only considers primarily the reduction of AU uncertainty. The authors recognise the benefits in further investigation of TUM in conjunction with the theory of optimisation, when a trade-off can be computed based on the classification's precision and accuracy. At the moment, our proposed concept does not take into account the sensor information based on human originated data. It is certainly an exciting future research topic, if the proposed concept is to be extended to cover identification systems where human originate information is employed.

## 9. References

Bastiere, A. (1997). Fusion methods for mltisensor classification of airborne targets, *Aerospace Science and Technology* 1: 83–94.

B.Ristic & P.Smets (2005). Target classification approach based on the belief function theory, *IEEE Transactions On Aerospace And Electronics Systems* 41(2).

Carson, R. Meyer, M. & Peters, D. (26-30 Oct 1997). Fusion of iff and radar data, *16th AIAA/IEEE Digital Avionics Systems Conference (DASC)* 1: 5.3–9–15.

Deng, H. Chao, P. & Liu, J. (2011). Entropy flow-aided navigation, *The Journal of Navigation*, Vol. 64, The Royal Institute of Navigation, pp. 109–125.

Dezert, J. & Smarandache, F. (2006). Dsmt: A new paradigm shift for information fusion, *Cogis ' 06 Conference, Paris* .

Hall, H. & Llinas, J. (2001). *Handbook of Multisensor Data Fusion*, CRC.

Harmanec, D. (1996). *Uncertainty in Dempster-Shafer Theory*, PhD Dissertation, State University of New York.

Hartley, R. (n.d.). Transmission of information, *The Bell System Technical Journal* 7(3): 535–563.

Klir, G. (2006). *Uncertainty and Information, Fundations of Generalised Information theory*, Wiley Interscience.

Klir, G. & Smith, R. (2001). On measuring uncertainty and uncertainty based information: Recent developments, *Annals of Mathematics and Artificial Intelligence* 32: 5–33.

Leung, H. & Wu, J. (2000). Bayesian and dempster-shafer target identification for radar surveillance, *IEEE Transactions on Aerospace and Electronic Systems* 36(2): 432–447.

Perlovsky, L. Chernick, J. & Schoendorf, W. (1995). Multi-sensor atr and identification of friend or foe using mlans, *Neural Networks* 8(7/8).

Porretta, M. Schuster, W. & Ochieng, W. (2010). Strategic conflict detection and resolution using aircraft intent information, *The Journal of Navigation*, Vol. 63, The Royal Institute of Navigation, pp. 61–88.

Schuster, W. & Ochieng, W. (2011). Airport surface movement - critical analysis of navigation system performance requirements, *The Journal of Navigation*, Vol. 64, The Royal Institute of Navigation, pp. 281–294.

Shafer, G. (1976). *A mathematical theory of evidence*, Princeton University Press.

Yu, B. & Sycara, K. (2006). Learning the quality of sensor data in distributed decision fusion, *Proceeding of the 9th International Conference on Information Fusion* .

# Permissions

The contributors of this book come from diverse backgrounds, making this book a truly international effort. This book will bring forth new frontiers with its revolutionizing research information and detailed analysis of the nascent developments around the world.

We would like to thank Ramesh K. Agarwal, for lending his expertise to make the book truly unique. He has played a crucial role in the development of this book. Without his invaluable contribution this book wouldn't have been possible. He has made vital efforts to compile up to date information on the varied aspects of this subject to make this book a valuable addition to the collection of many professionals and students.

This book was conceptualized with the vision of imparting up-to-date information and advanced data in this field. To ensure the same, a matchless editorial board was set up. Every individual on the board went through rigorous rounds of assessment to prove their worth. After which they invested a large part of their time researching and compiling the most relevant data for our readers. Conferences and sessions were held from time to time between the editorial board and the contributing authors to present the data in the most comprehensible form. The editorial team has worked tirelessly to provide valuable and valid information to help people across the globe.

Every chapter published in this book has been scrutinized by our experts. Their significance has been extensively debated. The topics covered herein carry significant findings which will fuel the growth of the discipline. They may even be implemented as practical applications or may be referred to as a beginning point for another development. Chapters in this book were first published by InTech; hereby published with permission under the Creative Commons Attribution License or equivalent.

The editorial board has been involved in producing this book since its inception. They have spent rigorous hours researching and exploring the diverse topics which have resulted in the successful publishing of this book. They have passed on their knowledge of decades through this book. To expedite this challenging task, the publisher supported the team at every step. A small team of assistant editors was also appointed to further simplify the editing procedure and attain best results for the readers.

Our editorial team has been hand-picked from every corner of the world. Their multi-ethnicity adds dynamic inputs to the discussions which result in innovative outcomes. These outcomes are then further discussed with the researchers and contributors who give their valuable feedback and opinion regarding the same. The feedback is then collaborated with the researches and they are edited in a comprehensive manner to aid the understanding of the subject.

Apart from the editorial board, the designing team has also invested a significant amount of their time in understanding the subject and creating the most relevant covers. They scrutinized every image to scout for the most suitable representation of the subject and create an appropriate cover for the book.

The publishing team has been involved in this book since its early stages. They were actively engaged in every process, be it collecting the data, connecting with the contributors or procuring relevant information. The team has been an ardent support to the editorial, designing and production team. Their endless efforts to recruit the best for this project, has resulted in the accomplishment of this book. They are a veteran in the field of academics and their pool of knowledge is as vast as their experience in printing. Their expertise and guidance has proved useful at every step. Their uncompromising quality standards have made this book an exceptional effort. Their encouragement from time to time has been an inspiration for everyone.

The publisher and the editorial board hope that this book will prove to be a valuable piece of knowledge for researchers, students, practitioners and scholars across the globe.

# List of Contributors

**Juraj Belan**
University of Žilina, Faculty of Mechanical Engineering, Department of Materials Engineering,
Žilina, Slovak Republic

**Robert D. Vocke III, Benjamin K.S. Woods, Edward A. Bubert and Norman M. Wereley**
University of Maryland, College Park, MD, USA

**Curt S. Kothera**
Techno-Sciences, Inc., Beltsville, MD, USA

**Giorgio Cavallini and Roberta Lazzeri**
University of Pisa-Department of Aerospace Engineering, Italy

**Melih Cemal Kushan**
Eskisehir Osmangazi University, Turkey

**Fehmi Diltemiz**
1st Air Supply and Maintenance Base, Turkey

**Sinem Cevik Uzgur**
Ondokuz Mayis University, Turkey

**Yagiz Uzunonat**
Anadolu University, Turkey

**I. A. Boguslavsky**
State Institute of Aviation Systems, Moskow Physical Technical Institute, Russia

**Matko Orsag and Stjepan Bogdan**
LARICS-Laboratory for Robotics and Intelligent Control Systems, Department of Control
and Computer Engineering, Faculty of Electrical Engineering and Computing, University
of Zagreb, Zagreb, Croatia

**Luca De Filippis and Giorgio Guglieri**
Politecnico di Torino, Italy

**Peter J. G. Teunissen**
Curtin University of Technology, Austalia
Delft University of Technology, The Netherlands

**Gabriele Giorgi**
Technische Universität München, Germany

**Andrea L'Afflitto and Wassim M. Haddad**
Georgia Institute of Technology, USA

**Jozsef Rohacs**
Budapest University of Technology and Economics, Hungary

**Peter Pong**
Jacobs Australia / University of Melbourne, Australia

**Subhash Challa**
NICTA Victoria Research Laboratory / University of Melbourne, Australia

Printed in the USA
CPSIA information can be obtained
at www.ICGtesting.com
JSHW011500221024
72173JS00005B/1153